Traditional Herbal Medicines for Modern Times

Juzen-taiho-to (Shi-Quan-Da-Bu-Tang)

Scientific Evaluation and Clinical Applications

Traditional Herbal Medicines for Modern Times

Each volume in this series provides academia, health sciences and the herbal medicines industry with in-depth coverage of the herbal remedies for infectious diseases, certain medical conditions or the plant medicines of a particular country.

Edited by Dr. Roland Hardman

Volume 1
Shengmai San, edited by Kam-Ming Ko

Volume 2
Rasayana, by H.S. Puri

Volume 3
Sho-Saiko-To and Related Formulations, by Yukio Ogihara and Masaki Aburada

Volume 4
Traditional Medicinal Plants and Malaria, edited by Merlin Willcox, Gerard Bodeker, and Philippe Rasoanaivo

Volume 5
Juzen-taiho-to (Shi-Quan-Da-Bu-Tang): Scientific Evaluation and Clinical Applications, edited by Haruki Yamada and Ikuo Saiki

Traditional Herbal Medicines for Modern Times

Juzen-taiho-to (Shi-Quan-Da-Bu-Tang)

Scientific Evaluation and Clinical Applications

Edited by
Haruki Yamada
Ikuo Saiki

CRC Press
Taylor & Francis Group
Boca Raton London New York

CRC Press is an imprint of the
Taylor & Francis Group, an **informa** business
A TAYLOR & FRANCIS BOOK

CRC Press
Taylor & Francis Group
6000 Broken Sound Parkway NW, Suite 300
Boca Raton, FL 33487-2742

First issued in paperback 2019

© 2005 by Taylor & Francis Group, LLC
CRC Press is an imprint of Taylor & Francis Group, an Informa business

No claim to original U.S. Government works

ISBN-13: 978-0-415-30830-4 (hbk)
ISBN-13: 978-0-367-39292-5 (pbk)

Library of Congress Cataloging-in-Publication Data

Catalog record is available from the Library of Congress

Visit the Taylor & Francis Web site at
http://www.taylorandfrancis.com

and the CRC Press Web site at
http://www.crcpress.com

Preface

As medical science has progressed, the profile of illness has changed greatly: disease related to aging and inappropriate lifestyles is increasing. Because it is difficult to cure patients with these diseases using only modern medicines, Kampo (Japanese herbal) medicine, which tends to regulate homeostasis, has an increasing role in medical practice in Japan.

Although Kampo medicine originated from Chinese medicine, it was modified and established as Kampo medicines in the 17th century in Japan. Japanese traditional medicine mainly involves Japanese herbal medicine, so-called Kampo medicine, and acupuncture. Modern medicines are used in an analytical way and tend to have a specific action against target molecules, whereas Kampo medicines have a holistic action and appear able to regulate the whole body system. Because of their holistic actions, Kampo medicines are suitable for the treatment of several diseases of the elderly, such as chronic diseases, multiorgan diseases, malaise, decreasing physical strength, senile dementia, and osteoporosis. Kampo medicines have also been thought to function as preventive medicine. Because Kampo medicine has been used for the treatment of a wide range of diseases with many reports of clinical effectiveness, it now plays a very important part in modern-day therapy in Japan.

The number of higher plant species on Earth is about 250,000, and it is estimated that 35,000 to 70,000 species have, at one time or another, been used in some cultures for medical purposes (Chen Ken, WHO Regional Office for the Western Pacific, 2002). Herbal medicines, traditionally used in Asia, have gained popularity globally. In the U.S., where the boom spread from Europe, people show great interest in complementary/alternative medicine, using various herbal medicine preparations as dietary supplements. Also in the U.S., huge research budgets are being spent at the National Institutes of Health (NIH) for scientific research into the safety and efficacy of these medicines.

The World Health Organization (WHO) published its traditional medicines strategy 2002–2005: objectives, components, and expected outcomes. It incorporates four objectives relating to policy: safety, efficacy and quality, access, and national use. In Australia, Europe, and the U.S., complementary and alternative medicine is increasingly used in conjunction with allopathic medicine, particularly for treating and managing chronic diseases. Concern about the adverse effects of chemical medicines, a desire for more personalized health care, and greater public access to health information fuel this increased use (Traditional Medicine—Growing Needs and Potential, WHO, Geneva, 2002).

Kampo medicines have traditionally been used as the "formulation," and now 148 kinds of Kampo formulations are allowed to be covered by the national health insurance system in Japan, starting in 1976, and 210 kinds of formulations have been recognized as useful formulations by the Ministry of Health, Labour and Welfare of Japan. Among such formulations, Juzen-taiho-to (Shi-Quan-Da-Bu-Tang in Chinese) is an important formulation derived from ten component herbs. It has been administered traditionally to patients with anemia, anorexia, or a debilitated general condition caused by surgery, chronic disease, or childbirth. Juzen-taiho-to has been widely studied in nonclinical and clinical research.

Although Kampo medicines have a long clinical experience to support their efficacies and safety, the efficacy of Kampo medicines requires the clarification of the mechanism of action and active ingredients in nonclinical studies and objective clinical evaluations as an evidence-based medicine. Because Kampo medicines contain many active ingredients due to the several component herbs, clarification of such combination effects and standardization are also very important in order to understand the mechanism of action and to supply quality-controlled materials. Attempted with

Juzen-taiho-to, these studies supply very useful examples of research methodologies and new ideas for the study not only of Kampo medicines but also of other natural source medicines.

In this book, we give a general introduction to Kampo medicine; studies of the taxonomy, cultivation, quality assurance, chemical constituents, and pharmacological actions of plant raw materials as the component herbs are described, followed by the therapeutic indications and traditional uses of Juzen-taiho-to. Recent progress in pharmacological studies of Juzen-taiho-to and the elucidation of its active ingredients, its important antitumor and antimetastatic properties, and its clinical use are described and discussed. Its toxicology and side effects, together with the use of Juzen-taiho-to and the other related formulations, are described by contributors who are experts involved in the study of Kampo medicines in general.

The editors believe that the scientific approaches described in this volume will be applicable to other herbal medicines and could provide a new strategy for the development of new medicines in the 21st century.

I wish to thank Dr. Shinyu Nunome for his helpful assistance in editing this book and for confirming the names of the plant raw materials given in the chapters.

Finally, I offer heartfelt thanks to Dr. Roland Hardman for his critical reading and valuable advice in the editorial process.

Haruki Yamada, Ph.D.

Editors

Haruki Yamada is the director of and a professor at the Kitasato Institute for Life Sciences and dean of the Graduate School of Infection Control Sciences, Kitasato University, Japan. He is also the executive director for research at the Oriental Medicine Research Center and the director of the WHO Collaborating Center for Traditional Medicine at the Kitasato Institute. He received the Li-Fu Academic Award for Chinese Medicine in 1999. He is well known in the fields of bioactive polysaccharides and the scientific elucidation of Kampo medicines.

Ikuo Saiki is a professor in the Department of Pathogenic Biochemistry, Institute of Natural Medicine, Toyama Medical and Pharmaceutical University. He has investigated the control of cancer invasion and metastasis and immunological diseases by various treatments, including Kampo medicines and their molecular mechanism of action. In 1990, he received the Incitement Award of the Japanese Cancer Association.

Contributors

Shigeru Abe
Institute of Medical Mycology
Teikyo University
Tokyo, Japan

Toshihiko Hanawa
Oriental Medicine Research Center
The Kitasato Institute
Tokyo, Japan

Hiroko Hisha
First Department of Pathology
Kansai Medical University
Osaka, Japan

Osamu T. Iijima
Department of Pharmacology
Tokyo Medical University
Tokyo, Japan

Susumu Ikehara
First Department of Pathology
Kansai Medical University
Osaka, Japan

Hiroaki Kiyohara
Kitasato Institute for Life Sciences
Kitasato University and Oriental Medicine
  Research Center
Tokyo, Japan

Katsuko Komatsu
Research Institute for WAKAN-YAKU
Toyama Medical and Pharmaceutical
  University
Toyama, Japan

Yasuhiro Komatsu
Department of Serology
Kanazawa Medical University
Ishikawa, Japan

Teruhiko Matsumiya
Department of Pharmacology
Tokyo Medical University
Tokyo, Japan

Tsukasa Matsumoto
Kitasato Institute for Life Sciences
Kitasato University and Oriental Medicine
  Research Center
Tokyo, Japan

Shinyu Nunome
Kitasato Institute for Life Sciences
Kitasato University and Oriental Medicine
  Research Center
Tokyo, Japan

Takashi Okamoto
Kanagawa Cancer Center
Yokohama, Japan

Ikuo Saiki
Department of Pathogenic Biochemistry
Institute of Natural Medicine, Toyama Medical
  and Pharmaceutical University
Toyama, Japan

Kiyoshi Sugiyama
Department of Clinical Pharmacokinetics
Hoshi University
Tokyo, Japan

Hiroshi Takeda
Department of Pharmacology
Toyko Medical University
Toyko, Japan

Tadahiro Takeda
Kyoritsu College of Pharmacy
Tokyo, Japan

Shigeru Tansho
Department of Microbiology and Immunology
Teikyo University School of Medicine
Tokyo, Japan

Kazuo Tarao
Kanagawa Cancer Center
Yokohama, Japan

Haruki Yamada
Kitasato Institute for Life Sciences
Kitasato University and Oriental Medicine
  Research Center
Tokyo, Japan

Hideyo Yamaguchi
Institute of Medical Mycology
Teikyo University
Tokyo, Japan

Nobuo Yamaguchi
Department of Serology
Kanazawa Medical University
Ishikawa, Japan

Toshihiko Yanagisawa
Medical Evaluation Research Laboratory
Tsumura & Co.
Ibaraki, Japan

Contents

1 Introduction: What Is Kampo Medicine?

Haruki Yamada

Chinese traditional herbal medicine was introduced into Japan from China between the 5th and 6th centuries and was established as a medical system in the 7th century. It was developed as Kampo (Japanese herbal) medicine with some modifications. When Japan opened its door to Western countries in the middle of the 19th century, the Japanese government refused to give medical doctors a license unless they had a license for Western medicine. As a result, the use of Kampo medicine declined at this stage.

Despite such an unfavorable period, Kampo medicine continued to thrive through the efforts of a few medical leaders who recognized its benefits. With the progress of modern science and technology, modern medicine has greatly improved; however, some chronic disorders of an endogenous nature surfaced during the second half of the 20th century. Furthermore, pressing medical problems such as nonspecific, constitutional, or psychosomatic diseases have also increased. Disillusion with modern medicine has on occasion been brought about by the severe adverse effects of synthetic compounds and environmental pollution. Against these social backgrounds, use of Kampo medicines in Japan emerged as an alternative. Consequently, Kampo medicine now plays an important role in medical treatment in Japan.

Kampo medicines benefit patients with a disease that affects physical functions and thus prevents mobility and ease of operation. They help those who have responded poorly to modern medical treatment because of its side effects or who have improved on clinical examination, yet remained ill. Similarly, those who are normal on clinical examination, but remain affected; those who are expected to show an improvement of the constitution or have a tendency toward psychosomatic disorders; the aged; or those who have decreased physical strength have seen benefits. Kampo medicines are often used for the treatment of:

- Hepatitis
- Menopausal disorders such as autonomic-nervous and hormonal manifestations
- Autonomic imbalances
- Bronchial asthma
- Cold syndrome
- Digestive disorders
- Atopy
- Dermatitis
- Eczema
- Hypersensitivity to low temperatures
- Allergic rhinitis
- Complaints of general malaise

These diseases are still difficult to treat using Western medicine; however, by correcting the imbalance of the whole body, Kampo medicines are relatively effective for their treatment. In Japan, now more than 80% of practicing physicians with a medical doctor's license for Western medicine have experienced the use of Kampo medicines. Therefore, Kampo medicines are a very important part of modern-day therapy in Japan.

Kampo medicines originated more than 2000 years ago in China and extensive knowledge and cumulative experience regarding their use have been acquired. These medicines have been traditionally used as a formulation consisting of several component herbs. Each formulation has its own name and has been used traditionally for a particular clinical application. One of the Kampo formulations, Juzen-taiho-to (Shi-Quan-Da-Bu-Tang in Chinese), consists of ten different component herbs, which are shown in Appendix 1.

Kampo medicines are generally administered in the form of a decoction. When patients need such a decoction for treatment, they must prepare it. In order to adapt Kampo medicines into modern-day therapy, several Kampo extract pharmaceutical preparations, which have the same efficacy as the decoction, have been developed. They are manufactured as granules by freeze-drying the decoction prepared on a large scale. Consequently, these drug forms are like Western medicines. Since 1976, the Ministry of Health and Welfare of Japan has approved several Kampo extract pharmaceutical preparations, and now 148 kinds of these Kampo formulations are allowed to be covered by Japan's national health insurance system. The medicinal plants used in the manufacture of these ethical Kampo prescriptions are also allowed to be covered by the insurance. Because these pharmaceutical preparations are very convenient to use and highly portable, their current uses are popular and extensive.

Kampo extract pharmaceutical preparations, such as the granule form, have been quality controlled by good manufacturing practice (GMP) guidelines since 1988. Because natural plant, animal, and mineral materials have been used as the raw material of ingredients in the pharmaceutical preparation, the quality control of these raw materials is also required. This control includes the selection of high-quality herbs and their quality control including, for example, limits of contamination by pesticides and heavy metals, which are evaluated by morphological and physicochemical methods. Quality control of herbs must be conducted by a qualified person who has special knowledge of and experience with herbs. Preservation of herbs is also very important for quality control. Process control and quality control for the manufacturing of the final pharmaceutical preparation are also carried out in Japan.

Western medicines and Kampo medicines have several different characteristics (Table 1.1). Western medicine is used in an analytical way but Kampo medicine is holistic. Western medicine uses synthetic drugs or purified drugs isolated from natural sources, whereas Kampo medicine uses natural herbal mixtures combined in a defined recipe. Western medicine is concerned with causes, such as bacteria, virus, and cancer, and their effect on a specific organ. Kampo medicine enhances the homeostasis and aids the recovery of the digestive, immune, endocrine, circulation, and neural systems as needed by the whole body.

Pharmacological studies based on clinical effects enable tests to be conducted on the quality control and reassessment aspects of Kampo extract pharmaceutical preparations. Some examples follow.

Iwama et al. (1987b) have investigated the effects of the Kampo medicine Sho-saiko-to (Xiao-Chai-Hu-Tang in Chinese) derived from seven herbs* on antibody responses to anti-sheep red blood cells (SRBC) by evaluating hemolytic plaque-forming cells (HPFC). Sho-saiko-to and levamisole are administered orally to ICR mice for 7 days (from 2 days before to 4 days after antigenic stimulation). When Sho-saiko-to is administered (1.2 g/kg, p.o.), a decrease in the number of SRBC-induced HPFC is observed, suggesting that Sho-saiko-to may have elicited

---

* *Sho-Saiko-To Scientific Evaluation and Clinical Applications*, edited by Yukio Ogihara and Masaki Aburada and published in 2003 in the book series, *Traditional Herbal Medicines for Modern Times*.

TABLE 1.1
Comparison of Western and Kampo Medicines

	Western Medicine	Kampo Medicine
Method	Analytical	Holistic
Diagnosis	Name of disease	Kampo diagnosis responds to patient's pathophysiological condition "Sho"
Pharmaceutical approach	Synthetic drugs or purified drugs from natural products	Herbal combination (formulation)
Principle	Chemotherapy	Homeostasis
	Correspond to causes	Recover whole body:
	Effect on organ	Digestive system
		Immune system
		Endocrine system
		Circulation system
		Neural system
		Effect on symptom in background of disease

little effect on the immune response of a normal individual mouse. However, when prednisolone (0.01 g/kg, i.p.) is administered to mice, the HPFC counts are reduced remarkably; Sho-saiko-to administrations (1.2 g/kg, p.o.) reverse the immunosuppression. These results reveal that Sho-saiko-to exhibits immunopotentiating activity when it is administered to immunocompliant mice. In a similar manner, pharmacological studies of Kampo medicine on a particular animal model of clinically related symptoms need to be conducted.

The action of a Western medicine aims specifically at the nature and functions of a disease, whereas the action of a Kampo medicine attempts to harmonize the set holistic pattern of a patient's symptoms that indicate the appropriate Kampo prescription (so-called "Sho" [Zheng in Chinese] clinically) required to recover a normal physiological environment in the system. For example, the Sho for the Kampo medicine Kakkon-to (Ge-Gen-Tang in Chinese) describes fever, stiff neck and back, little or no perspiration, aversion to wind, and a pulse diagnosis categorized as "floating" and "powerful." Although it is difficult to clarify Sho scientifically, pharmacological studies based on clinical effects are constructive for the evaluation of Kampo medicines.

If Western medicines and Kampo medicines are compared by using a globe, Western medicines affect diseases divided by longitude, whereas Kampo medicines affect several whole body systems, which are common in several diseases and are divided by latitude (Figure 1.1). Thus, the targets are different in both medicines.

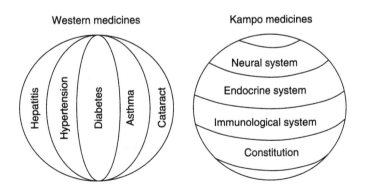

FIGURE 1.1 Concepts for Western medicines and Kampo medicines.

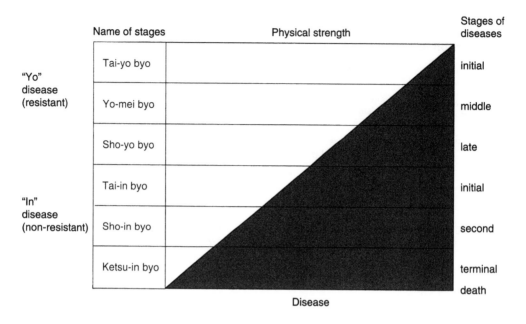

FIGURE 1.2 Physical strength and disease.

In classical literature, Shokanron (Shang Han Lun in Chinese), written in the late Han Dynasty in China for Kampo medicines, diseases are classified into six stages. Three correspond to "Yo" (Yang in Chinese) which has physical strength for resistance. The other three correspond to "In" (Ying in Chinese) which has no physical strength for resistance (Figure 1.2). In the initial stage, physical strength is still enough; however, as they proceed to middle and late stages in Yo and other In stages, diseases become serious and finally lead to death. Therefore, Kampo medicines must be used appropriately, according to the stage of the disease.

Kampo medicines (Figure 1.3) are known to regulate the disturbed harmony of "Ki" (Qi in Chinese)—the vital energy/circulation within the body. "Ketsu" (Xue in Chinese) refers to the regulation of blood–blood circulatory function and "Sui" (Shui in Chinese) refers to the regulation of body fluid/circulation of bodily fluids. Because Ki, Ketsu, and Sui are considered to correspond

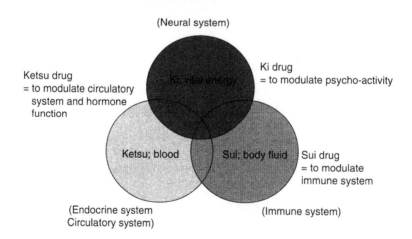

FIGURE 1.3 Harmony of Ki, Kestu, and Sui.

respectively to neural, endocrine, and immune systems in modern science, Kampo medicines and certain of their active ingredients are considered to regulate these body systems.

Efficacy of Kampo medicines cannot be explained by the pharmacological activity of any one of the active ingredients. Several active ingredients may affect the multiple systems of the whole body by several combined effects. Some of these active ingredients also work through structural modification to the actual active compounds by endogenous factors such as intestinal bacteria or gastric juice. Because Kampo medicines are generally administered orally, "inactive" compounds may be activated by endogenous factors such as gastric secretions, intestinal enzymes, and bacteria. For example, although sennoside A or B in Rhei Rhizoma (rhizome of *Rheum palmatum* Linne) is known to elicit a cathartic activity, such an effect is not produced when the sennoside is injected into mice. However, when either sennoside is administered orally, the cathartic activity is observed. In fact, rhein–anthrone, a metabolite formed by the action of intestinal bacteria, has been identified as the active component responsible for the catharsis (Hattori et al., 1982, 1988; Sasaki et al., 1979). In addition, rhein–anthrone is further metabolized to rhein, which displays antibacterial activities against intestinal anaerobic bacteria (Cyong et al., 1987). As such, the action of intestinal bacteria affects the activities of Rhei Rhizoma, and a certain feed-back mechanism may be involved in mild catharsis.

From such findings, studies on the postadministration products or byproducts of orally administered Kampo medicines are important to elucidate the mechanism(s) of action of a particular Kampo formulation. By the metabolism of rat and human livers, orally administered glycyrrhizin of Glycyrrhizae Radix produces a high concentration of its aglycone, glycyrrhetic acid, and a low level of glycyrrhizin in the circulating blood (Hattori et al., 1983). Therefore, glycyrrhetic acid is an important active principle when Kampo medicines containing Glycyrrhizae Radix are administered orally.

Shimizu et al. (1985) have demonstrated the complexity of metabolic changes involved in the study of saikosaponin a. This substance is transformed to at least eight compounds in the gut. Because each Kampo prescription contains many constituents derived from the component herbs, the action of the Kampo medicine becomes very complicated when the preparation is administered orally. When the effects of sera taken from orally prednisolone-treated mice 1 hour after administration are compared with those of the control mice, the former inhibit lipopolysaccharide (LPS)-induced mitogenic responses dose dependently. These *in vitro* findings are similar to *in vivo* effects of prednisolone in rats. Although the approach is indirect, it is useful to employ sera obtained from animals administered orally with the test samples for *in vitro* investigations. When spleen cells are treated with Sho-saiko-to without LPS, mitogenic responses of spleen cells of mice treated orally with Sho-saiko-to are not observed and the addition of sera from similarly treated mice does not affect the responses (Iwama et al., 1987). However, only the direct addition of Sho-saiko-to to the culture medium reveals a mitogenic effect (Iwama et al., 1987). Therefore, an indirect *in vitro* effect may reflect the *in vivo* action of the formulation. Because the serum samples contained active principles of orally administered Kampo medicines, which are absorbed from the gastrointestinal tract, Tashiro (1985) advocated a similar approach for the *in vitro* evaluation of Kampo medicines. He further proposed this method to be categorized as serum pharmacology and serum pharmacochemistry (Tashiro, 1985, 1991).

Because a Kampo medicine contains many active ingredients, several combination effects are involved in their efficacy. These combination effects can be classified as pharmacological effects and pharmaceutical effects. The former affect synergic, additive, and antagonistic actions, new pharmacological activity, reduction of adverse reaction, and so on. The pharmaceutical effects affect modulations of taste and pH, absorption, enhancement of extraction, interaction among constituents, and so on. The Kampo medicine Makyo-kanseki-to (Ma-Xing-Gan-Shi-Tang in Chinese), which is derived from Ephedra Herba (herb of *Ephedra sinica* Stapf), Armeniacae Semen (seed of *Prunus armeniaca* Linne), Glycyrrhizae Radix, and Gypsum Fibrosum (gypsum), has been used for the treatment of asthma and has a potent antitussive effect.

If Ephedre Herba is omitted from the formulation, the remaining herbal mixture shows no effect (Hosoya et al., 1988). If other herbs are omitted, the activity also decreases significantly. These results show that Ephedra Herba is the exclusive antiasthmatic component in this formulation; however, other component herbs also affect the activity. If each component herb was decocted separately and each of the extracts was mixed in the same ratio, the efficacy of the mixed extract shows a weaker effect than the extract of the original prescription. The content of the antitussive ingredient, ephedrine, in the original prescription was much higher than that in the mixture of the extracts. Therefore, this observation suggests that the combination of component herbs found in a Kampo medicine will affect the amount of the active ingredient in its decoction.

Some Western medicines are known to have potent adverse reactions. For example, antitumor agents generally cause anemia, decrease of leucocytes, systemic weariness, and vomiting depletion. Steroids have other side effects such as moon face (swelling), thrombus, infection, and menopausal disorder. The use of a Western medicine with a Kampo medicine, such as Juzen-taiho-to, can reduce adverse reactions to Western anticancer drugs in patients and increase their quality of life. Other Kampo medicines can reduce the adverse reactions of steroids. It is also possible to decrease the daily dose of steroid and even to stop its use. Other examples of the benefit of using the Kampo medicine, Juzen-taiho-to, with a Western medicine or as an alternative to it, will be found in this volume.

REFERENCES

Cyong, J.C., Matsumoto, T., Arakawa, K., Kiyohara, H., Yamada, H., and Otsuka, Y., Anti-*Bacteroides fragilis* substance from rhubarb. *J. Ethnopharmacol.,* 1987, 19: 279–283.

Hattori, M., Kim, G., Motoike, S., Kobashi, K., and Namba, T., Metabolism of sennosides by intestinal flora. *Chem. Pharm. Bull.,* 1982, 30: 1338–1346.

Hattori, M., Sakamoto, T., Kobashi, K., and Namba, T., Metabolism of glycyrrhizin by human intestinal flora. *Planta Medica,* 1983, 48: 38–42.

Hattori, M., Namba, T., Akao, T., and Kobashi, K., Metabolism of sennoside by human intestinal bacteria. *Pharmacology,* 1988, 36 Suppl. 1: 172–179.

Hosoya, E., Scientific evaluation of Kampo prescriptions using modern technology. In: Hosoya E. and Yamamura Y., Eds. *Recent Advances in the Pharmacology of Kampo (Japanese Herbal) Medicines.* Amsterdam: Excerpta Medica, 1988, 17–29.

Iwama, H., Amagaya, S., and Ogihara, Y., Effect of shosaikoto, a Japanese and Chinese traditional herbal mixture, on the mitogenic activity of lipopolysaccharide: a new pharmacological testing method. *J. Ethnopharmacol.,* 1987a, 21: 45–53.

Iwama, H., Amagai, S., and Ogihara, Y., Effect of Kampo-hozai (traditional Chinese medicine) on immune responses, *in vitro* studies of Sho-saiko-to and Dai-saiko-to on antibody responses to sheep red blood cells and lipopolysaccharide. *J. Med. Pharm. Soc. WAKAN-YAKU,* 1987b, 4: 8–19.

Sasaki, K., Yamauchi, K., and Kuwano, S., Metabolic activation of sennoside A in mice. *Planta Medica,* 1979, 37: 370–378.

Shimizu, K., Amagaya, S., and Ogihara, Y., Structural transformation of saikosaponins by gastric juice and intestinal flora. *J. Pharmacobio–Dynamics,* 1985, 8: 718–725.

Tashiro, S., Growth inhibition of Sai-Rei-To on cultured fibroblast cells (in Japanese). *J. Med. Pharm. Soc. WAKAN-YAKU,* 1985, 2: 108–109.

Tashiro, S., Problems in experimental studies of WAKAN-YAKU—problem of *in vitro* study (in Japanese). *J. Med. Pharm. Soc. WAKAN-YAKU,* 1991, 8: 218–221.

2 Crude Drugs of Juzen-taiho-to

Shinyu Nunome, Hiroaki Kiyohara, Katsuko Komatsu,
and Tadahiro Takeda

CONTENTS

2.1 ASTRAGALI RADIX

- *Definition.* Astragali Radix (astragalus root) is the root of *Astragalus mongholicus* Bunge or *A. membranaceus* Bunge (Leguminosae).
- *Vernacular name.* English: Astragalus root, milk vetch, milk vetch root; Chinese: Huang-Qi; Japanese: Ogi
- *Plant description.* (1) *Astragalus mongholicus* Bunge (synonym: *A. membranaceus* Bunge var. *mongholicus* (Bunge) Hsiao)
  - Perennial herb
  - Stem erect, 50 to 150 cm in height, angular, glabrous or pilose, upper part branching
  - Stipules lanceolate-triangular
  - Leaves imparipinnately compound, alternate
  - Leaflets 8 to 16 pairs, ellipsoidal, 5 to 10 mm in length, 3 to 5 mm in width
  - Flowers raceme, axillary, yellow, 10 to 25 mm, June to July
  - Ovary pubescent sparse
  - Legume black short hair
  - Seeds five to six, black, reniform
  - Root long axial, 15 to 30 mm in diameter
  - Epidermis whitish brown to brown
  - Distribution: northeastern China (Hei-long-jiang, Ji-lin, Liao-ning, He-bei, Shan-dong, Shan-xi, Nei-meng-gu, Ning-xia, Gan-su, Qing-hai, Xin-jiang, Si-chuan, Yun-nan, etc.), Mongolia, Russia

(2) *A. membranaceus* (Fischer) Bunge differs from *Astragalus mongholicus* Bunge as follows:

- Leaves 7 to 30 cm in length
- Leaflets 6 to 17 pairs, oblong to long-ovate, 7 to 30 mm in length, 4 to 10 mm in width
- Flowers, axillary, white–yellow, 10 to 20 mm
- Root 15 to 25 mm in diameter,
- Epidermis whitish brown to yellow
- Distribution: northeastern China (Hei-long-jiang, Ji-lin, He-bei, Shan-xi, Nei-meng-gu, etc.), Mongolia

- *Cultivation*. A cool and dry climate and well-drained, sandy soil with a deep and mellow layer are suitable for the cultivation of this plant. The field is cultivated to a depth of 60 cm, and weeds and stones are removed. The soil is pounded and made into ridges of 15 to 20 cm in height, 1 to 1.4 m in width. The well-selected seeds are sown on the rows or in the holes in April and May or September and October. The method of cultivation by rows is generally used; the seed is sown uniformly in groves of 3 cm depth on the ridges at intervals of 45 to 60 cm and covered with 2 cm of soil. When the soil is dry, water is sprinkled on it to assist seed sprouting.

  Using the method of the cultivation by holes, the rows are made at intervals of 60 cm and the holes dug at intervals of 30 cm in the rows. Five to ten seeds are sown in every hole and covered with 3 to 4 mm of soil. Pest and disease control: the plant mainly suffers from mildew and root rot, so careful drainage is needed. As antimicrobials, the agricultural chemicals, thiophanate (diluted 1000 times) or BO-10 biologics, are effective. Cultivation places: *A. mongholicus*: China (He-bei, Nei-meng-gu, Ji-lin, Shan-dong), *A. membranaceus*: China (Dong-bei, He-bei, Shan-xi, Si-chuan, etc.), Japan (Hokkaido).

- *Harvesting and processing*. This root is collected in spring and autumn and removed from the rootlets, the rhizome, and foreign matter. After drying in the sun, it is cut into thick slices. Storage: preserve in a ventilated dry place, protected from moisture and moths.

- *Macroscopy* (Figure 2.1). The plant has a nearly cylindrical root, 30 to 100 cm in length, 0.7 to 2 cm in diameter, with small bases of lateral root dispersed on the surface, twisted,

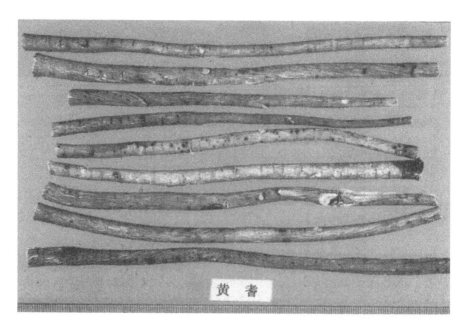

FIGURE 2.1 Astragali Radix.

nearly brown, externally light grayish yellow to light yellow–brown, and covered with irregular, dispersed longitudinal wrinkles and horizontal lenticel-like patterns. It is difficult to break; a fractured surface is fibrous. Under a magnifying glass, a transverse section reveals an outer layer composed of periderm; the cortex is light yellowish white and the xylem light yellow, with the zone near the cambium somewhat brown in color. Cortex thickness is from about one third to one half of the diameter of xylem; white medullary rays appear from xylem to cortex in thin root, but often appear as radiating cracks in thick root. The pith is usually unobservable.

- *Microscopy.* The transverse section shows cork consisting of many rows of cells. Phelloderm: 3 to 5 rows of collenchyma cells. Outer part of phloem rays is often curved and fissured; fibers are in bundles and walls are thick and lignified or slightly lignified, arranged alternately with sieve tube groups. Stone cells are sometimes visible near phelloderm. Cambium is in a ring. Xylem vessels are scattered singly or two to three are aggregated in groups; wood fibers among vessel stone cells appear singly or two to four in groups and are sometimes visible in rays. Parenchyma cells contain starch grains.

- *General test.* Purity of root of Hedysarum species and others: under a microscope, a vertical section of Astragali Radix reveals no crystal fiber containing solitary crystals of calcium oxalate outside the fiber bundle. Loss on drying is not more than 13.0% (6 h) and total ash not more than 5.0%. Acid-insoluble ash is not more than 1.0%; it has a slight odor and a sweet taste.

- *Chemistry* (Table 2.1a and Table 2.2). As major constituents, saponins, flavonoids, and polysaccharides are well known.

- *Pharmacology* (Table 2.2). The known biologically active constituents of Astragali Radix are represented by two major classes of chemical compounds: polysaccharides and saponins. Astragalus polysaccharides are already known to have anticancer and immune-enhancing properties in *in vitro* and *in vivo* experiments. Kajimura et al. (1996) reported the effects of those compounds in Astragali Radix against Japanese encephalitis virus (JEV) infection in mice. They studied the differences in the protective effect of four different Astragali Radix samples. Peritoneal exudate cells (PEC) play an important role in this resistance to JEV infection, e.g., by inducing the production of active oxygen (AO). A number of cycloartane-type triterpenoidal glycosides (saponins), named astragolosides (astragaloside I (1), II (2), III (3)), have been isolated and found to exert biological activities (e.g., anti-inflammatory, analgesic, diuretic, hypotensive, and sedative effects). Such saponins were evaluated to show immunomodulating activity by lymphocyte stimulation tests (Bedir et al., 1998). Sodium L-malate isolated from *A. membranaceus* provided significant protection against *cis*-diammine-dichloroplatinum (II) (CDDP)-induced nephrotoxity and bone marrow toxicity without reducing the antitumor activity.

- *Traditional use and application.* Astragali Radix has been used to reinforce Ki (Qi in Chinese), vital energy or functional activity, to strengthen superficial resistance, and to promote the discharge of pus and growth of new tissue in the body. Frequently, this drug is used in many prescriptions together with Ginseng Radix to reinforce the action. It has also been used for the treatment of diabetes mellitus, nephritis, leukemia, and uterine cancer. It has been used in China to improve a naturally weak constitution and to adjust unbalanced nutrition.

- *Supplementary.* In China, in addition to these two species of *Astragalus* spp., the root of *Hedysarum polybotrys* Handel–Mazzetti (Leguminosae) is used similarly as an alternative to Astragali Radix in the *Pharmacopoeia* of the People's Republic of China. As a classical quality evaluation test of Astragali Radix, whether the botanical origin is *A. mongholicus* or *A. membranaceus*, it is of good quality if the root is soft and cotton like to the touch.

TABLE 2.1a
Ingredients of Crude Drugs

2.1 Astragali Radix		2.2 Cinnamomi Cortex	2.3 Rehmanniae Radix
Astragalus Mongholicus	Astragalus Membranaceus	Cinnamomum Cassia	Rehmannia Glutinosa (nonsteamed)
2'-Hydroxy-3',4'-dimethoxyiso-flavone-7-O-β-glucoside	Acetylastragaloside I–VIII	3'-O-Methyl-(–)-epicatechin	Leonulide
3-9-O-di-methylnissolin	Astramembrannin I, II	4'-O-Methyl-(+)-epicatechin	Ajugol
3-Hydoxy-2-methylpyridine	γ-Aminobutylic acid	5,7-Dimethyl-3',4'-dimethylene-(±)-Epicatechin	Aucubin
3-O-β-xylopyranosyl-25-O-β-glucopyranosylcycloastragenol	Calyxosin	Benzaldehyde	Catalpol
5'-Hydroxyisomucronulatol-2',5'-di-O-glucoside	Coumarin	Benzyl benzoate	Rehmannioside A–D
7,2'-Dihydroxy-3',4'-dimethyliso-flavone-7-O-β-glucopyranoside	Daucosterol	Calamenene	Melittoside
7-O-Methylisomucronulatol	7,2'-dihydroxy-3',4'-dimethyliso-flavane-7-O-β-glucopyranoside	Cassioside	Rehmaglutin A–D
9,10-Dimethoxypterocarpan-3-O-β-glucoside	Formononetin	Cinnamaldehyde	Acteoside
Astragaloside I–IV	Formononetin-7-o-β-glucoside	Cinnamic acid	Isoacteoside
Calycosin-7-O-β-glucoside	3'-Hydroxyformononetin	Cinnamoside	Monomelittoside
Calyxosin	Isoastragaloside I, II	Cinnamtannin A2–4	Glutinoside
Coriolic acid	Isoliquiritigenin	Cinnamyl acetate	Ajugoside
Daucosterol	Kumatakenin	Cinncasside A,B,C1–3,D1–4,E	6-O-E-Feruloyl ajugol
Dimethyl-4-4'-dimethoxy-5.6,5',6'-dimethylene-dioxybiphenyl-2,2'-dicarboxylate	Lupeol	Cinncassiol A,B,C1–3,D1–4,E	Jioglutin D,E
Formononetin	β-Sitosterol	Cinnzeylamine	Jioglutolide
Isomucronulatol	Soyasaponin-I	Cinnzeylanol	Catalpolgenin
Isomucronulatol-7,2'-di-o-glucoside		Coumarin	Catalpogenin-α-L-arabinofranoside
Isomucronulatol-7-o-glucoside		Ethylcinnamate	Grardoside
Lariciresinol		Epicatechin	Rehmaionoside A–C
Lopeol		Lyoniresinol-3α-O-β-glucopyranoside	Jionosede A1,B1
Lupenone		Phenyl propyl acetate	Purpureaside C

Soyasaponin i
Syringaresinol

Procyanidin A2,B1,B2,B5,B7,C1
Protocatechuic acid
Salicylaldehyde
Syringaresinol
trans-cinnamaldehyde
β-Cadinene
β-Elemane

Rehmapicroside
Cistanoside
Echinacoside
Cinnanic acid
Daucosterol
Geniposide
Jioglutoside A
Mioporosidegenin
β-Sitosterol

2.2 CINNAMOMI CORTEX

- *Definition.* Cinnamomi Cortex (cinnamon bark) is the bark of the trunk of *Cinnamomum cassia* Blume (Lauraceae), or such bark from which a part of the periderm has been removed.
- *Vernacular name.* English: cinnamon bark, cassia bark, Chinese cinnamon bark, Vietnam cinnamon; Vietnamese: Que; Chinese: Rou-gui, Jun-gui, Mu-gui, Dong-xing-gui-pi, Quang-nan-gui-pi, Xi-jiang-gui-pi, Huang-yao-gui, Gui-tong, Qi-bian-gui, Ban-gui, Gui-sui, Japanese: Keihi, Toko-keihi, Kannan-keihi, Seiko-keihi, Betonamu-Keihi, Keitsu, Kihenkei
- *Plant description.Cinnamomum cassia* Blume (synonym: *C. aromaticum* Nees), English: Cassia bark, Chinese cinnamon
  - Evergreen tree, up to 10 m in height
  - Leaves alternate, petiolate, oblong or elliptical–oval, 8 to 15 cm in length by 3 to 4 cm in width, tip acuminate, base rounded, entire, three-nerved, coriaceous, lower side lightly pubescent
  - Petiole 10 mm in length, lightly pubescent
  - Inflorescence a densely hairy panicle as long as the leaves
  - Flowers yellowish white, small, in cymes of two to five
  - Perianth six-lobed, no petals; stamen six, pubescent; ovary free, 1-celled
  - Fruit a globular drupe, 8 mm in length, red
  - Distribution: southern China, Vietnam, Laos, Sumatra, etc., mostly cultivated
- *Cultivation.* The plants grow in tropical regions and prefer a warm and humid environment. They need an average temperature of over 21°C with the minimum temperature not less than –2.5°C. The mature plants need full sunlight, whereas the seedling cannot be exposed in strong sunshine. A rich and sandy or red-clay loam with a drainage system is suitable for cultivation. The plants are bred by sowing and, after 1 year, young plants are set to the field.

  Pest and disease control: brown lesion damages leaves. Charbons develop year round and harm seedlings, especially in spring and autumn. Bordeaux mixture or 1/1000 dilution of ambam is sprayed in early stages of both diseases, once every 7 to 10 days. Leaf curler bugs damage tender sprouts or leaves in spring, against which 1/1500 dilution of 50% dichlorvos emulsion is sprayed. When Capricorn beetles pester the inside back or branches, in addition to catching pests and eliminating eggs manually, it is effective to cram the hole with cotton soaked in 80% dichlorvos and seal it with mud to kill larvae.

  Cultivation places: China (two main areas: from Southwest of Guang-dong Province to Southeast of Guang-xi Zhuang Autonomous Region (Guang-nan-gui-pi or Xi-jiang-gui-pi) and in the south of Guang-xi (Dong-xing-gui-pi)], Vietnam [Quang Ninh, Yen Bai, Thanh Hoa and Quang Nam Provinces]).
- *Harvesting and processing.* In China, Cinnamomi Cortex in merchandise can be divided into twisted cinnamon (Gui-tong), channeled form of cinnamon (Qi-bian-gui), plank cinnamon (Ban-gui), and broken cinnamon (Gui-sui) according to the difference in growing year, varieties, and processing. Twisted cinnamon is made up of simple quills composed of the bark of young trunks and thick branches, 30 to 40 cm in length and 3 to 10 cm in width. After 5 to 6 years of growing, when plants are 3 to 4 m in height and the diameter is over 10 cm, Cinnamomi Cortex is collected during two seasons: February to March (spring cinnamon) and July to August (summer cinnamon).

  Summer cinnamon has a lower yield and is of better quality than spring cinnamon. The tree is cut down and decorticated. When the tree has reached 60 cm in height and is left uncut, it grows continuously and can be decorticated after 4 to 5 years. The bark for the channeled-form of cinnamon is collected in July to August after 10 years of cultivation. For processing, bark of 40 to 50 cm in length and 4.5 to 6 cm in width is rolled on both sides and stretched inside with the aid of bamboo canes during drying. "Huang-yao-gui" is collected in July to August after 30 to 40 years of growing.

In Vietnam, after at least 10 years of cultivation, collection of bark takes place in the first season from February to May and in the second from July to August or October. The bark of the early collection is better in quality than that of the later one. Bark for the channeled form of cinnamon is collected after more than 15 years (generally 30 years of cultivation) and for simple quills after 10 to 15 years. In Quang Nam Province, the channeled form of cinnamon is decorticated in 45 to 50 cm in length and 30 to 50 cm in width. After it has been inside a storehouse for 3 to 10 days, the piece of bark is cut on the top and bottom tips and channeled by a special method that takes place for 3 to 7 days. On the other hand, in North Vietnam the piece of bark is pressed, fitting in a model. Storage conditions: preserve in well-closed containers and store in a cool, dry place.

- *Macroscopy* (Figure 2.2). The plant usually consists of semitubular or tubularly rolled pieces of bark, 0.1 to 0.5 cm in thickness, 5 to 50 cm in length, and 1.5 to 5 cm in diameter. The outer surface is dark red–brown, and the inner surface red–brown and smooth; the plant is brittle. The fractured surface is slightly fibrous, red–brown, and exhibits a light brown, thin layer.
- *Microscopy.* A transverse section of Cinnamomi Cortex reveals a primary cortex and a secondary cortex divided by an almost continuous ring consisting of stone cells; nearly round bundles of fibers are present in the outer region of the ring. The wall of each stone cell is often thickened in a U-shape. The secondary cortex lacks stone cells and has a small number of sclerenchyma fibers coarsely scattered. Parenchyma are scattered with oil cells, mucilage cells, and cells containing starch grains; medullary rays are present with cells containing fine needles of calcium oxalate.

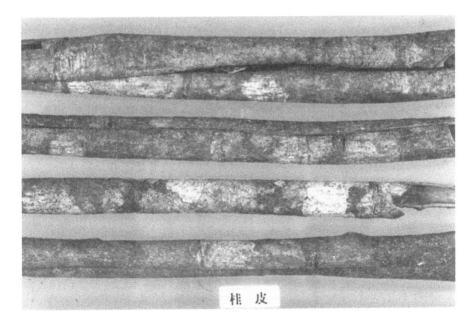

桂 皮

FIGURE 2.2 Cinnamomi Cortex from Vietnam (Vietnam cinnamon).

- *General test*. Identification: to 2.0 g of pulverized Cinnamomi Cortex, add 10 mL of ether, shake for 3 min, filter, and use the filtrate as follows. Spot 10 μL of the filtrate on a plate of silica gel with a fluorescent indicator for thin-layer chromatography. Develop the plate with a mixture of hexane, ethyl acetate, and chloroform (4 : 1 : 1) to a distance of about 10 cm, and air-dry the plate. Examine under ultraviolet light (main wavelength: 254 nm); a purple spot develops at the Rf value of about 0.4. Spray 2,4-dinitrophenylhydrazine TS upon the spot; a yellow–orange color then develops. Loss on drying is not more than 15.5% (6 h) and total ash not more than 5.0%. For essential oil content, perform the test with 50.0 g of pulverized Cinnamomi Cortex and 1 mL of silicon resin in the flask for determination of essential oil content; the volume of essential oil is not less than 0.5 mL. Odor is a characteristic aroma and the taste is sweet and pungent at first, but later rather mucilaginous and slightly astringent.
- *Chemistry* (Table 2.1a and Table 2.2). Cassia oil contains cinnamic aldehyde (4), 80 to 95%; the remainder consists of terpenes, such as limonene, *p*-cymene, (-)-linalool, β-caryophyllene, and other compounds such as eugenol.
- *Pharmacology* (Table 2.2). Cinnamomi Cortex is also known as a spice and a medicine for colds, flu, and digestive problems.
- *Traditional use and application in traditional Chinese medicine*. Cinnamomi Cortex has been used for warming up and invigorating spleen Yang (in Chinese) and kidney Yang, expelling cold to relieve pain, and promoting the circulation of vital energy and blood. Cinnamon oil is used as a flavoring agent. It is also a carminative and pungent aromatic and it has antiseptic properties.
- *Supplementary*. Cinnamon may be from the Arabic *kinnamon*; the cassia is from the Greek *kassia*, meaning to strip off the bark. Cinnamon is named as a spice in the biblical books of Moses, by the ancient Greek and Latin historians, and in Chinese herbals as early as 2700 B.C. (Newll et al., 1996). Its cultivation in Sri Lanka probably dates from 1200 A.D. Practically all commercial cinnamon is now obtained from cultivated trees in Sri Lanka, southwestern China, Vietnam, Laos, Indonesia, the West Indies, the Seychelles, Madagascar, and many other localities. Cinnamomi Cortex produced from southern China is often called "cassia bark." Although the cortices of *C. iners* Reinwardt ex Blume, *C. obutussifolium* Nees, *C. burmannii* Blume, etc. produced from Southern Asia are sold as a kind of Cinnamomi Cortex in markets, these qualities are said to be inferior to that of *C. cassia* for taste and odor.

  Varieties and grading: In China, twisted cinnamon (Gui-tong) is a main export variety, generally called "Guang-nan-guipi (Xi-jiang-gui-pi)" or "Dong-xing-gui-pi" in the market. The former bark is obtained from younger trees and is of lower quality than the latter one. "Huang-yao-gui" is of good quality and used for medicinal purpose under the name "Rou-gui." It is now produced from cultivated trees growing in southern Guang-xi.

  In Vietnam, the Cinnamomi Cortex is of better quality than that of China. The best one is from Quang Nam Province, from a Tra My variety, called "MN" in Japan. The next best grade is that of Yen Bai Province, called "YB," and then that of Quang Ninh Province, called "QN" in Japan. MN, YB, and QN have various subgrades such as special, first to fourth, and broken, depending on the weight, thickness of bark, and thickness of the oil layer (Stuart et al., 1987).

  In the last 20 years, the global export from Sri Lanka has been markedly reduced. The U.S. Food, Drug and Cosmetic Act permits the term "cinnamon" to be used for Ceylon (Sri Lanka) cinnamon (*C. verum* J.S. Presl), Chinese cassia, and Indonesian casssia (*C. burmannii* C. G. Th. Nees) (Ravindran et al., 2004).

2.3 REHMANNIAE RADIX

- *Definition.* Rehmanniae Radix (rehmannia root) is the root of *Rehmannia glutinosa* Liboschitz or *Rehmannia glutinosa* Liboschitz var. *purpurea* Makino (Scrophulariaceae), with or without the application of steaming.
- *Vernacular name.* English: Rehmannia root, Chinese foxglove root, dried Rehmannia root, prepared Rehmannia root; Chinese: Di-huang, Xian-di-huang, Sheng-di-huang, Sheng-di, Shu-di-huang, Shu-di, Ti-huang; Japanese: Jio, Kan-Jio, Juku-Jio
  Plant description. Rehmannia glutinosa Liboschitz or *R. glutinosa* Liboschitz var. *purpurea* Makino (synonyms (*R. glutinosa*): *Digitalis glutinosa* Gaertner, *R. chinensis* Liboschitz ex Fischer and C.A. Meyer, *R. glutinosa* Liboschitz var. *hemsleyana* Diels, *R. glutinosa* var. *huechingensis* Chao and Shih, *R. glutinosa* f. *huechingensis* (Chao and Shih) P.G. Hsiao, *R. glutinosa* f. *purpurea* Matsuda)
- Perennial herb, 15 to 50 cm in height, densely villous with glandular and eglandular hairs
- Root fleshy
- Stem erect, purplish, branching at the base
- Basal leaves usually rosulate, superior leaves alternate, oval, or spatulate, tapering to a short petiole, coarsely dentate, pubescent, the underside often reddish, 7 to 26 cm in length and 3 to 9 cm in width
- Axillary flowers solitary or in cymes
- Terminal flowers in few-blossomed cymes, violet–orange tinted with purple
- Calyx tube inflated in five segments, bent, oval; corolla 3 to 4.5 cm in length, white villous, tube compressed, limb broad, bilabiate, the superior lip bilobate, the inferior trilobate
- Fruit a globular capsule
- Appearances of two species similar to each other, but root of *R. glutinosa* is thicker than that of *R. glutinosa* var. *purpurea*
- Distribution: northern and northeastern China
- *Cultivation.* In China, rehmannia root used in ancient time is assumed to have been relatively slender. The root of cultivated *R. glutinosa* is coarse and thicker at present (Xie et al., 1994). The plant selection might be undertaken through long-term cultivation. The plant line cultivated in Huaiqing, Henan Province, is said to be of high yield and good quality. This line had been marked off from other *R. glutinosa* as *R. glutinosa* Liboschitz var. *huechingensis* Chao and Shih. However, now this variety is included in *R. glutinosa* (Wu et al., 1998).
  A temperate climate with sufficient sunlight is much preferred by this plant. The root can grow rapidly in temperatures of 25 to 28°C. It has no special requirement for soil, but does better in fertile, sandy loam. It is resistant to cold and drought, but water retention damages the crops. Repeated cultivation in the same field is not used. The plants are bred by vegetative propagation or sowing. Fresh and healthy roots of 0.8 to 1.0 cm in diameter are cut into 5- to 6-cm pieces and set to field in April or from late May to late June. Sowing from late March to early April and plantation are undertaken.
  Pest and disease control: the main infestations are spot-blight and dapples. Spot-blight in its initial stage is yellowish–green spots, then expands and the leaves turn into yellowish–brown or join together and are withered and twisted. A 1/500 dilution of 65% zineb is sprayed on the plants two to three times in early days, once in 7 days. Dapples damage the top of leaves and present significant concentric dapples, later appearing as cracks and perforation of the leaves. The prevention is the same as in spot-blight. The main pest is nymphalid, which eats leaves. For control, 1/800 dilution of trichlorfon is sprayed in its larva stage. Cultivation place: Henan, Zhejiang, Hebei, Liaoning and Shandong.

Three plant lines are present in Japan: *R. glutinosa* var. *purpurea*; *R. glutinosa*; and "Fukuchiyama 1." Fukuchiyama 1 was developed by clonal selection after making the hybrid groups between two taxa and selecting the superior strain as an indication of root shape. *R. glutinosa* var. *purpurea* has a low yield of its slender root, but better efficacy, whereas *R. glutinosa* has a high yield of its thick fusiform-shaped root. Fukuchiyama 1 has fusiform root thickened around the base and an increased number of thick roots; it reproduces easily through vegetative propagation and is resistant to a disease caused by species of *Pseudomonas*. Among these three plants, *R. glutinosa* var. *purpurea* is cultivated for commercial scale.

A temperate climate and fertile soil with a drainage system are suitable for these plant lines. The plants are bred by vegetative propagation. The stored string-like roots are cut into 5- to 10-cm pieces and set to field from late April to early May. Drainage on the field is very important because water retention causes root putrefaction. The seed roots are strictly chosen after harvesting and separating the thickened part of the root and are stored in sawdust until the following April. Cultivation: Hokkaido, Nara and Ohita.

- *Harvesting and processing.* In China, this drug is collected in October. After the fibrous roots are removed, the tuberous root is washed with water, dried in sunlight or by a stove, and then turned over frequently to dry uniformly until it becomes black and is soft inside but rigid outside. Radix Rehmanniae Preparata (Shu-di-huang in Chinese, Juku-jio in Japanese) is stewed with millet wine until it has absorbed the wine entirely, or it is steamed until it becomes blackish and shiny and then dried in the sun.

    In Japan, this drug is collected in November and December. When the aerial parts begin to wither, the plants are dug up and the aerial parts removed. After washing, the tuberous root is dried in the shade until it becomes black inside (Kan-jio). The hot-air drying makes the process faster. After half-drying, the root is rubbed down by hand. For storage, it is preserved in a ventilated dry place and protected from mold and moths.

- *Macroscopy* (Figure 2.3). The plant is usually thin and long with a fusiform root, 5 to 10 cm in length and 0.5 to 1.5 cm in diameter, and often broken or markedly deformed

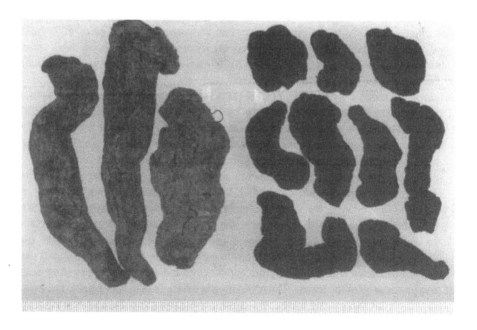

FIGURE 2.3 Dried Rehmanniae Radix and prepared Rehmanniae Radix.

in shape. Externally, it is yellow–brown to blackish brown, with deep, longitudinal wrinkles and constrictions; it is soft in texture and mucilaginous. A cross section is yellow–brown to blackish brown, and the cortex is a darker color than the xylem. The pith is barely observable.

- *Microscopy.* A transverse section reveals 7 to 15 layers of cork; the cortex is composed entirely of parenchyma cells. The outer region of the cortex has scattered cells containing brown secretes; the xylem is practically filled with parenchyma. The vessels are radially lined and mainly reticulate.
- *General test.* Total ash is not more than 6.0% and acid-insoluble ash not more than 2.5%. It has a characteristic odor and tastes slightly sweet at first, followed by a slight bitterness.
- *Chemistry* (Table 2.1a, Table 2.1b, and Table 2.2). Many constituents, such as iridoids, iridoid glucosides, ionone glucosides, a monoterpene glucoside, acteoside, and a cerebroside, have been isolated (Kitagawa et al., 1986).
- *Pharmacology* (Table 2.2). Immunosuppressive effects were assessed by using a hemolytic plaque-forming cell (HPFC) assay. Phenylethanoid glycosides (acteoside (5), jionoside A_1(6), B_1(7)) suppressed the induction of HPFC in mice. Immunosuppressive action of phenylethanoid glycosides (PAS) may be mainly due to the phenetyl alcohol moiety in the molecule. It is well known that PASs have some biological activities, such as antibacterial and inhibitory activities against 5-lipoxygenase (Sasaki et al., 1989).
- *Traditional use and application.* Rehmanniae Radix is one of the most effective blood and kidney tonics in traditional Chinese medicine and is said to prolong life, "quiet the soul, and confirm the spirit." Rehmanniae Radix is used as a tonic, an antianemic, and an antipyretic.
- *Supplementary.* In China, in addition to the dried root of *R. glutinosa*, the fresh root (fresh Radix Rehmanniae) is used as impairment of Yin in febrile disease marked by deep red tongue and thirst, skin eruption, spitting of blood, sore throat, etc.

2.4 PAEONIAE RADIX

In Japan, three types of Paeoniae Radix, called Shin-shakuyaku, Kiboshi-shakuyaku, and Kawatsuki-syakuyaku, are available and all are used in the same application as white peony root. The following descriptions are mainly for white peony; those of red peony are added in parenthesis.

- *Definition.* Paeoniae Radix (peony root) is the root of *Paeonia lactiflora* Pallas (Paeoniaceae). (Red peony, *P. veitchii* Lynch, is also used.)
- *Vernacular name.* English: peony root, white peony root (red peony root); Chinese: Shaoyao, Baishao (Chishao); Japanese: Shakuyaku, Shin-shakuyaku, Kiboshi-shakuyaku, Awatsuki-shakuyaku
- *Plant description. Paeonia lactiflora* Pallas (synonyms: *P. lactiflora* Pallas var. *trichocarpa* Stern; *P. albiflora* Pallas; *P. edulis* Salisbury; *P. officinalis* Thunberg); English: peony, paeony, white peony, white-flowered peony, red peony
  - Perennial herb, 50 to 80 cm in height, with a stout root
  - Stem fascicular, erect, and branched in upper part
  - Leaves alternate, biternately compound, the ultimate segments red veins
  - Leaflets narrowñovate, lanceolate, or ellipcal, 8 to 12 cm in length and 2 to 4 cm in width
  - Flowers terminal, solitary, large, white, red or purple color
  - Petals five to ten, large, wide; stamens numerous
  - Fruit coriaceous, three to five of follicles, several seeds included
  - Distribution: northeastern and northwestern China, Mongolia, Siberia and North Korea
- *Cultivation.* In Japan, there are three lines in original sources: peony of Japanese lineage that was cultivated from the Edo era in Japan; peony of Occidental lineage

TABLE 2.1b
Ingredients of Crude Drugs

2.3 Rehmanniae Radix	2.4 Paeoniae Radix	2.5 Cnidii Rhizoma	2.5 Szechwan Lovage Rhizoma
Rehmannia Glutinosa (steamed)	Paeonia Lactiflora	Cnidium Officinale	Ligusticum Chuanxiong
Acteoside	Albiflorin	Cnidilide	Tetramethylpyrazine
Leonulide	Benzoic acid	Neocnidilide	Perlolyrine
Aucubin	Benzoylpaeoniflorin	Ligustilide	Ligustilide
Catalpol	Catechin	Butyliphthalide	Wallichilide
Rehmannioside A–D	Daucosterol	Butylidenephthalide	3-Butylidenephthalide
Melittoside	Gallic acid	Senkyunolide	Butylphthalide
Rehmaglutin A,D	Galloylpaeoniflorin	Senkyunolide B–J	3-Butylidene-4-hydoxyphthalide
Glutinoside	Hederagenin	Pregnenolone	Neocnidilide
Jioglutin A–C	1,2,3,4,5,6-Hexa-O-galloyl-β-glucose	Vanillin	Pregnenolone
Jioglutolide	Lactoflorin	Coniferyl ferulate	Sedanolide
Jiofran	Oxybaeoniflorin	Ferulic acid	Senkyunolide
Rehmapicrogenin	Paeoniflorigenone	Ligustilidiol	Senkyunolide B–Q
Trihydroxy-β-ionone	Paeoniflorin	Sedanoic acid anhydride	Senkyunone
Dihydroxy-β-ionone	Paeonilactone A–C	Sedanolide	2-Methoxy-4-(3-methoxy-1-propenyl)phenol
Aeginetic acid	Paeonol	Tetramethylpyrazine	
5-c-Hydoxyaeginetic acid	1,2,3,4,6-Penta-O-galloyl-β-glucose		2-(1-Oxopentyl)-benzoic acid
5-Hydroxymethylfurfural	z-1s,5R-β-pinen-10-yl-β-vicianoside		methyl ester
5-Hydroxymethylfuroic acid	β-Sitosterol		4-Hydroxybenzoic acid
Succinic acid	1,2,3,6-*tetra*-O-galloyl-β-glucose		Vanillic acid
			Vanillin
			Tetramethylpyrazine
			Caffeic acid
			Protocatechuic acid
			Ferulic acid
			Chrysophanic acid
			Sedanoic acid anhydride
			L-Isoleucyl-L-vaniline anhydride
			L-Valyl-L-valine anhydride

created from Siberian and Chinese peony in the West; and peony of Dutch lineage that made a horticultural improvement from European peony. Nowadays, the hybrids are increasing among these lines. Of these hybrids, peony of Japanese lineage is cultivated in general for traditional medicinal use and because of chemical differences between Japanese and Occidental lines. Peony of Japanese lineage cultivated in Nara Prefecture, called "Yamato-shakuyaku" in Japanese, is said to be the best in quality.

A temperate and humid climate with full sunshine is suitable for peony. It can resist cold and endure drought. For cultivation, a drained area with sandy or clay loam, without a sunken place filled with water, is suitable. Repeated cultivation is not productive. The plants are bred by sowing or separating the rhizomes, each with two to three sprouts, after removing all roots. The cultivation takes 4 to 5 years. Bud picking or flower picking is performed to enlarge peony roots.

Pest and disease control: in Japan, gray mold damages this plant badly. The plants show symptom of flower rot and a brown fleck occurs in leaves, on which gray mold is developed and the infected parts die. Susceptibility depends on the plant line. Leaf spot, rust, and powdery mildew are also found in this plant. The invasion of the caterpillar of the swift moth causes the stem to wilt and parasitism of the root-knot nematode forms a knob in the root, both causing a poor crop. For effective control of diseases, infected stems and leaves are separated and destroyed by fire. Before they are planted, the rhizomes are disinfected with 1/100 dilution benomyl for 1 h.

In China, rust is the main infestation. Yellow or yellowish-brown pellets are found in the lower surface of leaves of infected plants, and a dark brown, splinter-like hair will grow up from spores in a later stage. For prevention, a 1/500 dilution of carboxin or 0.3 to 0.4 dilution of lime sulfur is sprayed in the early stage. Cultivation place: Japan (mainly in Hokkaido, Niigata, Nagano and Nara); China (white peony: mainly in Zhe-jiang, An-hui, and Si-chuan (Xie et al., 1994); red peony: mainly collected in Inner Mongolia and Si-chuan).

- *Harvesting and processing*. In Japan, this drug is collected from September to November after 4 to 5 years of cultivation. The underground parts are dug up by hand using a mattock or spade, or by a machine such as a tractor with digger, and divided into rhizomes for the next cultivation and roots for medicinal uses. The roots are stored without drying till processing. The root is washed, removed from the lower part and rootlet, and then dried (Kawatsuki-shakuyaku) or, after peeling the outer skin, drying with wind and hot air takes place (Kiboshi-shakuyaku), or peeled after boiling in water for 5 to 10 min or boiled after peeling and drying with warm air (Shin-shakuyaku). For storage, the roots should be preserved in a dry place and protected from moths.

  In China, the plant is collected from August to October after 4 to 5 years of seed cultivation or 3 to 4 years of cultivation of peony sprouts. Red peony root is collected in spring and autumn. After rhizomes and rootlets are removed, the root is peeled by scraping after or before slightly boiling and softening in hot water, then drying in the sun. In the case of red peony root, the root is directly dried in the sun after rhizomes and rootlets are removed.

- *Macroscopy* (Figure 2.4). The plant has a cylindrical root, 7 to 20 cm in length and 1 to 2.5 cm in diameter; externally, it is brown to light grayish brown, with distinct longitudinal wrinkles, warty scars of lateral roots, and laterally elongated lenticels. The fractured surface is dense in texture and light grayish brown, with light brown radiating lines in the xylem.

- *Microscopy*. A transverse section reveals 4 to 10 layers of cork, with several layers of collenchyma inside the layer; the cortex exhibits many oil canals surrounded by secretary cells; often large hollows appear. The boundary of phloem and xylem is distinct. In the xylem, numerous vessels radiate alternately with medullary rays; vessels in the outer part of the xylem are single or in several groups and disposed rather densely in a cuneiform pattern. However, vessels in the region of the center are scattered very sparsely. Starch grains are simple grains, not more than 20 μm in diameter, and rarely 2- to 5-compound grains, up to 25 μm in diameter; starch grains often gelatinized.

- *General test*. Identification: (1) Shake 0.5 g of pulverized Paeoniae Radix with 30 mL of ethanol for 15 min and filter. Shake 3 mL of the filtrate with 1 drop of ferric chloride TS: a blue–purple to blue–green color is produced, and it changes to dark blue–purple to dark green.

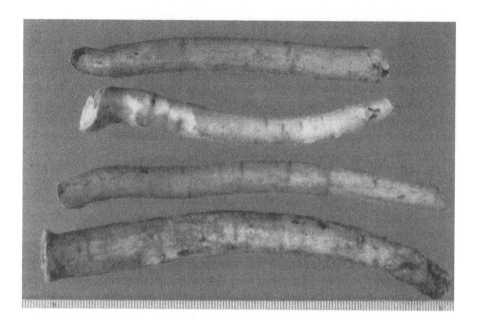

FIGURE 2.4 Paeoniae Radix produced in China.

(2) To 2 g of pulverized Paeoniae Radix add 10 mL of methanol, warm on a water bath for 5 min, cool, filter, and use the filtrate as the sample solution. Separately, dissolve 1 mg of paeoniflorin for thin-layer chromatography in 1 mL of methanol, and use this solution as the standard solution. Spot 10 μL each of the sample solution and the standard solution on a plate of silica gel for thin-layer chromatography. Develop the plate with a mixture of acetone, ethyl acetate and glacial acetic acid (10 : 10 : 1) to a distance of about 10 cm and air-dry the plate. Spray *p*-anisaldehyde-sulfuric acid TS evenly upon the plate and heat at 105°C for 5 min: one spot among the spots from the sample solution and the purple–red spot from the standard solution show the same color tone and the same Rf value.

Component determination: peony root contains 2 to 6% of paeoniflorin. Weigh about 0.5 g of pulverized Paeoniae Radix accurately, add 50 mL of diluted methanol (1 in 2), heat under a reflux condenser on a water bath for 30 min, cool, and filter. To the residue add 50 mL of diluted methanol (1 in 2) and proceed in the same manner. Combine the filtrates, add diluted methanol (1 in 2) to make exactly 100 mL, and use this solution as the sample solution. Separately, accurately weigh about 0.01 g of paeoniflorin for component determination, after previously drying in a desiccator (in vacuum, phosphorus pentoxide, 80°C) for 8 h; dissolve in diluted methanol (1 in 2) to make exactly 100 mL and use this solution as the standard solution. Pipet 2 μL each of the sample solution and the standard solution and perform the test as directed under the liquid chromatography according to the following conditions. Determine the peak areas, AT and AS, of paeoniflorin in each solution:

> Amount (mg) of paeoniflorin
> = amount (mg) of paeoniflorin for component determination × AT/AS

High-pressure liquid chromatography operating conditions: the detector is an ultraviolet absorption photometer (wavelength: 230 nm). The column is a stainless steel column of 4 to 6 mm inside diameter and 15 to 25 cm in length, packed with octadecylsilianized silica gel (5 to 10 μm in diameter). Column temperature is a constant temperature of about 20°C. In the mobile phase, mix water and acetonitrile (4 : 1). Adjust the flow rate so that the retention time of paeoniflorin is about 10 min. To select the column, dissolve 1 mg each of paeoniflorin for component determination and *p*-hydroxyacetophenone in diluted methanol (1 in 2) to make 10 mL.

Perform the test with 20 mL of this solution under the preceding operation conditions. Use a column giving elution of paeoniflorin and *p*-hydroxyacetophenone in this order with the resolution between their peaks of not less than 3. Reproducibility: repeat the test five times with the standard solution under the preceding operation conditions; the relative standard deviation of the peak area of paeoniflorin is not more than 1.5%.

Loss on drying is not more than 14% (6 h) and total ash not more than 6.5%. Acid-insoluble ash is not more than 0.5%. The odor is characteristic and the taste is slightly sweet at first, followed by astringency and a slight bitterness.

- *Chemistry* (Table 2.1b and Table 2.2). The plant contains no less than 2.0% of paeoniflorin, calculated on the dried basis. Seven triterpenoids were isolated from the roots of this plant: oleanolic acid, hederagenin, and betulinic acid, and, as a new triterpene, 11α, 12α-epoxy- 3β, 23-dihydroxy- 30-norolean-20 (29)-en- 28, 13β-olide was isolated. The drug is derived from the dried roots of a plant known to contain gallic acid, oxypaeonifloron, albiflorin, paeoniflorin, benzoic acid, pentagalloyl glucose, paeonol, and benzoylalbiflorin as its bioactive constituents.

- *Pharmacology* (Table 2.2). The pharmacological effects of this plant, whose crude extracts are used therapeutically in Asia and Europe, are reported to include anti-inflammatory effects, immunomodulating effects, and the restoration of memory deficit. Paeoniflorin and paeony root extract have been mentioned to attenuate the scopolamine-induced deficit in radial maze performance in rats (Ohta et al., 1993). Mediation of glucose is suggested as the possible mechanism of paeoniflorin to meet the hypothesis that glucose may improve memory. An i.v. injection of paeoniflorin lowered the blood sugar markedly in streptozotocin (STZ)-induced diabetic rats. Plasma glucose levels decreased gradually with the prolongation of treatment time in STZ-diabetic rats that received i.v. injections of paeoniflorin. Similar results were also observed in rats that received an i.v. injection of 8-debenzoylpaeoniflorin (Hsu et al., 1997).

  Paeoniflorin (8) has the ability to lower blood sugar like 8-debenzoylpaeniflorin (9) in the rats. Gallotannin, pentagalloylglucose, hexagalloylglucose, heptagalloylglucose, and octagalloylglucose have been extracted from paeony root and have exhibited an endothelium-dependent vasodilator effect on isolated rat aorta (Goto et al., 1996). Lactinolide (10), 6-O-β-D-glucopyranosyl-lactinolide (11), 1-O-β-D-glucopyranosyl-paeonisuffrone, oxybenzoyl-paeoniflorin, and paeonilactinone inhibited twitch contraction of guinea pig ileum induced by electric field stimulation. On the other hand, 1-O-β-D-glucopyranosyl-paeonisuffrone inhibited histamin release from rat peritoneal exudate cells induced by the antigen–antibody reaction. This plant has long been known to improve blood flow (Terasawa, 1990).

- *Traditional use and application.* White peony root is used as a remedy for nourishing the blood, regulating menstruation, and reinforcing "Yin." Red peony root is used for removing pathogenic heat from the blood, dissipating blood, and relieving pain.

- *Supplementary.* In China, peony produced in Zhe-jiang (called "Hang-bai-shao" in Chinese) is said to be the best in quality.

2.5 CNIDII RHIZOMA

- *Definition.* Cnidii Rhizoma (cnidium rhizome) is the rhizome of *Cnidium officinale* Makino (Umbelliferae), usually passed through hot water.
- *Vernacular name.* English: Cnidium rhizome; Chinese: Chuan-xiong; Japanese: Senkyu

- *Plant description. Cnidium officinale* Makino
  - Perennial herb, 30 to 60 cm in length, with a massive rhizome
  - Stem erect and branched in upper
  - Leaves pale green, alternate, bipinnate, or tripinnate
  - Leaflet, oval–lanceolate, pinnatifid, or pinnatipartite, with dentate lobes
  - Radical leaves with long petiole; petiole sheathed
  - Compound umbels terminal, multiflorus
  - Flower small, white; petals five, incurvated
  - No fructification
  - Distribution: all cultivated in northern part of Japan
- *Cultivation.* This is a typical temperate plant with strong resistance to cold and weak tolerance of heat. A fertile soil with a drainage system is suitable for its cultivation. The plants are bred by vegetative propagation. Generally, in autumn, the seed rhizomes are selected from harvested rhizomes and planted by hand or with a machine such as a potato planter. Pest and disease control: diseases such as downy mildew, black root rot, leaf blight, and black rot, as well as pests such as the two-spotted spider mite, seedcorn maggot, false melon beetle, and larva of common yellow swallowtail, are known in this plant. Downy mildew occurs in the leaf blade from June to September; control is carried out in high growing stage. To treat against black root rot, in which seed rhizomes are putrefied by infection with *Phoma* spp., use of diluted benomyl on seed rhizomes is effective. Cultivation place: Japan (mainly in Hokkaido Prefecture, as well as Iwate, Gunma and Niigata).
- *Harvesting and processing.* From September to November after 1 year of cultivation, whole plants are dug up by hand using a mattock or by a machine such as a potato digger. After residual soil is removed, the underground part is cut longitudinally into two to four pieces and washed. The underground parts are kept in water for 15 to 20 min at 60 to 80°C and the bunches of plants are dried outside while hanging on a rack. After drying, the aerial parts are removed and the rhizomes are polished. On a large scale, the aerial parts are cut off with a forage chopper and then their underground parts are dug up by a potato digger. In the processing factory, these parts are washed, cut into several pieces, and then washed again by brushing on the conveyor system. Washed rhizomes are dried with warm or hot air after boiling with water or heating with steam. For storage, they should be preserved in a ventilated and dry place and protected from moths.
- *Macroscopy* (Figure 2.5). Cnidii Rhizoma is an irregular massive rhizome, occasionally cut lengthwise; 5 to 10 cm in length and 3 to 5 cm in diameter. Externally, it is grayish brown to dark brown, with gathered nodes and knobbed protrusions on the node; the margin of the vertical section is irregularly branched. Internally, it is grayish white to grayish brown, translucent, and occasionally has hollows that are dense and hard in texture.
- *Microscopy.* A transverse section reveals cortex and pith with scattered oil canals; in the xylem, thick-walled and lignified xylem fibers appear in groups of various sizes. Starch grains are usually gelatinized, but on rare occasions remain as grains of 5 to 25 μm in diameter. Crystals of calcium oxalate are not observable.
- *General test.* Total ash is not more than 6.0% and acid-insoluble ash is not more than 1.0%. It has a characteristic odor and a slightly bitter taste.
- *Chemistry* (Table 2.1b and Table 2.2). The rhizome contains a variety of volatile nonpolar alkylphthalide derivatives, ligustilide, senkyunolide A-J, butylidenephthalide, butylphthalide, cnidilide, neocnidilide, which have been shown to have antifungal and smooth-muscle relaxing activities (Kobayashi et al., 1984). Volatile alkylphthalide derivatives are the common and unique constituents of the two Umbelliferae crude drugs "Senkyu,"

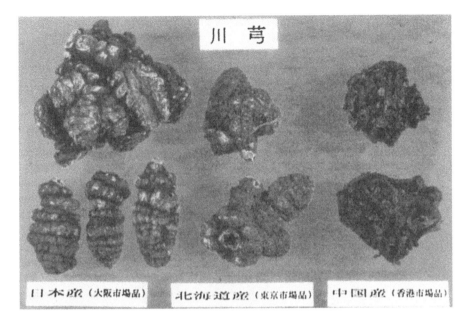

FIGURE 2.5 Cnidii Rhizoma and Szechwan Lovage rhizome.

the dried rhizomes of *Cnidium officinale*, and "Toki," the dried roots of *Angelica acutiloba*, although the content in the latter is less than one tenth that of the former. The predominant compound is ligustilide in both plants.

- *Pharmacology* (Table 2.2). The polar and nonvolatile components have analgesic activity. Pregnenolone, coniferylferulate, tetramethylpyrazine, falcarinol, farcarinolone, and falcarindiol are reported. It is known that orally administered pregnenolone is mostly excreted through the enterohepatic circulation in a short period (Kobayashi et al., 1987). Tetramethylpyrazine and ferulic acid showed an inhibitory effect on uterine movement when given perorally and intravenously. The combination of both compounds, at doses individually insufficient to inhibit, synergistically inhibited uterine contraction (Ozaki and Ma, 1990).
- *Traditional use and application.* Since ancient times, this has been an important herb, especially as a female tonic used to invigorate blood circulation for the treatment of irregular menstruation, menopause syndrome, and coronary heart diseases, and to relieve headache, body pain, and arthritis.
- *Supplementary. Rhizoma Chuanxiong* is derived from the rhizome of *Ligusticum chuanxiong* Hortorum (Umbelliferae). It is official in the Chinese *Pharmacopoeia* but not in the Japanese one.
  - *Vernacular name*: Rhizoma Ligustici Chuanxiong, Ligustici Chuanxiong Rhizoma. English: Szechwan Lovage rhizome, Ligusticum Chuanxiong rhizome; Chinese: Chuan-xiong, Xiong-qiong.
  - *Plant description. Ligusticum chuanxiong* Hortorum (synonyms: *L. wallichi* auct. sin. non Franch)
    - Flavorful, perennial herb, 40 to 70 cm in height, with an irregularly tuberous rhizome
    - Stem erect, lightly striate and branched in upper part; nodes in the middle and base swelling like a disk and those in the base with aerial roots
    - Leaves alternate, bipinnate, or tripinnate

- - Leaflet three to five pairs, oval–deltoid, pinnatisect, or pinnatipartite with narrow lobes, glabrous except on the vein
  - Petiole sheathed
  - Compound umbels terminal, multiflorus; peduncle and bract hairy
  - Flower small, white
  - Calyx teeth almost wanting
  - Petals five, elliptical, incurvated
  - Sterile cremocarp oboid, edged with narrow wings
  - Distribution: western China, all by cultivation
- *Cultivation.* The plants prefer to grow in temperate and humid environments with sufficient rain. The soil should be deep, loose, and fertile, with the good drainage. Cultivation in the same soil for continuous years is not used. Seed stems are obtained from the mountain area in Si-chuan Province. The immature rhizome is first cultivated in the plains area, and then transplanted to a mountain area in early February. These plants are dug up in July and left inside the room or in a dark, cool cave for a month, and then the stem is cut into 3- to 4-cm pieces with one swelling node. These cut stems are used for cultivation in the plains (Xie et al., 1994).

  Pests and disease control: infected rhizomes putrefy from inside and have a yellowish brown paste-like form with a special smell, leading to the aerial parts becoming yellow and withered. For prevention, the rhizomes and stems for cultivation are strictly chosen and infected rhizomes are removed and burned. The main pest is the stem moth, whose larva first causes damage on the top of stem and then gets inside and kills the plant. A 1/1500 dilution of 50% phosphorus emulsion is sprayed before or shortly after the larva gets inside the stem. Cultivation: Mainly in Si-chuan, also Hu-bei, Shan-xi, Gui-zhou, and Yun-nan.
- *Harvesting and processing.* After 2 years of cultivation, in late May on the plains or from August to September in the mountain area, the plants are harvested. After gathering rhizomes, stems, leaves, and residual soil are removed. The rhizomes are washed, dried in sunlight or by use of a stove, and then put into a bomboo cage, which is shaken to get rid of fibrous roots. For storage plants should be preserved in a ventilated and dry place and protected from moths.
- *Macroscopy.* Plants are in irregular knotty and fist-like masses, 2 to 7 cm in diameter. Externally, they are yellowish brown, rough, and shrunken, with many patrolled and raised annulations showing dented and subrounded stem scars on the crown and numerous tuberculous fibrous root scars beneath the crown and at the annulations.
- *Microscopy.* A transverse section reveals cork of over ten rows of cells, a narrow cortex, scattered with root-trace vascular bundles, and distinct interfascicular cambium. The vessels in the xylem are polygonal or subrounded and mostly uniseiate; V-shaped, wood fibers in bundles are occasionally found. Yellowish brown oil cavities are scattered throughout the parenchyma, the cells of which contain starch grain.

2.6 ATRACTYLODIS LANCEAE RHIZOMA AND ATRACTYLODIS RHIZOMA

Autractylodis Lanceae Rhizoma and Atractylodis Rhizoma were originally described without distinction as the same crude drug, "Jutsu," in the Chinese ancient literature. Now, in the pharmacopoeias of Japan and China, they are treated as separate crude drugs; "Sojutsu" is Atractylodis Lanceae Rhizoma (Rhizoma Atractylodis in China) and "Byakujutsu" is Atractylodis Rhizoma (Rhizoma Atractylodis Macrocepharae in China). In this section, we describe these crude drugs separately.

2.6.1 Atractylodis Lanceae Rhizoma

- *Definition.* Atractylodis Lanceae Rhizoma (Atractylodes Lancea Rhizome) is the rhizome of *Atractylodes lancea* De Candolle or of *Atractylodes chinensis* Koidzumi (Compositae).
- *Vernacular name.* English: Atractylodes Lancea Rhizome, Atractylodis Lanceae Rhizoma (thistle type); Chinese: Cang-zhu; Japanese: Sojutsu
- *Plant description.* (1) *Atractylodes lancea* (Thunberg) De Candolle (synonym: *Atractylodes lancea* De Candolle)
  - A perennial herb, glabrous, 30 to 80 cm in height
  - Stem erect, simple
  - Cauline leaves, lobes three, alternate, ovate–lanceolate, elliptical, simple lobate–pinnate
  - Terminal leaves elliptical, 4 cm in length, 1 to 1.5 cm in width, entire or three to seven partite, pinnatilobate, margin serrate and spinulescent
  - Inflorescence a terminal head: involucre with scariose bracts, firm, imbricate, acuminate, ciliate
  - Floral head, August to October, white, 2 cm in diameter
  - Corolla of fertile flowers tubular, white, slightly tinged with purple
  - Fruit, September to December, an achene crowned with a silky pappus
  - Rhizome irregularly moniliform or knobby-cylindrical, slightly curved, occasionally branching
  - Distribution: China (Jiang-su, Zhe-jiang, An-hui, Jiang-xi, Hu-bei, Si-chuan, etc.)

  (2) *Atractylodes chinensis* (De Candolle) Koidzumi (synonym: *A. chinensis* De Candolle, *A. lancea* De Candolle var. *chinensis* Kitamura) differs from (1) as follows:
  - Herb, 30 to 50 cm in height
  - Cauline leaves, pinnatiparted lobes three to five
  - Floral head, July to August, 1 cm in diameter
  - Fruit, August to September
  - Rhizome knotty-lumpy or knobby-cylindrical
  - Distribution: China (Hei-long-jiang, Ji-lin, Liao-ning, He-bei, He-nan, Shan-dong, Shan-xi, Ning-xia, Gan-su, etc.)
- *Cultivation.* In China, a somewhat cool, dry climate and well-drained sandy mellow soil with a deep and mellow layer are suitable for the cultivation. The best temperature for growth is 15 to 22°C. The plants are bred by sowing or division.

  Method of direct sowing in a bed: in the middle of March to the beginning of April, the holes are dug at the size of 20 × 40 cm. Four to five grain seeds are sown in every hole and sprinkled with water. After germination, a layer of rice grass is used to cover the seedlings and it is sprinkled with water. The plants are bedded, 1 to 2 years later, at the beginning of March in holes at the depth size of 20 × 20 cm and the width size of 6 to 8 cm.

  Method of division of rhizome: after harvesting, the rhizome is cut and separated into pieces with shoots. Each piece is planted in a 20- × 20-cm hole and covered with soil. Pest and disease control: root rot in May to June; it is necessary to remove the damaged root and to take care of the drainage. Chemicals are used against plant lice—for example, 50% of thiophanate, diluted 800 times. Cultivation places: China (Zhe-jiang, Hu-nan, An-hui, etc.)
- *Harvesting and processing.* The rhizome is dug up in winter when the lower leaves of the plants turn yellow and the upper leaves become fragile; they are freed from soil, baked or sunned to dryness, and then removed from the fibrous roots. They should be stored in a ventilated dry place and protected from moisture.
- *Macroscopy* (Figure 2.6-1). This plant is an irregularly curved, cylindrical rhizome, 3 to 10 cm in length and 1 to 2.5 cm in diameter. Externally, it is dark grayish brown to dark yellow–brown; a transverse section is nearly orbicular, with light brown to red–brown secretes as fine points. White cotton-like crystals are often produced on its surface.

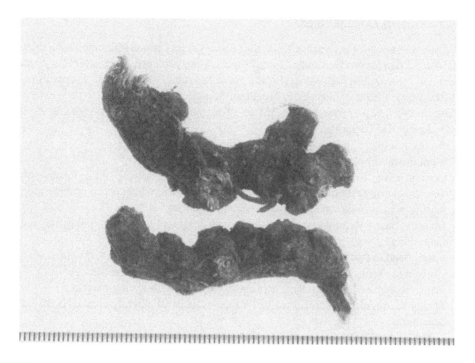

FIGURE 2.6-1 Atractylodis Lanceae Rhizoma from China.

- *Microscopy.* A transverse section usually reveals periderm with stone cells; parenchyma of cortex, usually without any fiber bundle; and oil sacs containing light brown to yellow–brown substances located at the end region of medullary rays. Xylem exhibits vessels surrounded by fiber bundles and arranged radially on the region adjoining the cambium; pith and medullary rays exhibit the same oil sacs as in the cortex. Parenchyma cells contain spherocrystals of inulin and fine needle crystals of calcium oxalate.
- *General test.* To test purity, macerate 0.5 g of pulverized Atractylodes Lancea Rhizome with 5 mL of ethanol by warming in a water bath for 2 min, and then filter. To 2 mL of the filtrate add 0.5 mL of vanillin-hydrochloric acid TS, and shake immediately: no red to red–purple color develops within 1 min (difference from Atractylodis Rhizoma). For essential oil content, perform the test with 50.0 g of pulverized Atractylodis Lanceae Rhizoma as directed in the essential oil content under the crude drugs; the volume of essential oil is not less than 0.7 mL. Total ash is not more than 7.0%. Acid-insoluble ash is not more than 1.5%. It has a characteristic odor and slightly bitter taste.
- *Chemistry* (Table 2.1c and Table 2.2). The rhizome of *A. lancea* is rich in essential oil (3.5 to 7.0%). Essential oil components, i.e., elemol, atractylon (12), hinesol (13), β-eudesmol, selina-4(14),7(11)-dien-8-one, and atractylodin, are commonly contained in *A. lancea* and *A. chinensis.*
- *Pharmacology* (Table 2.2). The essential oil has been reported as important (atractylodin, atractylon). The biotransformation compound of hinesol has spasmolytic activity (Takeda et al., 1996). Sesquiterpenes (+)-eudesma-4(14),7(11)-dien-8-one and atractylenolide I (14) are active compounds in anti-inflammatory assays. Atractylenolide I and atractylon showed anti-inflammatory activity in the rat cotton-pellet-granuloma assay. Atractylon inhibited 12-*O*-tetradecanoylphorbol 13 acetate (TPA) induced ear inflammation in mice (Resch et al., 1998).

TABLE 2.1c
Ingredients of Crude Drugs

2.6.1 Atractylodis Lanceae Rhizoma		2.6.2 Atractylodis Rhizoma	
Atractylodes Lancea	Atractylodes Chinensis	Atractylodes Ovata	Atractylodes Japonica
1,9-Aristolodiene	Acetylatractylodinol	12-Senecioyl-2E,8E,10E-atractylentriol	(4E,6E,12E)-Tetradecatriene-8,10-diyne-1,3-diol diacetate
2-Carcene	Atractylodin	12-Senecioyl-2E,8Z,10E-atractylentriol	(6E,12E)-Tetradecatriene-8,10-diyne-1,3-diol diacetate
3β-Hydroxyatractylon	Atractylodinol	12α-Methylbutyryl-14-acetyl-2E,8E,10E-atractylentriol	2-Furaldehyde
Acetylatractylodinol	Atractylon	12α-Methylbutyryl-14-acetyl-2E,8Z,10E-atractylentriol	3β-Acetoxyatractylon
Atractylodin	Elemol	14-Acetyl-12-senecioyl-2E,8E,10E-atractylentriol	3β-Hydroxyatractylon
Atractylodinol	Hinesol	14-Acetyl-12-senecioyl-2E,8Z,10E-atractylentriol	5α-10β-Selina-4(14),7(11)-diene-8-one
Atractylon	Selina-4(14),7(11)-diene-8-one	14α-Methylbutyryl-14-acetyl-2E,8E,10E-atractylentriol	Acetaldehyde
Atractyloside A–I	α-Bisabolol	14α-Methylbutyryl-14-acetyl-2E,8Z,10E-atractylentriol	Atractan A–C
	β-Eudesmol	3β-Acetoxyatractylon	Atractylenolide I–III
Butenolide B	β-Selinene	5β-10β-Selina-4(14),7(11)-diene-8-one	Atractylon
Caryophyllene		8β-Ethoxyatractylenolide-II	Diacetylatractylodiol
Chamigrene		Atractylenolide I–III	Furfural
Elem013mene		Atractylon	
Elemol		Butenolide A	
Furaldehyde		Hinesol	
Guaiene		Humulene	
Guaiol		Scopoletin	
Hinesol		α-Curcumene	
Humulene		β-Elemol	
Patchoulene		β-Eudesmol	
Selina-4(14),7(11)-diene-8-one		β-Selinene	
α-Bisabolol		γ-Caudinene	
β-Eudesmol		γ-Patchoulene	
β-Maaliene			
β-Selinene			

- *Traditional use and application.* For treatment of rheumatic diseases, digestive disorders, mild diarrhea, and influenza, it is said to be diaphoretic.
- *Supplementary.* The plant origin of Atractylodis Lanceae Rhizoma is regulated as *A. lancea* and *A. chinensis* in Japanese *Pharmacopoeia*, and most of them are imported from China. Although the same two species are described in Chinese *Pharmacopoeia*, the best quality is said to be Bozan-sojutsu (Mao-shan-cang-zhu in Chinese) or Bo-sojutsu (Mao-cang-zhu in Chinese), which are derived from *A. lancea* produced in Jiangsu in China and often occur with flocculent crystals. It is said that the good quality of Atractylodis Lanceae Rhizoma is characterized by its massive size, oily feeling to the touch, strong odor, and flocculent crystals on the exterior. In Sojutsu, deposits of white cotton-like crystals are found frequently on its surface like a mold; these consist of hinesol and β-eudesmol and traditionally denote a high quality of the crude drug (Takeda et al., 1996).

2.6.2 ATRACTYLODIS RHIZOMA

- *Definition.* The Atractylodis Rhizoma (Atractylodes root) is the rhizome of *Atractylodes ovata* (Thunberg) De Candolle or *Atractylodes japonica* Koidzumi ex Kitamura (Compositae).
- *Vernacular name.* English: Atractylodes root, Chinese thistle daisy, largehead Atractylodes Rhizoma; Chinese: Bai-zhu; Japan: Byakujutsu, Wa-byakujutsu (derived from *A. japonica*), Kara-byakujutsu (derived from *A. ovata*)
- *Plant description.* (1) *Atractylodes ovata* De Candolle (synonym: *A. macrocephara* Koidzumi); Japanese: Obanaokera
  - Perennial herb, glabrous, 30 to 80 cm in height
  - Stem erect, simple, lignified at the base
  - Cauline leaves alternate, simple or lobate–pinnate, base cuneiform, lobes three to five, acuminate
  - Terminal leaves elliptical, 5 to 8 cm in length, entire, margin ciliate
  - Petiole short, slightly winged
  - Floral heads 3 to 4 in corymbs, pale yellow, September to October
  - Peduncle 2.5 cm in diameter. Involucre with scariose bracts, firm, imbricate, acuminate
  - Corolla of neuter flowers nearly radiant, corolla of fertile flowers tubular
  - Fruit an achene crowned with a silky pappus
  - Rhizome irregularly thick mass
  - Distribution: China (An-hui, Zhe-jiang, Jiang-xi, Hu-nan, Hu-bei, Shan-xi, etc.)

  (2) *Atractylodes japonica* Koidzumi ex Kitamura (Japanese: Okera) differs from (1) as follows:
  - Height: 30 to 100 cm
  - Cauline and radical leaves alternate, simple or lobate–pinnate
  - Terminal leaves smaller, often simple, subsessile, elliptical, entire, margin short, spinulose
  - Petiole 3 to 8 cm in length
  - Floral heads white–pink, 1.5 to 2 cm, August to September
  - Corolla of neuter flowers nearly radiant, whitish, 10 to 12 mm in length; corolla of fertile flowers tubular
  - Fruit an achene crowned with a pappus
  - Rhizome irregular masses or irregularly curved cylinder
  - Distribution: Japan, Korea, northeastern China (Hei-long-jiang, Ji-lin, Liao-ning)
- *Cultivation.* Most of Atractylodis Rhizoma derived from *A. ovate* are cultivated in China. Somewhat cool and mild climate is suitable. The soil should be well drained, deep layered, and of sand-like consistency. The field is cultivated to a depth of 60 cm, and weeds and stones are removed. The soil is pounded and made into ridges of 15 to 20 cm in height,

1 to 1.4 m in width. In the spring of the first year, the young plants are raised. In the spring of the next year, the rhizome is planted in the field and harvested in autumn.

(1) Seedling culture: the well-selected seeds are sown on the rows in spring. The rows are made at intervals of 30 to 40 cm and the seeds are sown in them. During May to July, the seedlings are thinned out at intervals of 10 cm. In late autumn, the rhizome, which has many eyes or buds, is dug out and many eyes or buds sprouted from rhizome are cut off into two to three buds. These rhizomes are reserved in boxes in the shade.

(2) Planting a seedling: In the next spring, the eyes and buds from the rhizome are cut off again into two to three buds and dibbled at intervals of 30 cm into the rows. The fertilizer has been put on the field before the spring planting. Pest and disease control: the plant suffers damage from *Corticium rolfsii*, meloidogyne, cut worm, or by wilt disease. Chemicals are used against them. Cultivation: China (Zhe-jiang, Hu-nan, An-hui, etc.)

- *Harvesting and processing.* The rhizome is dug up in late autumn to winter, washed in the water, dried in the sun, and freed from soil and roots by polishing. Storage should preserve the rhizomes in a ventilated, dry place, protected from moisture.
- *Macroscopy* (Figure 2.6-2). (1) Wa-byakujutsu (*A. japonica*) periderm-removed rhizome is an irregular mass or irregularly curved cylinder, 3 to 8 cm in length and 2 to 3 cm in diameter. Externally, it is light grayish yellow to light yellowish white, with scattered grayish brown parts. The rhizome covered with periderm is externally grayish brown, often with node-like protuberances and coarse wrinkles. It is difficult to break, and the fractured surface is fibrous. A transverse section can be observed to have fine dots of light yellow–brown to brown secretions.

  (2) Kara-byakujutsu (*A. ovata*) is an irregularly enlarged mass, 4 to 8 cm in length and 2 to 5 cm in diameter. Externally, it is grayish yellow to dark brown, with sporadic, knob-like small protrusions. It is difficult to break; the fractured surface has a light brown to dark brown xylem that is remarkably fibrous.
- *Microscopy.* (1) Wa-byakujutsu (*A. japonica*): a transverse section usually reveals periderm with stone cell layers and fiber bundles in the parenchyma of the cortex, often adjoined to the outside of the phloem. Oil sacs containing light brown to brown substances

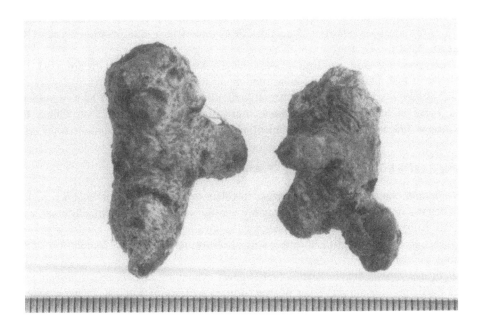

FIGURE 2.6-2 Atractylodis Rhizoma from Japan and China.

are situated at the outer end of medullary rays. In the xylem, radially lined vessels surround a large pith and a distinct fiber bundle surrounds the vessels. In pith and in medullary rays, oil sacs similar to those in the cortex can be found and, in parenchyma, crystals of inulin and small needle crystals of calcium oxalate.

(2) Kara-byakujutsu (*A. ovata*): A transverse section usually reveals periderm with stone cells and absence of fibers in the cortex. Oil sacs containing yellow-brown contents in phloem rays are found and also at the outer end of it. Xylem with radially lined vessels surround a large pith, and a distinct fiber bundle surrounds the vessels. Pith and medullary rays exhibit oil sacs as in the cortex; parenchyma contains crystals of inulin and small needle crystals of calcium oxalate.

- *General test.* For identification, macerate 0.5 g of pulverized Atractylodis Rhizoma with 5 mL of ethanol by warming in a water bath for 2 min, and then filter. To 2 mL of the filtrate add 0.5 mL of vanillin-hydrochloric acid TS and shake immediately: a red to red–purple color develops and persists. To test purity, add exactly 5 mL of hexane to 2.0 g of pulverized Atractylodis Rhizoma, shake for 5 min, filter, and then use this filtrate as the sample solution. Spot 10 μL of this solution on a plate of silica gel for thin-layer chromatography. Develop the plate with a mixture of hexane and acetone (7 : 1) to a distance of about 10 cm and air-dry the plate. Evenly spray *p*-dimethylaminobenzaldehyde TS for spraying on the plate, and heat at 100°C for 5 min; no green to grayish green spot appears between Rf 0.3 and 0.6 (difference from Atractylodis Lancea Rhizoma).

  Total ash is not more than 7.0% and acid-insoluble ash not more than 1.5%. To determine essential oil content, perform the test as directed in the essential oil content under the crude drugs, with 50.0 g of pulverized Atractylodis Rhizoma: the volume of essential oil is not less than 0.5 mL. (1) Wa-byakujutsu has a characteristic odor and somewhat bitter taste. (2) Kara-byakujutsu has a characteristic odor and tastes somewhat sweet at first, but that is followed by slight bitterness.

- *Chemistry* (Table 2.1c and Table 2.2). The rhizomes of *A. lancea* contain essential oil (1.5 to 3.5%). Sesquiterpenes with the furosesquiterpene (atractylon) as the main compound and acetylenic compounds have been described.

- *Pharmacology* (Table 2.2). Atractylenolide I(14), II, III and selina-4(14),7(11)-dien-8-one show as active compounds in anti-inflammatory assays. Atractylon inhibited 12-*O*-tetradecanoylphorbol 13 acetate (TPA) induced ear inflammation in mice (Resch et al., 1998).

- *Traditional use and application.* Atractylodis Rhizoma has been recommended as a digestive and diuretic, and as anhidrotic medication.

- *Supplementary.* Plant origins of Atractylodis Rhizoma are regulated as *A. japonica* and *A. ovata* in Japanese *Pharmacopoeia*; most of these are imported from China. In the Chinese *Pharmacopoeia*, the botanical origin of Atractylodis Rhizoma is *A. ovata*.

2.7 ANGELICAE RADIX (JAPANESE ANGELICA ROOT)

- *Definition.* Angelicae Radix (Japanese angelica root) is the root of *Angelica acutiloba* Kitagawa or *Angelica acutiloba* Kitagawa var. *sugiyamae* Hikino (Umbelliferae), usually after being passed through hot water.

- *Vernacular name.* English: Japanese Angelica root; Japanese: Toki, Yamato-toki, Obuka-toki, Hokkai-toki

- *Plant description.* *Angelica acutiloba* Kitagawa (synonyms: *Ligusticum acutilobum* Siebold et Zuccarini, *A. acutiloba* Kitagawa var. *sugiyamae* Hikino), English: Japanese angelica
  - Fragrant, perennial herb, 40 to 90 cm in height
  - Stem erect, reddish to purplish, glabrous, striate
  - Leaves alternate, one or two ternately pinnate, shining deep green in upper surface

- Leaflet two to three partite
- Lobes lanceolate, dentate incised, the teeth obtuse
- Petiole 10 to 30 cm in length, sheathed
- Compound umbels terminal, multiflorus; rays 15 to 45, 1 to 10 cm in length
- Petals white
- Cremocarp ellipsoidal, slightly winged
- Distribution: all cultivated in Japan
- *Cultivation.* There are two plant-lines: "Yamato-toki," derived from *A. acutiloba*, and "Hokkai-toki" from *A. acutiloba* var. *sugiyamae.* The latter is supposed to be selected from the hybrids between *A. acutiloba* cultivated in Nara Prefecture and other wild angelica. It has a bolting character, yellowish green petioles, and stronger resistance to cold. *A. actiloba* is lower in stem height and has deep purplish green petioles, parted terminal leaflet, more branched root, and fewer flowers per umbel. It is generally said that this (Yamato-toki) is of better quality than the variety *sugiyamae* (Hokkai-toki).

  A somewhat cool climate and well drained land with deep and mellow soil are suitable for the cultivation of this plant. The sowing season is March to April in the Honsyu region and April to May in Hokkaido Prefecture. In early spring of the second year, the young plant with less than 8 mm of root head is set to field. In the case of "Yamato-toki," bud cutting is undertaken to avoid bolting. Pest and disease control: Diseases like downy mildew and cottony rot and pests like two-spotted spider mite, aphid, cabbage army-worm, and larva of the common yellow swallowtail are known in this plant. Cultivation place: *A. acutiloba* is cultivated in Nara, Wakayama, Ehime, Kohchi, Miyazaki, and Toyama and *A. acutiloba* var. *sugiyamae* in Hokkaido.
- *Harvesting and processing.* After 2 years of cultivation, the drug is collected from November to December in the Honshu area or in October in Hokkaido Prefecture. On sunny days when leaves turn yellow, the plants are dug up, left on the ground for 2 to 3 days, and then tied up into small bunches after removing the soil and drying outside hanging on a rack. When the plants are moderately dry, the roots are put into hot water (70 to 80°C) and rubbed down by hand to arrange the shape. After they are dryied, the aerial parts are removed. In Hokkaido Prefecture, after rack-drying and removal of the aerial parts, the root is dried again with hot air. Storage conditions should preserve the plants in a cool, dry place, protected from moisture and moths.
- *Macroscopy* (Figure 2.7). The herb has a thick and short main root, with numerous branched roots nearly fusiform, 10 to 25 cm in length. Externally, it is dark brown to red–brown, with longitudinal wrinkles and horizontal protrusions composed of numerous scars of fine rootlets. The fractured surface is dark brown to yellow–brown in color and smooth, with little remains of leaf sheath at the crown.
- *Microscopy.* A transverse section reveals four to ten layers of cork, with several layers of collenchyma inside the layer. The cortex exhibits many oil canals surrounded by secretary cells; often large hollows appear. The boundary of phloem and xylem is distinct in the xylem and numerous vessels radiate alternately with medullary rays. Vessels in the outer part of the xylem appear singly or in several groups and are disposed rather densely in a cuneiform pattern, but vessels in the region of the center are scattered very sparsely. Starch grains are simple grains, not more than 20 μm in diameter and rarely two- to five-compound grains, up to 25 μm in diameter; starch grains are often gelatinized.
- *General test.* To test purity of (1) leaf sheath, the amount of leaf sheath contained in Angelicae Radix does not exceed 3.0%; for (2) foreign matter, the amount of foreign matter other than leaf sheath contained in Angelicae Radix does not exceed 1.0%. Total ash is not more than 7.0% and acid-insoluble ash not more than 1.0%. Extract content (dilute ethanol-soluble extract) should be not less than 35%. The herb has a characteristic odor and slightly sweet taste, followed by slight pungency.

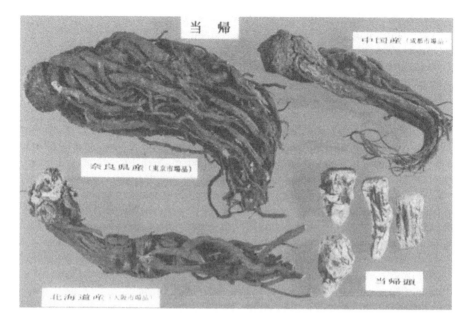

FIGURE 2.7 Angelicae Radix and Chinese angelica root. Angelicae Radix: Yamato-tohki and Hokkai-tohki in Japanese; Chinese angelica root: Quan-danggui and Danggui-tou in Chinese.

- *Chemistry* (Table 2.1d and Table 2.2). These plants are a variety of volatile alkylphthalide derivatives, ligustilide, butylphthalide, and as shown under Cnidii Rhizoma.
- *Pharmacology* (Table 2.2). More than 30 constituents, mostly coumarins, have been isolated and identified. Scopoletin (15), umbelliferone, and bergapten have been regarded as the major active principles showing anti-inflammatory and antiproliferatory activities. The also exhibit relaxant effects on the trachealis and thoracic aorta and inhibitory effects on thromboxane formation in platelets and phosphoinositides breakdown.
- *Traditional use and application.* This herb is used to treat various gynecological diseases and as a crude drug in Kampo medicine.
- *Supplementary.* Dang-gui (Kara-toki in Japanese) used in China is derived from the root of *A. sinensis* (Oliver) Diels, which is not listed in the *Pharmacopoeia* of Japan. The synopsis of Dang-gui is as follows:
  - *Vernacular name.* English: Chinese angelica root; Chinese: Dang-gui, Tang Kuei, Quan-dang-gui, Dang-gui-tou, Dang-gui-shen, Dang-gui-wei
  - *Plant description. Angelica sinensis* (Oliver) Diels (synonyms: *A. polymorpha* Maximowicz var. *sinensis* Oliver), English: Chinese angelica
    - Perennial herb, 40 to 100 cm in height
    - Stem erect, purplish, glabrous, lightly striate
    - Leaves bipinnate or tripinnate
    - Leaflet three pairs oval or oval–lanceolate, dentate-incised, the teeth obtuse
    - Petiole 3 to 11 cm in length, sheathed
    - Compound umbels terminal, multiflorus; rays irregular
    - Pedicels slender, longer than the fruit
    - Calyx five, slender, oval
    - Petals five, white, elliptical, incurvated
    - Cremocarp ellipsoidal, five edged, with membranous wings
    - Distribution: China (Gan-su), mainly cultivated

TABLE 2.1d
Ingredients of Crude Drugs

2.7 Angelicae Radix Angelica Actiloba	(2.7 Chinese Angelica) Angelica Actiloba	2.8 Ginseng Radix Panax Ginseng	2.9 Poria Polia Cocos
Angeloylsenkyunolide F	Angeloylsenkyunolide F	(8E)-1,8-Heptadecadiene-4,6-diyne-3,10-diol	Pachymic acid
Bergaptene	Bergaptene	20-Glucoginsenoside Rf	7,9(11)-Dehydropachymic acid
Butylidenephthalide	Butylidenephthalide	Acetyl panaxydol	Tumulosic acid
Butylphthalide	Butylphthalide	Caryophylene	Dehydrotumulosic acid
Caffeic acid	Caffeic acid	Caryophyllene alcohol	3β-Hydroxylanosta-7,9(11),24-trien-21-oic acid
Carvacrol	Carvacrol	Gensenol	Pachymic acid methyl ester
p-Cymene	p-Cymene	Ginsenoside Ra1–3,Rb1–3,Rc,Rd,Rg3	Tumulosic acid methyl ester
Falcarindiol	Falcarindiol	Ginsenoside Re,Rf,Rg1–2,Rh1	7,9,(11)-Dehydropachymic methyl ester
Falcarinol	Falcarinol	Ginsenoyne A–K	Polyperenic acid C methyl ester
Falcarinolone	Falcarinolone	Guaiene	Trametenolic acid
Ferulic acid	Ferulic acid	Gurjunene	Eburicoic acid
Folinic acid	Folinic acid	Heptadec-1-ene-4,6-diyn-3,9-diol	Dehydroeburicoic acid
Isopimpinellin	Isopimpinellin	Malonyl-ginsenoside Rb1–2,Rc,Rd	Poricoic acid A–D,DM,AM
Isosafrole	Isosafrole	Notoginsenoside R1,R4	β-Amyrin acetate
Levistolide A	Levistolide A	Panasinsanol A,B	Ergosterol
Ligustilide	Ligustilide	Panaxacol	Caprylate
Nicotinic acid	Nicotinic acid	Panaxydol	Caprylic acid
Safrole	Safrole	Panaxydol chlorohydrine	Dodecenoate
Scopoletin	Scopoletin	Panaxynol	Dodecenoic acid
Sedanoic acid lactone	Sedanoic acid lactore	Panaxytriol	Lauric acid
Senkyunolide E,F,H,I	Senkyunolide E,F,H,I	Quinquenoside R1–2	Palmitic acid
β-Sitosterol	β-Sitosterol	Selina-4(14),7(11)-diene	Undecanoic acid
Tokinolide A,B	Tokinolide A,B	α-Selinene, β-, γ-	
Umbelliferone	Umbelliferone	α-Humulene, β-	
O-Valerophenone carboxylic acid	O-Valerophenone carboxylic acid	α-Neoclovene, β-	
Vanillic acid	Vanillic acid	α-Panasinsene, β-	
Volic acid	Volic acid	β-Farmesene	
Xanthotoxin	Xanthotoxin	β-Sitosterol	

- *Cultivation.* It is known that the products from Min-xian of Gan-su Province and Li-jiang of Yun-nan Province are of high grade and high yield. The plants commonly grow in highland and cold mountain areas. The aerial parts and roots grow rapidly in the second year when the average temperature is more than 14°C, while the growth rate is slowed if the average temperature in August is 16 to 17°C. The soil should be deep, loose, fertile, and rich in humus. Sufficient rain will lead to high production, but overabundant water also can make roots putrefy. Repeated cultivation on the same land is not productive. Sowing is undertaken in June. Young plants are dug up in early October, stored with dry Chinese yellow soil in a cave, and set to field the following April (Xie et al., 1994).

  Pest and disease control: the plants are mainly infested by root putrefaction. The root twigs and tender roots of infected plants are watery-like, turn yellow, and break off; lately taproots also become yellow and rust-like putrefaction ensues. In early stages, lime is used to sanitize the infected hole of cultivation or 1/500 dilution of 50% car-bendaxol to perfuse the infected area. The main pest is larva of common yellow swallowtails, which eats leaves, thus causing them to be indented. A 1/800 dilution of 90% trichlofon is sprayed once a week during 2 weeks. Cultivation place: China (mainly in Gan-su, Si-chuan, Yun-nan, Shan-xi, Gui-zhou, and Hu-bei).

- *Harvesting and processing.* At the beginning of October after 2 years of cultivation, aerial parts are cut off when leaves turn yellow and the ground is exposed directly to sunlight to mature the roots. The roots are dug out in late October and residual stems and soil are removed. The roots are tied up into small bunches according to the size after a slightly evaporation, hung on a frame, and then treated with smoke and fire. The roots are scrubbed after drying and removing the fibrous roots. Storage should preserve the roots in a cool, dry place, protected from moisture and moths.

2.8 GINSENG RADIX

Ginseng Radix is the root of *Panax ginseng* C.A. Meyer, which is native to northern China, Korea, and northeastern Russia; it is one of the most famous and expensive crude drugs and is used as a tonic in Oriental medicine. Panax is derived from panacea, which means "cure-all" and "longevity." At present, Ginseng Radix sold in the markets is generally cultivated in China, Korea, and Japan, and only a little Ginseng Radix is derived from the wild type, which is extremely expensive.

The herb is classified roughly into two varieties by the difference of the processing. One is a white variety that is immediately dried after harvesting. Another is a red variety, which is repeatedly dried after scalding for several 10-min periods. Ginseng Radix is used in prescriptions that combine several crude drugs, although in the prescription "Dokujin-to," Ginseng Radix is used alone for gastric atony, dyspepsia, and feelings of cold. It is also used as a health food for a tonic, a stomachic, and for improvement of resistance against illness.

In addition to the root of *Panax ginseng*, three kinds of ginsengs exist: the rhizome of *Panax japonicus* C.A. Meyer; the root of *Panax quinquefolius* Linne; and the root of *Panax notoginseng* (Burkill) F.H. Chen. These three Ginsengs are seldom used in traditional prescriptions.

- *Definition.* Ginseng Radix (ginseng) is the dried root of *Panax ginseng* C.A. Meyer (Araceae) from which rootlets have been removed. The root has been quickly passed through hot water.
- *Vernacular name.* English: ginseng, ginseng root, sweet wood; German: Ginsengwurzel; France: racine de ginseng; Chinese: Ren-shen; Japanese: Ninjin
- *Plant description. Panax ginseng* C.A.Meyer (synonym: *P. schinseng* Nees)
  - Perennial herb 60 to 80 cm in height
  - Root often characteristic bifurcate from the middle of the main root in the form of a human figure, aromatic

- Stem erect, simple, not branching
- Leaves verticillate, compound, digitate, ovate–oblong, thin
- Leaflets five to seven, with the three terminal leaflets larger than the lateral ones, elliptical, acuminate, 5 to 15 cm in length, 2 to 6.5 cm in width
- Apex acuminate, base cuneate, margin serrulate or finely bidentate
- In general, one leaf in the 1st year with one leaflet, added annually until the 6th year
- Petiole short, slightly winged
- Inflorescence a small terminal umbel, hemispherical in early summer
- Peduncle 2.5 cm in length, involucre with scariose bracts, firm, imbricate, acuminate, reddish
- Flowers April to July, polygamous, pink
- Corolla of neuter flowers nearly radiant, corolla of fertile flowers tubular
- Fruit September to October, an achene crowned with a silky pappus, compressed legume, 3 to 4 cm in length, 5 to 6 cm in width, and red when ripe in autumn
- Distribution: northeastern to northwestern China, Korea, Mongolia, Siberia

- *Cultivation*. A somewhat cold climate is suitable for this plant, and strong sunshine, high temperature, and high moisture are not suitable. Well-drained and fertile soil is favorable for the growth. The seeds are treated to encourage germination before being sown on the farm for seedlings in spring or autumn. Then seedlings are planted on the field. A shelter that is higher on the north side than on the south side is necessary to avoid the direct rays of the sun and rainfall. No additional fertilizer is given. After the third summer, the flowers blossom and then bear the red fruits.

  Pest and disease control: the seedling of the first year is subject to damping off. As the disease rapidly spreads, the agricultural chemicals are stirred into soil before sowing. The disease, which is seen as spots on the terrestrial part or the root apex, is easy to break out; the agricultural chemicals should be sprayed from May to July. The noxious insects for Ginseng Radix are armyworms, cutworms, and mealy bugs; insecticide is used against them. Cultivation place: China (Ji-lin, Liao-ning, Hei-long-jiang), Korea, Japan (Fukushima, Nagano, Shimane)

- *Harvesting and processing*. The roots are dug up mostly in autumn and washed well. The rootlets are removed, then exposed to the sunshine or passed through hot water and drying at a low temperature (see sun-dried Ginseng Radix in Figure 2.8-1). In some cases, the cork layer of the root and stolon is peeled off. When the roots are steamed and then dried, they are known as red Ginseng Radix (Figure 2.8-2). Ginseng Radix is used usually as sun-dried Ginseng Radix in Kampo medicines. For storage, preserve in well-closed containers stored in a cool and dry place and protected from worms and moths.

- *Macroscopy* (Figure 2.8-1). The herb has a thin and long cylindrical to fusiform root, often branching two to five lateral roots from the middle. It is 5 to 20 cm in length and the main root is 0.5 to 3 cm in diameter. Externally, it is light yellow–brown to light grayish brown, with longitudinal wrinkles and scars of rootlets; sometimes the crown is somewhat constricted, with short remains of rhizome. The fractured surface is practically flat, light yellow–brown in color, and brown in the neighborhood of the cambium.

- *Microscopy*. The transverse section shows cork consisting of several rows of cells, with narrow cortex and phloem showing clefts in the outer part. Parenchyma cells are densely arranged and scattered with resin canals containing yellow secretions in the inner part; cambium is in a ring and xylem rays are broad. Vessels are singly scattered or grouped in an interrupted radial arrangement and occasionally are accompanied by nonlignified fibers; parenchyma cells contain abundant starch grains and a few clusters of calcium oxalate.

- *General test*. Indication: (1) To a section of Ginseng Radix, add dilute iodine TS dropwise: a dark blue color is produced on the surface. (2) To 2.0 g of pulverized Ginseng

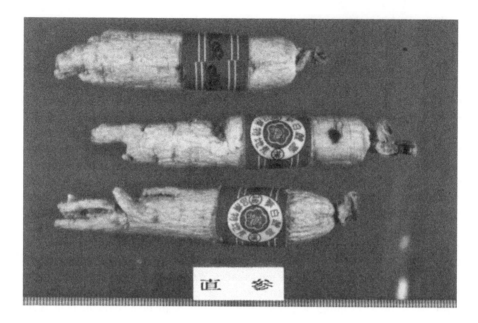

FIGURE 2.8-1 Ginseng Radix.

Radix, add 20 mL of methanol, boil gently under a reflux condenser on a water bath for 15 min, cool, filter, and use the filtrate as the sample solution. Separately, dissolve 1 mg of ginsenoside Rg_1 for thin-layer chromatography in 1 mL of methanol and use this solution as the standard solution. Spot 10 μL each of sample and standard solutions on a plate of silica gel for thin-layer chromatography. Develop the plate with the lower layer of a mixture of chloroform, methanol and water (13 : 7 : 2) to a distance of about 10 cm and air-dry the plate. Evenly spray dilute sulfuric acid on the plate and heat at 110°C

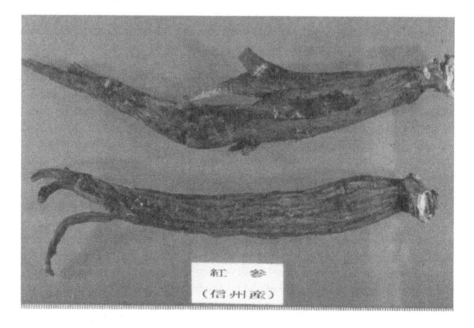

FIGURE 2.8-2 Red Ginseng Radix.

for 5 min: one of the spots from the sample solution and a red–purple spot from the standard solution show the same color tone and the same Rf value.

Purity: the amount of the stems and other foreign matter contained in Ginseng Radix does not exceed 2.0%. Total ash is not more than 4.2% and extract content (dilute ethanol-soluble extract) is not less than 14.0%. The herb has a characteristic odor and tastes slightly sweet at first, followed by a slight bitterness.

- *Chemistry* (Table 2.1d and Table 2.2). Ginseng contains a complex mixture of triterpenoid saponins that can be steroidal triterpenes or pentacyclic related to oleanoic acid. These glycocides have been categorized into three series: ginsenosides, panaxosides, and chikusetsusaponins. Ginsenoside Rg_1, one of the major saponins found in ginseng, is an example of a glycoside with a triterpene aglycone, 20(S)-protopanaxatriol. One or more of three groups of glycosides appear to account for the biological properties of ginseng.

  Systematic studies on *P. ginseng* cultivated in China have been carried out by many groups. Many kinds of saponins along with other compounds, including flavonoids, sterols, sesquiterpenoids, and saponins, have been isolated and characterized from the roots, rhizomes, stems, leaves, flowers, and fruits of this plant. Shibata et al. (1962, 1963) reported the isolation of saponins from ginseng and proposed a structure of panaxadiol as a sapogenin. Panaxa-diol and –triol are artifacts formed by the action of acid hydrolysis of saponins, and 20(S)-protopanaxadiol and -triol dammarane-type tetracyclic triterpene have been proved to be the genuine sapogenins of ginseng saponins. The saponins, named ginsenoside Rx ($x = o, a_1, a_2 b_1, b_2, b_3, c, d, e, f$, 20-gluco-f,$g_1, g_2$ and h_1), have been established. That of only ginsenoside Ro is a oleanane-type pentacyclic triterpenoid, which is a minor component in ginseng. The genuine sapogenins of ginsenosides Re, Rf, 20-gluco-Rf, Rg_1 and Rg_2 are 20(S)-protopanaxatriol and possess hydroxyl groups at 3β, 6α, 12β, and 20 positions. The sugar moieties are linked to 6 and 20-hydroxyls, and the 3-hydroxyl remains free (Nagai et al., 1971).

- *Pharmacology* (Table 2.2). The herb reportedly has adaptogenic (antistress) activity and also acts favorably on metabolism, the central nervous system, and endocrine secretions. It is employed in the Orient for the treatment of anemia, diabetes, insomnia, neurasthenia, gastritis, and, especially, sexual impotence. Western interest in the drug has vastly increased, and ginseng has become widely available in"health-food"outlets (Robbers et al., 1996). Widespread use has been accompanied by a veritable deluge of literature pertaining to the product and its purported activity. Ginseng is classified as an adaptogen because some studies in animals suggest that it may help the body adapt to stress and correct adrenal and thyroid dysfunctions. Ginseng is also heavily promoted as an aphrodisiac and a performance and endurance enhancer. The drug is administered in such forms as powders, extracts, and teas.

  Takagi et al. (1972, 1974) studied the pharmacological properties of Ginseng saponins. Slight central nervous system (CNS)-stimulating action and antifatigue action were observed in ginsenoside Rg_1 (16). CNS-suppressing effect and tranquillizing action were shown in ginsenoside Rb_1. The CNS-activating effect of ginsenoside Rg_1 was shown from increased discrimination in Y-maze and pole-climbing tests. The antifatigue action of ginsenoside Rg_1 was revealed in acceleration of recovery from depressed spontaneous and exploratory behaviors, motor incoordination, conditioned avoidance, and fighting behavior tests.

  Ginsenoside Rb_1 also inhibits stress ulcer in mice (Shibata et al., 1985). The polyacetylene compounds, panaxynol (17), panaxydol (18), and panaxytriol (19) strongly inhibit cholesteryl ester transfer protein (CETP). This is a hydrophobic glycoprotein with a molecular mass of 74 kDa, a lipid transfer protein found in plasma that mediates the transfer of cholesteryl ester (CE) and triglyceride (TG) between high-density lipoprotein

(HDL) and other low-density lipoproteins (VLDL, LDL). Feeding experiments with *P. ginseng* roots in human and monkeys, respectively, found that the total serum cholesterol and LDL-cholesterol were decreased, whereas the HDL-cholesterol level was increased (Kwon et al., 1996). Kitagawa et al. demonstrated *in vitro* that $20(R)$-ginsenoside Rg_3 inhibits cancer cell invasion and metastasis (Kitagawa et al., 1995). Recently, ginsenoside Rg_3 has been prepared as an antiangiogenic anticancer drug in China (Rg_3 Shenyi Jiaonang, *Health News*, Chinese Ministry of Health; May, 2000).

- *Traditional use and application.* Ginseng is a favorite remedy in Chinese medicine and is considered to have tonic, stimulant, diuretic, and carminative properties.
- *Supplementary.* Ginseng Radix has been an expensive and rare crude drug; therefore, some counterfeits and substitutes have been described in Chinese books. Well-known counterfeits or substitutes of Ginseng Radix in China include Radix Platycodi (*Platycodon grandiflorum* A. De Candolle, Campanulaceae); Radix Salviae Miltiorrhizae (*Salvia miltiorrhiza* Bunge, Labiatae); Radix Codonopsis (*Codonopsis pilosula* or *Codonopsis tangshen* Oliver, Campanulaceae); Radix Adenophorae (*Adenophora tetraphylla* (Thunberg) Fisher or *Adenophora stricta* Miquel, Companulaceae); Radix Dichroae (*Dichroa febrifuga* Loureiro, Saxifragaceae); and Radix Rhapontici (*Rhaponticum uniflorum* (Linne) De Candolle, Compositae). These root shapes are comparatively similar to that of Ginseng Radix. Now, in China, Radix Codonopsis is commonly used as a substitute for Ginseng Radix, and that derived from *Panax ginseng* is exported to foreign countries as a high value crop.

2.9 PORIA

- *Definition.* Poria is the sclerotium of *Poria cocos* Walf (Polyporaceae), from which usually the external layer has been mostly removed.
- *Vernacular name.* English: Poria, Hoelen, Indian bread, tuckahoe, Virginia truffle, orange Poria; Chinese: Fu-ling; Japanese: Bukuryo
- *Plant description. Poria cocos* (Fries) Wolf (synonym: *Poria cocos* (Schweinitz) Wolf, *Poria cocos* Fries, *Pachyma hoelen* Rumph, *Pachyma cocos* Wolf, *Wolfiporia cocos* (Wolf) Ryvarden et Ginilbertson)
  - Subterranean saprophytic fungus grown in association with the root of various conifers, particularly *Pinus densiflora* Siebold et Zuccarini, *Pinus thunbergii* Parlatore, *Pinus massoniana* Lambert, and *Pinus yunnanensis* Siebold et Zuccarini
  - Large, ovate, globular and amorphous, ponderous tuberiform bodies, 10 to 30 cm in length, with a reddish brown covering
  - Interior consisting of a compact mass of considerable hardness, varying in color from cinnamon brown to pure white
  - Bodies actually considered to be an altered state of the root of the tree, occasioned by the presence of a fungus whose mycelium penetrates the ligneous substance
  - Distribution: China (Guang-dong, He-nan, An-hui, Fu-jian, Guang-xi, Yun-nan, etc.), Korea, Japan, northern America
- *Cultivation.* Pine tree is cut down to pieces in autumn and lies during a half year. In early spring of the next year, holes are drilled in the pieces and these are spread with the spawn of Poria. After 1 month, the pieces are buried under a depth of 10 cm of soil in a sunny field. In December of next year, the Poria growing on the pieces of pine tree is gathered and processed. Cultivation place: China (Guang-xi, An-hui, Hu-bei, Fu-jian, Gui-zhou, etc.)
- *Harvesting and processing.* Poria is dug up mostly in July to September, immersed in water, and washed clean. After softening and removal of the surface from the Poria, it is cut into slices or pieces and then dried in the sun. For storage, preserve in a dry place protected from moisture.

Crude Drugs of Juzen-taiho-to

FIGURE 2.9 Poria.

- *Macroscopy* (Figure 2.9). The plant's mass is about 10 to 30 cm in diameter, up to 0.1 to 2 kg in weight; usually, it appears as broken or chipped pieces. The color is white, grayish white, or slightly reddish white; sclerotium with remaining outer layer is dark brown to dark red–brown in color, coarse, with fissures and hard in texture, but brittle.
- *Microscopy.* When cut into pieces, Poria reveals colorless and transparent hyphae strongly refracting light and fragments of false tissue consisting of granules and mucilage plates. It has thin hyphae, 2 to 4 μm in diameter; thick ones are usually 10 to 20 μm, up to 39 μm.
- *General test.* Identification: (1) Warm 1 g of pulverized Poria with 5 mL of acetone in a water bath for 2 min with shaking, and filter. Evaporate the filtrate to dryness, dissolve the residue in 0.5 ml of acetic anhydride, and add one drop of sulfuric acid: a light red color develops, which changes immediately to dark green. (2) To a section or powder of Poria, add one drop of iodine TS: a deep red–brown color is produced. Total ash is not more than 1.0%. Poria is almost odorless and is tasteless with slight mucous.
- *Chemistry* (Table 2.1d and Table 2.2). Water-insoluble polysaccharide (pachyman) makes up 90% of dry weight. Other compounds such as sterol (ergosterol) and triterpenoids (ebricoic acid, tumulosic acid, pachymic acid, etc.) also comprise the chemistry.
- *Pharmacology* (Table 2.2). Poria has been found to have a remarkable inhibitory effect on the secretion of the cytokines such as IL-1β, IL-6, TNF-α, and GM-CSF from human peripheral blood monocytes (Tseng and Chang, 1992). The six triterpenes from *Pinus cocos* (lanostane-type triterpenes, some of which reveal antiemetic activity [Tai et al., 1995b])—pachymic acid (20), dehydropachymic acid, dehydrotrametenolic acid, 3-β-p-hydroxy benzoyldehydrotumulosic acid, secolanostane-type triterpenes, and poricoic acid A and B—have showed inhibitory activity against 12-*O*-tetradecanoylphorbol-13-acetate (TPA)- and arachidonic acid (AA)- induced ear inflammation in mice when isolated from *Pinus cocos* (Yasukawa et al., 1998).

  The two lanostane-type triterpenes, pachymic acid and dehydrotumulosic acid (21), showed significant phospholipase A2 (PLA2) inhibitory activity (Cuellar et al., 1997). The higher potency of dehydrotumulosic acid could be attributed to the presence of a free 3-OH that is blocked by acetylation in pachmic acid. Moreover, dehydrotumulosic acid presents an heteroanular 7-8,9-11-dien group, which is thought to increase the planarity of the molecule and could be one reason for its greater activity. Poricoic acid A, B, 3,4-secolanostane-type triterpenes isolated from *P. cocos* has shown the inhibitory activity against emetic action induced by oral administration of copper sulfate pentahydrate to leopard frog. The results indicate that some triterpenes with an exo-methylene group at C-24 in their side chain reveal antiemetic activity (Tai et al., 1995a).

- *Traditional use and application.* Poria is used as a diuretic and sedative in the treatment of edema with oliguria, dizziness, and palpitation caused by retained fluid; diminished function of the spleen marked by anorexia, loose stools, or diarrhea; restlessness; and insomnia. It is often an ingredient in Kampo and Chinese prescriptions. In China, it is also used in confectioneries as an edible material. Regarding color, there are two types of Poria. The common kind is white and sometimes called "Shiro-buku." The other, "Aka-buku," is rarer and is somewhat red, and this is said to be of better quality than Shiro-buku, but without scientific justification.

2.10 GLYCYRRHIZAE RADIX

- *Definition.* Glycyrrhizae Radix (licorice) is the root and stolon, with (unpeeled) or without (peeled) the periderm of *Glycyrrhiza uralensis* Fisher or *Glycyrrhiza glabra* Linne (Leguminosae).
- *Vernacular name.* English: licorice, Glycyrrhiza, Glycyrrhiza root, sweet wood; German: Sussholzwurzel, Lakritzenwurzel, Lacrisse; France: racine de reglisse, bois doux, racine douce; Chinese: Gan-cao; Japanese: Kanzo
- *Plant description.* (1) *G. uralensis* Fisher:
  - Perennial herb, 30 to 100 cm in height
  - Stems erect, with whitish hairs and echinate glandular hairs, lower part of stem woody
  - Leaves imparipinnate, alternate
  - Leaflets 5 to 17, elliptical–ovate, 1.5 to 5 cm in length by 1 to 3 cm in width, apex obtuse-rounded, base rounded, both surfaces covered with sticky glandular hairs and short hairs
  - Stipules lanceolate
  - Inflorescence an axillary cluster
  - Flowers June to July, raceme, axillary, purplish, papillionaceous, 10 to 24 mm in length
  - Fruit July to October, compressed legume, 3 to 4 cm in length, 5 to 9 mm in width, densely covered with brownish echinate glandular hairs
  - Root and stolon cylindrical, fibrous, flexible, with cork reddish, light yellow inside, main root comparatively long, coarse
  - Distribution: northern China, Mongolia, Siberia
  (2) *G. glabra* Linne differs from *G. uralensis* Fisher as follows:
  - Height of 1 m or more
  - Stem, pubescent
  - Leaflets 5 to 13, orbicular–ovate, oblong to elliptical lanceolate, 1.5 to 4 cm in length by 0.8 to 2.3 cm in width, lower surface covered with sticky glandular trichomes
  - Flowers June to August, 9 to 12 mm
  - Fruit July to September, 2 to 3 cm in length by 4 to 7 mm in width
  - Distribution: northwestern to middle China (Xin-jiang, Qing-hai, Gan-su), southern Russia, middle-southern Asia, Mediterranean area
- *Cultivation.* In China, a somewhat cool climate and well-drained, sandy-like, and deep layers of soil are suitable to enable easy root strikes with the growth of this plant. The planting season is spring or autumn. The soil should have a good staple of mold 0.8 to 1 m in depth and be manured if necessary.

(1) By seedling: well-selected seeds are sown on the rows or in the holes in spring or autumn. On the rows, the rows are made at intervals of 30 to 40 cm and the seeds are sown in them. In the holes, the rows are made at intervals of 30 cm, the holes are dug in the rows at intervals of 15 cm and a depth of 6 cm, and six to ten grains of the seeds are sown in every hole.

(2) By the division of rhizome or root: the root or rhizome, which has eyes or buds, is cut into sections of about 10 to 15 cm in length. These are dibbled in, in rows 50 cm apart, about 10 cm underneath the surface and about 15 cm apart in the rows. When the shoots are overcrowded, they are thinned out. Every year, the weeds are removed and fertilizer applied to the field. During the first 2 years, the growth is slight and plants do not rise above 1 foot. After the third year, the growing Glycyrrhizae Radix plants cover the soil.

Pest and disease control: the plant suffers mainly spider mite and rust damage from May to August. As the antimicrobial, the agricultural chemicals mixed with sulfur are useful. In May and June, the plant suffers also from the larva of the turnip moth; acariasides and insecticides are used against them. Cultivation place: China (Hei-long-jiang, Gan-su, Nei-meng-gu, Xin-jiang)

- *Harvesting and processing.* In autumn, the root and stolon are dug up and removed from rootlet and dried in the sun. In some cases, the cork layers of the root and stolon are peeled off in the process of drying. For storage, preserve in a ventilated dry place protected from moisture.
- *Macroscopy* (Figure 2.10). This herb has nearly cylindrical pieces, 0.5 to 3.0 cm in diameter and over 1 m in length. Glycyrrhizae Radix is externally dark brown to red–brown, longitudinally wrinkled, and often has lenticels, small buds, and scaly leaves; peeled Glycyrrhizae Radix is externally light yellow and fibrous. The transverse section reveals a rather clear border between phloem and xylem and a radical structure that often has radiating splits; pith is present in the stolon pieces, but absent from the root ones.
- *Microscopy.* The transverse section reveals several yellow–brown cork layers and a one- to three-cellular layer of cork cortex inside the cork layer. The cortex exhibits medullary

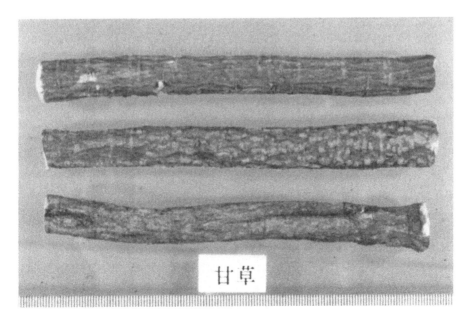

FIGURE 2.10 Glycyrrhizae Radix.

rays and obliterated sieve portions radiated alternately; the phloem exhibits groups of phloem fibers with thick but incompletely lignified walls and surrounded by crystal cells. Peeled Glycyrrhizae Radix sometimes lacks periderm and medullary rays in three to ten rows radiated alternately. The vessels are accompanied with xylem fibers surrounded by crystal cells and with xylem parenchyma cells; the parenchymatous pith originates from stolon. The parenchyma cells contain starch grains and often solitary crystals of calcium oxalate.

- *General test.* Identification: To 2.0 g of pulverized Glycyrrhizae Radix, add 10 ml of a mixture of ethanol and water (7 : 3), heat by shaking in a water bath for 5 min, cool, filter, and use the filtrate as the sample solution. Separately, dissolve 5 mg of glycyrrhizinic acid for thin-layer chromatography in 1 ml of a mixture of ethanol and water (7 : 3), and use this solution as the standard solution. Spot 2 µl each of the solutions on a plate of silica gel with fluorescent indicator for thin-layer chromatography. Develop the plate with a mixture of n-butanol, water, and glacial acetic acid (7 : 2 : 1) to a distance of about 10 cm, and air-dry the plate.

Examine under ultraviolet light (main wavelength: 254 nm): one spot among the spots from the sample solution and a dark purple spot from the standard solution show the same color tone and the same Rf value. Loss on drying is not more than 12.0% (6 h) and total ash not more than 7.0%. Acid-insoluble ash is not more than 2.0% and extract content (dilute ethanol-soluble extract) not less than 25%. A slight odor and a sweet taste are present.

- *Chemistry* (Table 2.1e and Table 2.2). Glycyrrhizae Radix contains 2 to 6% glycyrrhizinic acid. Accurately weigh about 0.5 g of pulverized Glycyrrhizae Radix in a glass-stoppered centrifuge tube, add 70 ml of dilute ethanol, shake for 15 min, centrifuge, and separate the supernatant liquid. To the residue, add 25 ml of dilute ethanol and proceed in the same manner. Combine all the extracts, add dilute ethanol to make exactly 100 ml, and use this solution as the sample solution.

Separately, accurately weigh about 0.025 g of glycyrrhizinic acid for component determination, after previously drying in a desiccator (in vacuum at a pressure not exceeding 5 mmHg, phosphorus pentoxide, 50°C) for not less than 12 h. Dissolve in dilute ethanol to make exactly 100 ml and use this solution as the standard solution. Proceed as directed in the component determination under Glycyrrhizae Radix extract. Glycyrrhiza (GL) contains a saponin-like glycoside, glycyrrhizin (22) (glycyrrhizic acid, 3-0-[β-D-glucuronopyranosyl- (1→2)-β-D-glucuronopyranosyl] glycyrrhetic acid), which is 170 times as sweet as sugar. Upon hydrolysis, the glycoside loses its sweet taste and is converted to the aglycone glycyrrhetic acid (23) (glycyrrhetinic acid) plus two molecules of glucuronic acid.

- *Pharmacology* (Table 2.2). Licorice is one of the most thoroughly studied herbs; clinical studies have been conducted on its antimutagenic activity, antiulcer effect, protective action for hepatotoxicity, antitumor-promoting activity, antimicrobial effect, etc. Glycyrrhizic acid (ammonium or potassium salt) and its derivative carbenoxolone (24) (INN, 3-carboxy-1-oxoprepoxyglycyrrhetic acid) are used in the treatment of peptic ulcers (Ottenjann and Rosch, 1970) and other gastrointestinal disorders and as an antiinflammatory agent. In Japan, a GL-preparation combined with L-cysteine and glycine has been used for more than 60 years under the trade name"Stronger Neo-Minophagen C (SNMC). It has been used as a drug for chronic hepatitis by i.v. administration.

"Glycyron" (GL + L-methionine, glycine) has been applied orally as a supplement drug for SNMC (Shibata, 2000). In treatment of chronic hepatitis and as a precaution against transfusion hepatitis, glycyrrhizic acid is also used by injection in large amounts. Glycyrrhizic acid and its derivative are antiulcer drugs but have serious side effects, such as edema, reduced serum aldosterone, elevated blood pressure, etc.

TABLE 2.1e
Ingredients of Crude Drugs

2.10 Glycyrrhizae Radix		
Glycyrrhiza Ularensis	Glycyrrhiza Glabra	Glycysrrhiza Infrata
18β-Glycyrrhetic acid	11-Deoxyglycyrrhetic acid	11-Deoxyglycyrrhetic acid
5,6,7,8-tetrahydro-2,4-dimethyl-quinoline	18α-Hydroxyglycyrrhetic acid	18β-Glycyrrhetic acid
5,6,7,8-Tetrahydro-4-methylquinoline	18β-Glycyrrhetic acid	4′,7-Dihydoxyflavone
5-O-Methyllicoricidin	21α-Hydroxyisoglabrolide	5′-Prenyllicodione
Apioisoliquiritin	24-Hydroxy-11-deoxyglycyrrhetic acid	Apioglycyrrhizin
Apioliquiritin	24-Hydroxyglycyrrhetic acid	Apioisoliquiritin
Clycyrol	7-Hydroxy-2-methylisoflavone	Apioliquiritin
Formononetin	7-Methoxy-2-methylisoflavone	Araboglycyrrhizin
Glycycoumarin	Deoxyglabrolide	Echinatin
Glycyrin	Glabranin	Glabrone
Glycyrrhizinic acid	Glabrene	Glycyrdione A,B
Isolicoflavonol	Glabridin	Glycyrrhizinic acid
Isoliquiritigenin	Glabrolide	Glyinflanin A–D
Isoliquiritin	Glabrolide	Isoliquiritigenin
Licobenzofuran	Glabrone	Isoliquiritin
Licocoumarone	Glazarin	Licochalcone A–D
Licopyranocoumarin	Glycyrrhetol	Licoflavone A
Licoricesaponin	Glycyrrhizinic acid	Licoricesaponin A3,G2,H2
A3,B2,C2,D3,E2,F3,G2,H2,J2,K2		
Licoricidin	Glyzaglabrin	Liquiritigenin
Licoricone	Isograbrolide	Liquiritin
Liquiritigenin	Isoliquiritogenin	Uralsaponin B
Liquiritin	Isoliquiritoside	β-Sitosterol
Neoglycyrol	Licuraside	
Neoglycyrol	Liquiritic acid	
Neoisoliquiritin	Liquiritogenin	
Neoliquiritin	Liquiritoside	
Ononin	Liquoric acid	
Uralsaponin A,B	Pinocembein	
	Prumetin	

About 50 other sapogenins and/or saponins have been isolated from Glycyrrhiza species. Glycyrrhetic acid (18β-olean-11-oxo-12-ene-3β-ol-30-oic acid) is a pentacyclic triterpene derivative of the β-amyrin type.

Other constituents include flavonoid glycosides, (liquiritin, isoliquiritin, liquiritoside, isoliquiritoside, rhamnoliquiritin, and rhamnoisoliquiritin) and coumarin derivatives (herniarin and umbelliferone). In 1967, Takagi et al. (Takagi and Ishii, 1967) reported that one of the flavonoid-rich (isoflavonoid, chalcone) fractions of licorice, which also included about 15% of glycyrrhiziec acid (FM-100 fraction), is effective in prevention of digestive ulcer by suppressing gastric secretion. The fraction has been developed as an antiulcer drug and has been supplied as Aspalon®. Thus far, about 300 kinds of phenolic compounds have been isolated from various species of Glycyrrhiza, about 150 of which are new and have been obtained from the underground parts.

Hatano et al. (1988) have reported that some licorice flavonoids, isolicoflavonol, glycycoumarin, licochalcones A (25) and B, glycyrrhisoflavone, and licopyranocoumarin, inhibit the cytopathic activity of HIV. Shibata et al. (1991) reported that licochalcone A shows antitumor-promoting activity in a two-stage mouse skin carcinogenesis

TABLE 2.2
Pharmacologically Active Polysaccharides Purified from Component Herbs of Juzen-Taiho-To

Component Herb	Original Plant	Polysaccharide	Type	Activity	Ref.
Astragali Radix	Roots of *Astragalus membranaceus* Bunge var. *mongholics*	FB	Crude PS fraction	Enhancement of GVH reaction (*in vivo*, i.p.)	Wang, 1989
		APS	Crude PS fraction	Activation of C3 in complement pathway Preventive effect on endotoxin-induced hepatotoxicity (*in vivo*, i.p.)	Wang et al., 1989 Wang and Han, 1992
		AMon-S	Acidic a-Arabino-β-3,6-galactan	Enhancement of phagocytosis of reticuloendotherial system (*in vivo*, i.p.)	Shimizu et al., 1991a
Cinnamomum Cortex	Bark of *Cinnamomum cassia* Blume	Cinnaman AX	Arabinoxylan	Reticuloendotherial system-enhancing activity (*in vivo*, i.p.)	Kanari et al., 1989
Rehmaniae Radix	Roots of *Rehmania glutinosa* Libosch. var. *purpure* Makino	LRPS	Low molecular weight polysaccharide	p53 Gene expression-enhancing activity in Lewis lung cancer cell (*in vivo*, i.p.)	Wei and Ru, 1997
		RGP-b		Enhancement of cytotoxic T-cell (*in vivo*, i.p.)	Chen et al., 1995
		Rehmannan FS-I Rehmannan FS-II Rehmannan SA	Pectic arabinogalactan Pectin Acidic arabino-3,6-galactan	Reticuloendotherial system enhancing activity (*in vivo*, i.p.)	Tomoda et al., 1994d
		Rehmannan FS-II Rehmannan SA	Pectin Acidic arabino-3,6-galactan	Reticuloendotherial system enhancing activity (*in vivo*, i.p.) Anticomplementary activity	Tomoda et al., 1994e,f
	Roots of *Rehmania glutinosa* Libosch. f. *hueichingensis* Hsiao	Rehmannan SB RG-WP	pectic arabinogalactan Crude polysaccharide fraction composed of pectin-like polysaccharide	Anticomplementary activity Hypoglycemic activity (*in vivo*, i.p.)	Tomoda et al., 1994e Kiho et al., 1992
Paeoniae Radix	Roots of *Paeonia lactiflora* Pullas	Paeonan PA	Acidic arabino-3,6-galactan	Reticuloendotherial system-enhancing activity (*in vivo*, i.p.)	Tomoda et al., 1994c

Crude drug	Plant source	Fraction	Structure	Activity	Reference
Cnidii Rhizoma	Rhizomes of *Cnidium officinal* Blume Makino	Paeonan SA	Heteroglycan composed of branched α-(1 → 4)-glucan	Anticomplementary activity	Tomoda et al., 1993a
		Paeonan SB	Acidic arabino-3,6-galactan		
		Cnidiran AG	Acidic arabinogalactan	Reticuloendotherial system-enhancing activity (*in vivo*) / Anticomplementary activity / Mitogenic activity	Tomoda et al., 1992
		Cnidiran SI	Branched β-(1 → 4)-glucan	Reticuloendotherial system-enhancing activity (*in vivo*) / Anticomplementary activity	Tomoda et al., 1994a
		Cnidiran SIIA	Heteroglycan composed mainly of β-(1 → 6)-glucan	Anticomplementary activity	Tomoda et al., 1994b
Atractylodis lancea Rhizoma	Rhizomes of *Atractylodes lancea* DC.	ALR-5IIa-1-1	Arabino-3,6-galactan	Intestinal immune system-modulating activity through Peyer's patch cells	Yu et al., 1998
		ALR-5IIb-2-2Bb	Unique polysaccharide		
		ALR-5IIc-3-1	Pectic polysaccharide		
Atractylodis Rhizoma Koidzumi	Rhizomes of *Atractylodes japonica*	Atractan A	Neutral heteroglycan	hypoglycemic activity (*in vivo*, i.p.)	Konno et al., 1985a
		Atractan B	Neutral heteroglycan		
		Atractan C	Acidic heteroglycan		
Angelicae Radix	Roots of *Angelica acutiloba* Kitagawa	AGIIa	Arabino-3,6-galactan	Complement-activating activity through classical and alternative pathways	Yamada et al., 1985 / Kiyohara et al., 1986, 1989
		AGIIb-1	Pectic arabinogalactan	Complement-activating activity through classical and alternative pathways	Kiyohara et al., 1988
		AR-2IIa	Pectin		
		AR-2IIb			
		AR-2IIc			
		AR-2IId		Antitumor activity (*in vivo*, i.p.)	Yamada et al., 1990
		AR-4E-2	Pectic arabinogalactan	Polyclonal B cell activation / Mitogenic activity	Kumazawa et al., 1985 / Ohno et al., 1983
		AIP	Crude PS fraction		
Angelicae Radix	Roots of *Angelica sinensis*	AP	Crude PS fraction	Enhancement of proliferation and differentiation of hematopoietic progenitor cell	Wang and Zhu, 1996

(*continued*)

TABLE 2.2
(Continued)

Component Herb	Original Plant	Polysaccharide	Type	Activity	Ref.
Ginseng Radix	Roots of *Panax ginseng* C.A. Meyer	F-2pc-A	Acidic polysaccharide	Antitumor activity (*in vivo*, i.p.)	Choy et al., 1994
		Panaxan A	α-(1 → 3,6)-glucan	Mitogenic activity NK cell induction activity (*in vivo*, i.p., i.v.) Hypoglycemic activity (*in vivo*, i.p.)	Konno et al., 1984 Tomoda et al., 1984
		Panaxan E Panaxan C Panaxan D Panaxan E Panaxan Q Panaxan R Panaxan S Panaxan T Panaxan U Panaxan I Panaxan J Panaxan K Panaxan L			Konno et al., 1985b Oshima et al., 1985
		GRA-4	Neutral polysaccharide Fraction composed of pectic polysaccharides	Anti-ulcer activity (*in vivo*, p.o.)	Sun et al., 1991
		Ginsanan PA	Pectic arabinogalactan	Reticuloendotherial system-enhancing activity (*in vivo*, i.p.)	Tomoda et al., 1993b
		Ginsanan PB	Pectic arabinogalactan	Anticomplementary activity Mitogenic activity	
		Ginsanan S-IA	Acidic α-(1 → 5)-arabino- β-(1 → 3,6)-galactan	Reticuloendotherial system-enhancing activity (*in vivo*, i.p.) Anticomplementary activity Induction of cytotoxic T cells	Tomoda et al., 1993c

Crude drug	Source	Compound	Chemical structure	Activity	Reference
Poria	*Poria cocos* Wolf	Ginsan		IL-8-inducing activity; Induction of cytotoxic T-cells	Sonoda et al., 1998; Lee et al., 1997; Kim et al., 1998
		Pachyman	β-(1 → 3)-glucan	Antinephritic activity (*in vivo*, i.p.)	Warsi and Whelan, 1957; Hattori et al., 1992; Kanayama et al., 1983
	Cultured mycelia of *Poria cocos* Wolf	H11	(1 → 3)(1 → 6)-β-D-glucan	Antitumor activity (*in vivo*, i.p.)	
Glycyrrhizae Radix	roots of *Glycyrrhiza ularensis* Fisch *et* DC	GR-2IIa; GR-2IIb	Pectic arabinogalactan	Anticomplementary activity	Zhao et al., 1991a
		GR-2IIc-1-2A	Pectic heteroglycan	Anticomplementary activity; Mitogenic activity	Zhao et al., 1991b
		GlycyrrhizanUA	Pectic arabino-3,6-galactan	Reticuoendotherial system-enhancing activity (*in vivo*, i.p.)	Tomoda et al., 1990
		GlycyrrhizanUB; GlycyrrhizanUC	Pectic arabinogalactan		Shimizu et al., 1990
		GlycyrrhizanGA	Arabino-3,6-galactan		Shimizu et al., 1991b

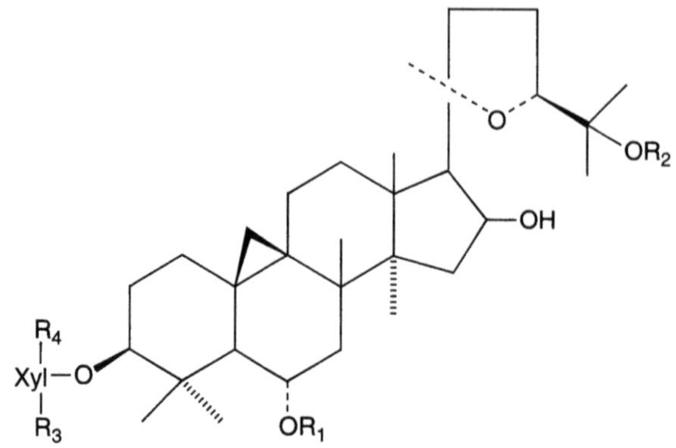

		R_1	R_2	R_3	R_4
1	astragaloside I	Glc	H	Ac	Ac
2	astragaloside II	Glc	H	Ac	H
3	astragaloside III	H	H	Glc	H

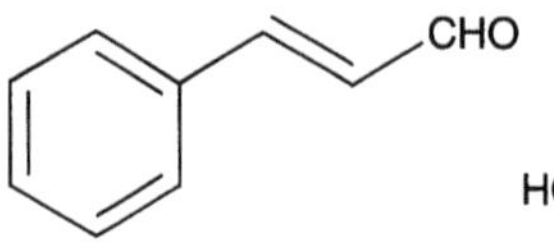

4 cinnamic aldehyde

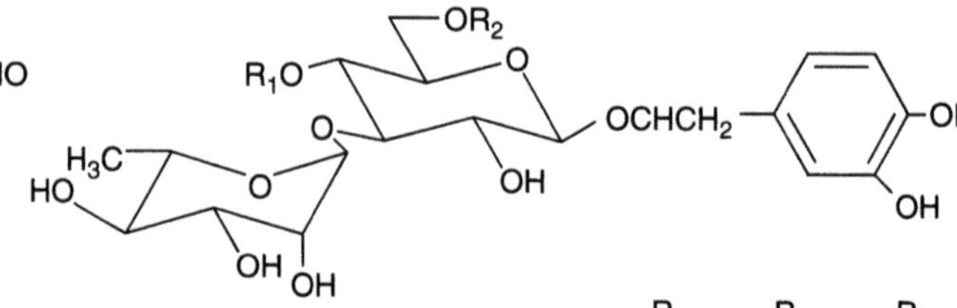

		R_1	R_2	R_3
5	acteoside	H	caffeoyl	H
6	jionoside A_1	H	feruloyl	H
7	jionoside B_1	OMe	feruloyl	H

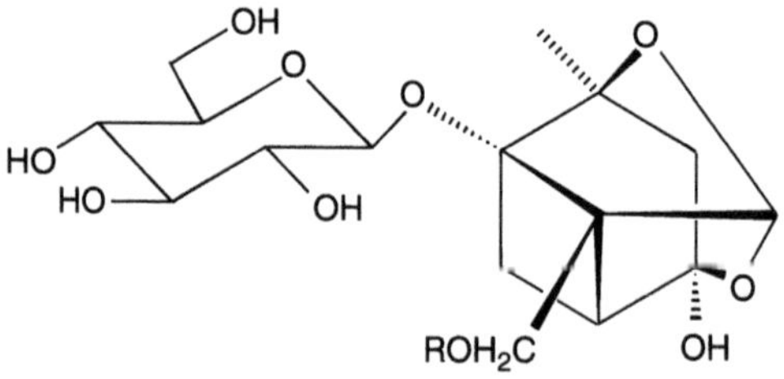

		R
8	paeoniflorin	Benzoyl
9	8-debenzoyl	H

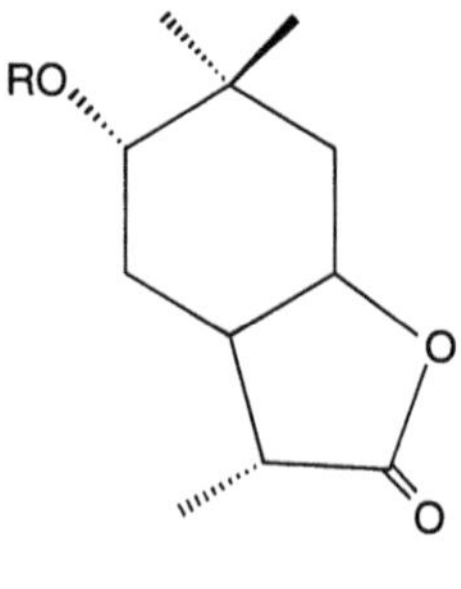

		R
10	lactinolide	H
11	lactinolide	Glc

FIGURE 2.11A

experiment. Flavonoids obtained from aerial parts of *G. glabra* are not used as drugs. They exhibit relatively high inhibitory activity for several kinds of enzymes such as xanthine oxidase, monoamine oxidase, aldose reductase, phosphodiesterase, and hyaluronidase (Nomura and Fukai, 1998).

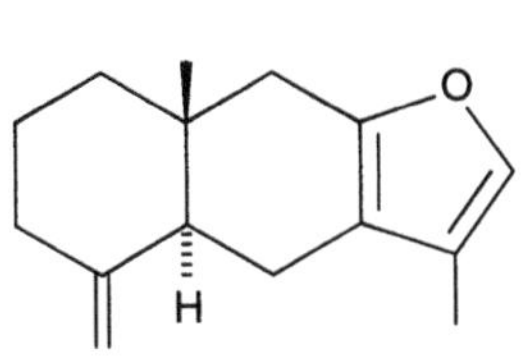

12 atractylon

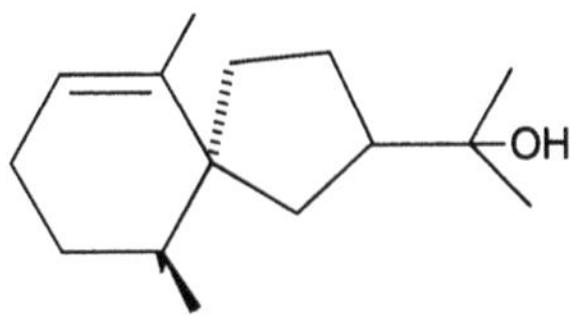

13 hinesol

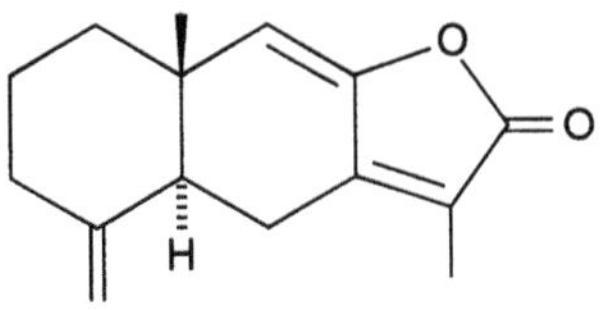

14 atractylenolide I

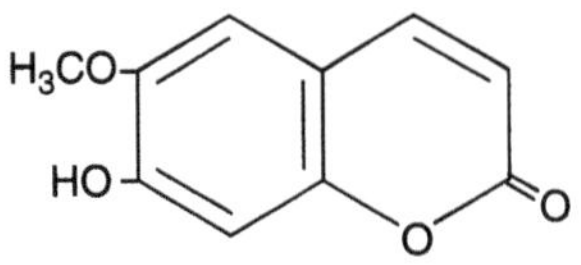

15 scopoletin

17 $CH_2=CHCH(C\equiv C)_2CH_2CH=CH(CH_2)_6CH_3$
OH
18 $CH_2=CHCH(C\equiv C)_2CH_2CH-CH(CH_2)_6CH_3$
$OH\qquad\quad O$
19 $CH_2=CHCH(C\equiv C)_2CH_2CH-CH(CH_2)_6CH_3$
$OH\qquad\quad OH\ OH$

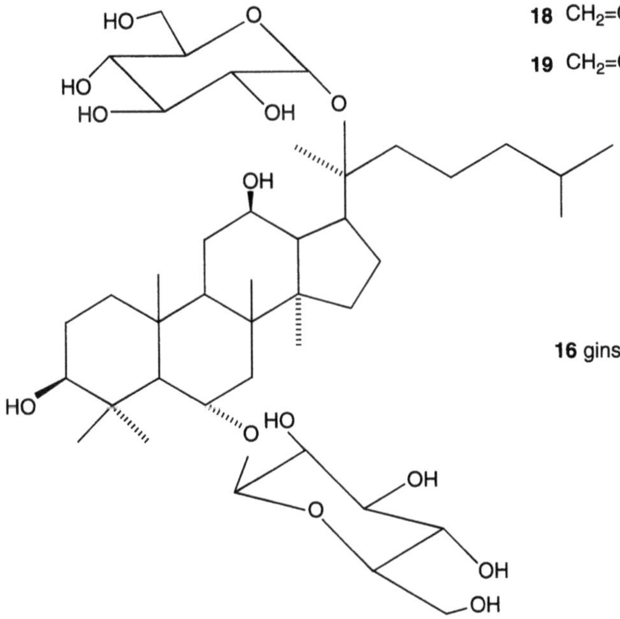

16 ginsenoside Rg$_1$

FIGURE 2.11B

- *Traditional use and application.* Licorice is described as belonging to the upper class and is recommended for its life-enhancing properties, for improving health, for cures for injury or swelling, and for its detoxification effect (Otsuka, 1973). It is used to relieve cough, remove heat, alleviate spasmodic pain, and reinforce the function of the spleen, and compounded in most kinds of traditional prescriptions in Japan and China.
- *Supplementary.* Glycyrrhiza is also the dried rhizome and roots of *Glycyrrhiza glabra*, known commercially as Spanish licorice, or of *Glycyrrhiza glabra* Linne var. *glandulifera*

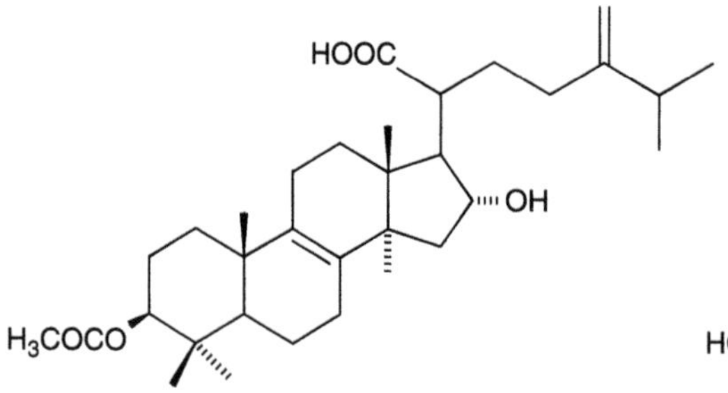

20 pachymic acid

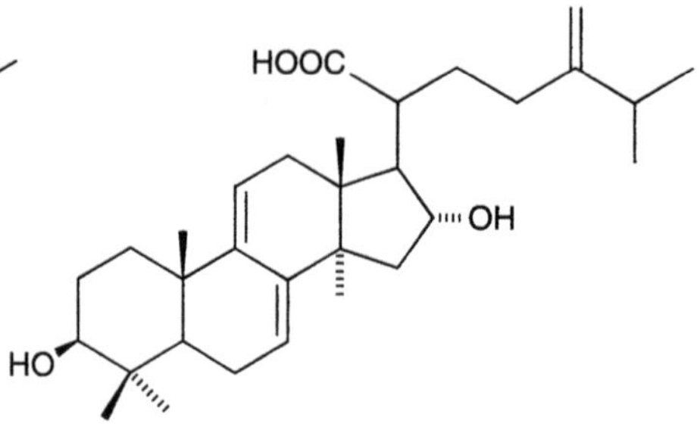

21 dehydrotumulosic acid

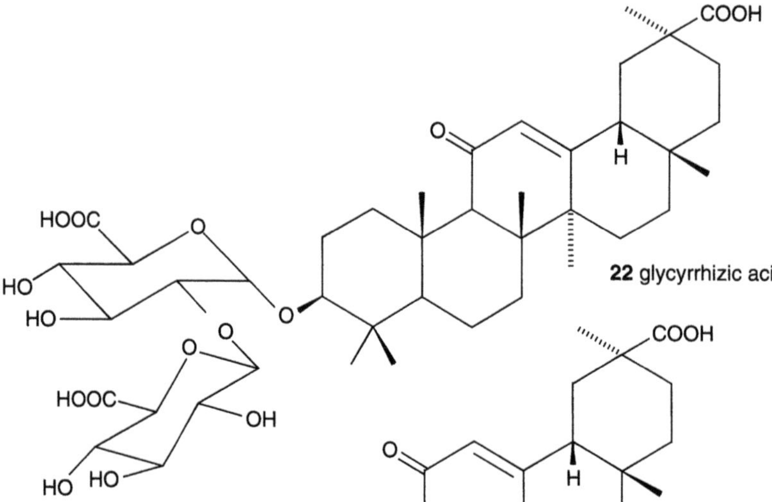

22 glycyrrhizic acid

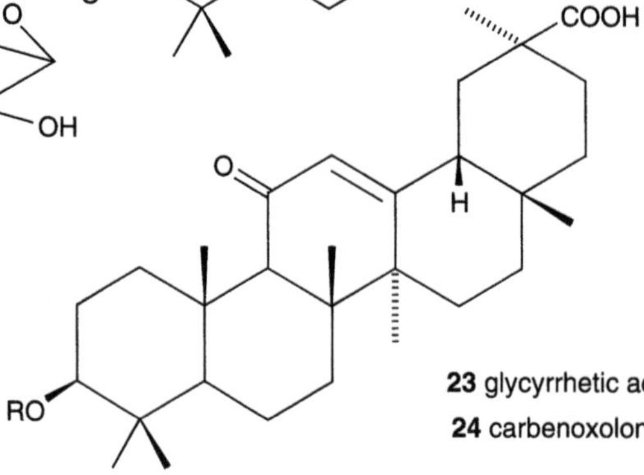

23 glycyrrhetic acid   R

24 carbenoxolone   H
                          COCH$_2$CH$_2$COOH

FIGURE 2.11C

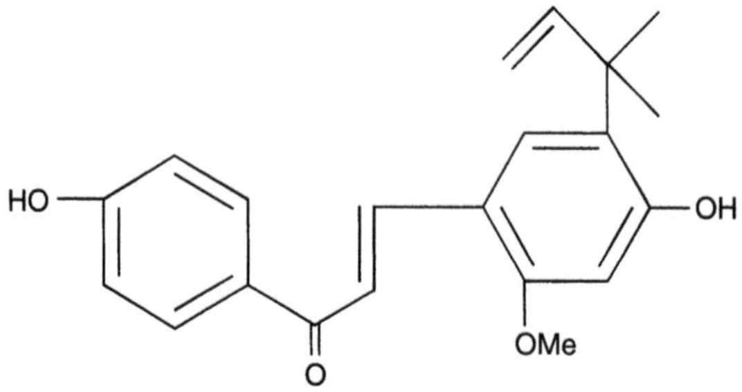

25 licochalcone A

FIGURE 2.11D

Waldstein et Kitaibel, known commercially as Russian licorice, or of other varieties of *Glycyrrhiza glabra*. Glycyrrhiza is of Greek origin and means "sweet root"; glabra means "smooth" and refers to the smooth, podlike fruit of this species (Wichtl, 1989).

Following extraction, Glycyrrhiza yields the licorice commercial products used as flavoring for American type tobaccos, chewing gums, candies, sweetening agents, and depigmentation agents in cosmetic and pharmaceutical products, e.g., in antiulcer drugs (Aspalon, Caved-SR, etc.) (Okabe and Ohtsuki, 1979).

In the Chinese *Pharmacopoeia*, three species of Glycyrrhizae Radix are described: *G. uralensis*, *G. glabra*, and *G. inflata*; these are used equally as Glycyrrhizae Radix in China. The characteristics of *G. inflata* differ from those of *G. uralensis* as follows:
- Height of 50 to 120 cm, glabrous
- Leaves compound
- Leaflets three to seven, ovate to oblong, 2 to 6 cm in length by 1 to 3 cm in width, upper and lower surface covered with sticky glandular trichome
- Flowers June to August, 9 to 12 mm
- Fruit July to September, 2 to 3 cm in length by 4 to 7 mm in width
- Distribution: northwestern to middle China (Xin-jiang, Gan-su). Glycyrrhizae Radix is widely distributed in the northern part of China and produced in large quantities. Counterfeits rarely occur.

ACKNOWLEDGMENTS

The overall aspects of the species are referred to in the references of Jiangsu New Medical College (1990), Keys (1981), Pharmaceutical Affairs Division (1992–1989) and Pharmacopoeia Commission of PRC (1997).

Photographs of the crude drugs were supplied by K. Komatsu.

REFERENCES

Bedir, E., Calis, I., Zerbe, O., and Sticher, O. (1998) Cyclocephaloside I: a novel cycloartane-type glycoside from *Astragalus microcephalus, J. Nat. Prod.,* 61, 503–505.

Chen, L.-Z., Feng, X.-W., and Zhou, J.-H. (1995) Effects of *Rehmannia glutinosa* polysaccharide b on T-lymphocytes in mice bearing sarcoma 180, *Chung Kuo Yao Li Hsueh Pao,* 16, 337–340.

Chevallier A. (1996) *The Encyclopedia of Medicinal Plants,* DK Publishing Inc., New York, p. 5 (Angelicae Radix); pp. 80 (Cinnamomi Cortex); p. 115 (Paeoniae Radix); p. 123 (Rehmanniae Radix).

Choy, Y. M., Leung, K.N., Cho, C.S., Wong, C.K., and Pang, P.K.T. (1994) Immunopharmacological studies of low molecular weight polysaccharide from *Angelica sinensis, Am. J. Chin. Med.,* XXII, 137–145.

Court, E.W. (2000) Ginsing: the genus Panax, in *Medicinal and Aromatic Plants—Industrial Profiles,* Vol. 15, Taylor & Francis Limited, London.

Cuellar, M.J., Giner, R.M., Recio, M.C., Just, M.J., Manez, S., and Rios, J.L. (1997) Effect of the basidiomycete *Poria cocos* on experimental dermatitis and other inflammatory conditions. *Chem. Pharm. Bull.,* 45, 492–494.

Deni Bown (1997) *Herbs,* Dorling Kindersley Limited, London, Seibundo Shinkosha, Publishing Co., Limited, pp. 246 (Astragali Radix); p. 289 (Glycyrrhizae Radix); pp. 321–322 (Ginseng Radix); pp. 372 (Poria).

Goto, H., Shimada, Y., Akechi, Y., Kohta, K., Hattori, M., and Terasawa, K. (1996) Endotherium-dependent vasodilator effect of extract prepared from the roots of *Paeonia lactiflora* on isolated rat aorta, *Planta Med.,* 62, 436–439.

Hasegawa, H. and Saiki, I. (2003) *Cancer Prevention by Ginseng via its Intestinal Bacterial Metabolites,* Art Village Inc., Tokyo.

Hatano, T., Yasuhara, T., Miyamoto, K., and Okuda, T. (1988) Anti-human immunodeficiency virus phenolics from licorice, *Chem. Pharm. Bull.,* 36, 2286–2288.

Hattori, T., Hayashi, K., Nagao, T., Furuta, K., Ito, M., and Suzuki, Y. (1992) Studies on antinephritic effects of plant components (3): effect of phachyman, a main component of *Poria cocos* Wolf on original-type anti-GBM nephritis in rats and its mechanisms, *Jpn. J. Pharmacol.*, 59, 89–96.

Hsu, F-L., Lai, C-H., and Cheng, J-T. (1997) Antihyperglycemic effects of paeoniflorin and 8-debenzoylpae-oniflorin, glucosides from the root of *Paeonia lactiflora, Planta Med.*, 63, 323–325.

Institute of Materia Medica, Chinese Academy of Medical Sciences et al. (1982) *A Record of Chinese Materia Medica* (Zhong yao zhi), The People's Health Publishing House, Beijing, Vol. I (1979): pp. 1–10 (Ginseng Radix); pp. 152–155 (Atractylodis Rhizoma); pp. 156–164 (Atractylodis Lanceae Rhizoma); pp. 182–185 (Paeoniae Radix); pp. 355–366 (Glycyrrhizae Radix); pp. 417–423 (Angelicae Radix); Vol. II (1982): pp. 187–198 (Astragali Radix); pp. 257–261 (Cnidii Rhizoma); Vol. V (1994): pp. 388–396 (Cinnamomi Cortex); pp. 799–803 (Poria).

Jiangsu New Medical College, Shogakkan, (1990), *Chinese Materia Medica Dictionary* (Zhong yao da ci dian), Shanghai Science and Techmology Press and Shogakkann Publishing Co., Ltd., pp. 121–126 (Astragali Radix); pp. 349–352 (Rehmanniae Radix); pp. 371–378 (Glycyrrhizae Radix); pp. 642–643 (Cinnamomi Cortex); pp. 1494–1496 (Cnidium Rhizime); pp. 1585–1589 (Atractylodis Lanceae Rizoma); pp. 1887–1891 (Angelicae Radix); pp. 2017–2026 (Ginseng Radix); pp. 2066–2070 (Paeoniae Radix); pp. 2247–2252 (Atractylodis Rhizoma); pp. 2312–2313 (Poria).

Keys, J.D. (1981), *Chinese Herbs,* Charles Tuttle Company, Inc., Tokyo, pp. 21–22, 2312–2313 (Hoelen), pp. 85–86, 2017–2026 (Ginseng Radix); p. 111 (Cinnamomi Cortex); pp. 114, 121–126 (Astragali Radix); pp. 120–121, 371–378 (Glycyrrhizae Radix); pp. 161 (Paeoniae Radix); p. 195 (Angelicae Radix); pp. 221–222 (Atractylodis Rhizoma, Atractylodis Lanceae Rhizoma); p. 273 (Rehmanniae Radix); pp. 1585–1589 (Atractylodis Lanceae Rhizoma); pp. 2247–2252 (Atractylodis Rhizoma).

Kajimura, K., Takagi, Y., Ueba, N., Yamasaki, K., Sakagami, Y., Yokoyama H., and Yoneda, K. (1996) Protective effect of Astragali Radix by intraperitoneal injection against Japanese encephalitis virus infection in mice, *Biol. Pharm. Bull.*, 19, 855–859.

Kanari, M., Tomoda, M., Gonda, R., Shimizu, N., Kimura, M., Kawaguchi, M., and Kawabe, C. (1989) A reticuloendotherial system-activating arabinoxylan from the bark of *Cinnamomum cassia, Chem. Pharm. Bull,* 37, 3191–3194.

Kanayama, H., Adachi, N., and Togami, M. (1983) A new antitumor polysaccharide from the mycelia of *Poria cocos* Wolf, *Chem. Pharm. Bull.*, 31, 115–118.

Kiho, T., Watanabe, T., Nagai, K., and Ukai, S. (1992) Hypoglycemic activity of polysaccharide fraction from rhizome of *Rehmannia glutinosa* Libosch. f. hueichingensis Hsiao and the effect on carbohydrate metabolism in normal mouse liver, *Yakugaku Zasshi,* 112, 393–400.

Kim, K.-H., Lee, Y.-S., Jung, I.-S., Park, S.-Y., Chung, H.-Y., Lee. I.-R., and Yun, Y.-S. (1998) Acidic polysaccharide from *Panax ginseng,* ginsan, induces Th1 cell and macrophage cytokines and generates LAK cells in synergy with rIL-2, *Planta Med.*, 64, 110–115.

Kitagawa, I., Fukuda, Y., Taniyama, T., and Yoshikawa, M. (1986) Absolute stereostructures of rehmaglutins A, B and D: three new iridoids isolated from Chinese Rehmanniae Radix, *Chem. Pharm. Bull.*, 34, 1399–1402.

Kitagawa, I., Kobayashi, M., Akedo, H., Tatsuta, M., Ishii, H., Shinkai, K., Mukai, M., and Imamura, F. (1995) Inhibition of tumor cell invasion and metastasis by ginsenoside Rg$_3$, *Ginseng Rev.,* 20, 41–46.

Kiyohara, H., Cyong, J.-C., and Yamada, H. (1988) Structure and anticomplemenatry activity of pectic polysaccharides isolated from the root of *Angelica acutiloba* Kitagawa, *Carbohydr. Res.,* 182, 259–275.

Kiyohara, H. and Yamada, H. (1989) Structure of an anticomplementary arabinogalactan from the roots of *Angelica acutiloba* Kitagawa, *Carbohydr. Res.,* 193, 173–192.

Kiyohara, H., Yamada, H., Cyong, J.-C., and Otsuka, Y. (1986) Studies on polysaccharides from *Angelica acutiloba*-V. Molecular aggregation of arabinogalactan from *Angelica acutiloba, J. Pharmacobio-Dyn.,* 9, 339–346.

Kobayashi, M., Fujita, M., and Mitsuhashi, H. (1984) Components of *Cnidium officinale* Makino, *Chem. Pharm. Bull.,* 32, 3770–3773.

Kobayashi, M., Fujita, M., and Mitsuhashi, H. (1987) Studies on the constituents of Umbelliferae plants. XV. Constituents of *Cnidium officinale, Chem. Pharm. Bull.,* 35, 1427–1433.

Konno, C., Sugiyama, K., Kano, M., Takahashi, M., and Hikino, H. (1984) Isolation and hypoglycemic activity of panaxans A, B, C, D and E, glycans of *Panax ginseng* roots, *Planta Med.,* 50, 424–436.

Konno, C., Suzuki, Y., Oishi, K., Munakata, E., and Hikino, H. (1985a) Isolation anad hypoglycemic activity of atractans A, B and C, glycans of *Atractylodes japonica* rhizomes, *Planta Med.*, 102–103.

Konno, C., Murakami, M., Oshima, Y., and Hikino, H. (1985b) Isolation and hypoglycemic activity of panaxans Q, R, S, T and U, glycans of *Panax ginseng* roots, *J. Ethnopharmacol.*, 14, 69–74.

Kumazawa, Y., Nakatsuru, Y., Fujisawa, H., Nishimura, C., Mizunoe, K., Otsuka, Y., and Nomoto, K. (1985) Lymphocyte activation by a polysaccharide fraction separated from hot water extracts of *Angelica acutiloba* Kitagawa, *J. Pharmacobio-Dyn.*, 8, 417–424.

Kwon, B.M., Nam, J.Y., Lee, S.H., Jeong, T.S., Kim, Y.K., and Bok, S.H. (1996) Isolation of cholesteryl ester transfer protein inhibitors from *Panax ginseng* roots, *Chem. Pharm. Bull.*, 44, 444–445.

Lee, Y.-S., Chung, I.-S., Lee, I.-R., Kim, K.-H., Hong, W.-S., and Yun, Y.-S. (1997) Activation of multiple effector pathways of immune system by the antineoplasmic immunostimulator acidic polysaccharide ginsan isolated from *Panax gnseng, Anticancer Res.*, 17, 323–331.

Liu, B.-C. and Fang, J.-N. (1991) Isolation, purification and chemical structure of a glucan from *Glycyrrhiza uralensis* Fisch, *Yao Hsueh Hsueh Pao*, 26, 672–675.

Nagai, Y., Tanaka, O., and Shibata, S. (1971) Chemical studies on the Oriental plant drugs. XXIV. Structure of ginsenoside Rg_1, a new saponin of ginseng root, *Tetrahedron*, 27, 881–883.

Namba, T. (1993) *The Encyclopedia of Wakan-Yaku (Traditional Sino-Japanese Medicines) with Color Pictures*, Hoikusha Publishing Co., Ltd., Osaka, Vol. I: pp.1–8 (Ginseng Radix); pp. 23–24 (Cnidii Rhizoma); pp. 42–45 (Glycyrrhizae Radix); pp. 47–55 (Atractylodis Lanceae Rhizoma, Atractylodis Rhizoma); pp. 58–61 (Angelicae Radix); pp. 63–64 (Rehmanniae Radix); pp. 102–104 (Paeoniae Radix); pp. 149–150 (Astragali Radix); Vol. II: pp. 140–142 (Cinnamomi Cortex); pp. 241–243 (Poria).

Newll, C.A., Anderson L.A., and Phillipson J.K. (1996) *Herbal Medicines, A Guide for Health-Care Professionals*, The Phrmaceutical Press, London, p. 63 (Cinnamomi Cortex).

Nomura, T. and Fukai T. (1998) *Phenolic Constituents of Licorice (Glycyrrhiza Species)*, Fortschritte der Chemie organischer Naturstoffe, 73, 1–140. Springer Wien, New York.

Ohta, H., Ni, J.W., Matsumoto, K., and Watanabe, H. (1993). Peony and its major constituents, paeoniflorin, improve radial maze performance impaired by scopolamine in rats, *Pharmacol. Biochem. Behav.*, 45, 719–723.

Ohno, N., Matsumoto, S., Suzuki, I., Miyazaki, T., Kumazawa, Y., Otsuka, Y., and Yadomae, T. (1983) Biochemical and physicochemical characterization of a mitogen obtained from an Oriental crude drug, Tohki (*Angelica acutiloba* Kitagawa), *J. Pharmacobio-Dyn.*, 6, 903–912.

Okabe, S. and Ohtsuki, H. (1979) New drugs of peptic ulcer, pirenzepine, aspalon, and cetraxate, *Farumashia (J. Pharm. Soc. Jpn.)*, 15, 802.

Oshima, Y., Konno, C., and Hikino, H. (1985) Isolation and hypoglycemic activity of panaxans I, J, K and L, glycans of Panax ginseng roots, *J. Ethnopharmacol.*, 14, 255–259.

Otsuka, Y. (1973) Introductory remarks on licorice from the viewpoint of medical history, *Taisha (Metabol. Dis., Jpn.)*, 10, 613–618.

Ottenjann, R. and Rosch, W. (1970) Therapie des ulcus ventriculi mit carbenoxolon-natrium, *Ergebnisse eines doppelblindversuch, Med. Klin.*, 65, 74–82.

Ozaki, Y. and Ma, J-P. (1990) Inhibitory effects of tetramethylpyrazine and ferulic acid on spontaneous movement of rat uterus *in situ, Chem. Pharm. Bull.*, 38, 1620–1623.

Ou, M. (1992) *Manual of Commonly Used in Herbs in Traditional Chinese Medicine*, Guandong Science and Technology Press, Hong Kong, p. 52 (Cnidii Rhizoma); pp. 160, 665 (Rehmannia Rhizome); p. 223 (Cinnamomi Cortex); p. 614 (Angelicae Radix).

Pharmaceutical Affairs Division, Ministry of Health and Welfare, *Medicinal Plants, Cultivation and Quality Evaluation*, Yakuji Nippo, Ltd., Tokyo, Part I (1992): pp. 15–26 (Rehmanniae Radix); pp. 39–50 (Angelicae Radix); Part II (1993): pp. 23–34 (Cnidii Rhizoma); Part III (1994): pp. 43–56 (Paeoniae Radix); Part V (1996): pp. 27–38 (Ginseng Radix); pp. 15–24 (Atractylodis Rhizoma); Part VIII (1999): pp. 55–64 (Poria).

Pharmacopoeia Commission of PRC (Engl. ed.) (1997), *Pharmacopoeia of the People's Republic of China*, vol. 1, Chemical Industry Press, Beijing, p. 28 (Cinnamomi Cortex); p. 138 (Angelicae Radix); p. 159 (Paeoniae Radix); p. 166 (Rehmanniae Radix); p. 186 (Cnidii Rhizoma).

Pharmacopoeia Commission of the Ministry of Public Health, PRC (2000) *Pharmacopoeia of the People's Republic of China* (2000 ed.), Chemical Industry Press, Beijing, Vol. 1, pp. 6–7 (Ginseng Radix); pp. 30–31 (Cnidium Rhizome); pp. 65–66 (Glycyrrhizae Radix); pp. 77–78 (Atractylodes Rhizome);

pp. 78–79, p. 125 (Paeoniae Radix); pp. 94–95 (Rehmanniae Radix); pp. 101–102 (Japanese Angelicae Radix); p. 104 (Cinnamomi Cortex); p. 131 (Atractylodis Rhizoma); pp. 127–128 (Atractylodis Lanceae Rhizoma); p. 193 (Poria); pp. 249–250 (Astragali Radix).

Ravindran, P.N., Babu Nirmal, K., and Shylaja, M. (2004) *Cinnamon and Cassia: the Genus Cinnamomum. Medicinal and Aromatic Plant Industrial Profiles,* vol. 36, CRC Press LLC, Boca Raton, FL.

Resch, M., Steigel, A., Chen, Z-L., and Bauer, R. (1998) 5-Lipoxygenase and cyclooxygenase-1 inhibitory active compounds from Atractylodes lancea, *J. Nat. Med.,* 61, 347–350.

Robbers, J.E., Speedie, M.K., and Tyler, V.E. (1996) *Pharmacognosy and Pharmaco-Biotechnology,* Williams & Wilkins, Baltimore,156–157.

Ron, T. (1984) *Chinese Tonic Herbs,* Japan Publications, Inc., New York, p. 85 (Angelicae Radix); p. 101 (Rehmanniae Radix); p. 128 (Paeoniae Radix); p. 138 (Cnidii Rhizoma).

Sasaki, H., Nishimura, H., Morita, T., Chin, M., Mitsuhashi, H., Komatsu, Y., Maruyama, H., Guo-rui, T., Wei, H., and Yulang, X. (1989) Immunosuppressive principles of *Rehmannia glutinosa* var. hueichingensis, *Planta Med.,* 55, 458–462.

Shibata, S., Fujita, M., Itokawa, H., Tanaka, O., and Ishii, T. (1962) The structure of panaxadiol, a sapogenin of Ginseng, *Tetrahedron Lett.,* 419–422.

Shibata, S., Tanaka, O., Sado, M., and Tsushima, S. (1963) On genuine sapogenin of ginseng, *Tetrahedron Lett.,* 795–800.

Shibata, S., Tanaka, O., Shoji, J., and Saito, H. (1985) Chemistry and pharmacology of Panax, in *Economic and Medicinal Plant Research,* Vol. 1, Wagner, H., Hikino, H., and Farnsworth, N.R., (Eds.), Academic Press Inc. Orlando, FL, 217–287.

Shibata, S., Inoue, H., Iwata, S., Ma, R.-D., Yu, L.-J., Ueyama, H., Takayasu, J., Hasegawa, T., Tokuda, H., Nishino, A., Nishino, H., and Iwashima, A. (1991) Inhibitory effects of licochalcone a isolated from *Glycyrrhiza inflata* root on inflammatory ear edema and tumour promotion in mice, *Planta Med.,* 57, 221–224.

Shibata, S. (2000) A drug over the millennia: pharmacognosy, chemistry, and pharmacology of licorice, *Yakugaku zasshi,* 120, 849–862.

Shimizu, N., Tomoda, M., Kanari, M., Gonda, R., Satoh, A., and Satoh, N. (1990) A novel neutral polysaccharide having activity on the reticuloendotherial system form the root of *Glycyrrhiza uralensis, Chem. Pharm. Bull.,* 38, 3069–3071.

Shimizu, N., Tomoda, M., Kanari, M., and Gonda, R. (1991a) An acidic polysaccharide having activity on the reticuloendotherial system from the root of *Astragalus membranaceus, Chem. Pharm. Bull.,* 39, 2969–2972.

Shimizu, N., Tomoda, M., Satoh, M., Gonda, R., and Ohara, N. (1991b) Characterization of a polysaccharide having activity on the reticuloendotherial system from the stolon of *Glycyrrhiza glabra* var. *glandulifera, Chem. Pharm. Bull.,* 39, 2082–2086.

Society of Japanese Pharmacopoeia (2001) *The Japanese Pharmacopoeia,* 14th ed , Hirokawa Publishing Co., Tokyo, D 135–138 (Astragalus root); D-242–251 (Glycyrrhiza); D-327–333 (cinnamon bark); D-480–482 (Rehmannia root); D-503–509 (peony root); D638–642 (Cnidium Rhizome); D-674–677 (Atractylodes Lanceae Rhizome); D-795–800 (Japanese Angelica root); D-862–873 (ginseng); D-980–985 (Atractylodes Rhizome); D-1012–1015 (Poria).

Sonoda, Y., Kasahara, T., Mukaida, N., Shimizu, N., Tomoda, M., and Takeda, T. (1998) Stimulation of interleukin-8 production by acidic polysaccharides from the root of *Panax ginseng, Immunopharmacol.,* 38, 287–294.

State Pharmaceutical Administration (1998) *Zhonghua Bencao,* Shanghai Science and Technology Press, Shanghai, pp. 187–193 (Poria); pp. 454–463 (Cinnamamomi Cortex); pp. 645–663 (Paeoniae Cortex); pp. 815–835 (Astragali Cortex); pp. 867–884 (Glycyrrhizae Radix); pp. 1269–1294 (Ginseng Radix); pp. 1341–1355 (Japanese Angelicae Radix); pp. 1397–1407 (Cnidii Rhizoma); pp. 1734–1750 (Rehmanniae Radix); pp. 1885–1893 (Atractylodis Lanceae Rhizome); pp. 1893–1902 (Atractylodis Rhizoma).

Stuart, M. (1987) *The Encyclopedia of Herbs and Herbalism,* Macdonald & Co. Ltd., London, p. 174 (Cinnamomi Cortex).

Sun, X.-B., Matsumoto, T., Kiyohara, H., Hirano, M., and Yamada, H. (1991) Cytoprotective activity of pectic polysaccharides from the root of *Panax ginseng, J. Ethnopharmacol.,* 31, 101–107.

Tai, T., Akita, Y., Kinoshita, K., Koyama, K., Takahashi, K., and Watanabe, K. (1995a) Antiemetic principles of *Poria cocos, Planta Med.,* 61, 527–530.

Tai, T., Shingu, T., Kikuchi, T., Tezuka, Y., and Akahori, A. (1995b) *Phytochemistry*, 39, 1165–1169.

Takagi, K. and Ishii, Y. (1967) Peptic ulcer inhibiting properties of a new fraction from licorice root (FM 100). I. Experimental peptic ulcer and general pharmacology, *Arzneim.-Forsch*, 17, 1544–1547.

Takagi, K., Saito, H., and Nabata, H. (1972) Pharmacological studies of Panax ginseng root, *Jpn. J. Pharmacol.*, 22, 245–250.

Takagi, K., Saito, H., and Nabata, H. (1974) Effect of *Panax ginseng* on continuous movement and exercise in mice, *Jpn. J. Pharmacol.*, 24, 41–47.

Takasu, S. (2000) *Today's Therapy 2000*, IGAKU-SHOIN Ltd., Tokyo, Vol. 42, pp. 1080.

Takeda, O., Miki, E., Terabayashi, S., Okada, M., Lu, Y., and He, H-S. (1996) A comparative study on essential oil components of wild and cultivated *Atractylodes lancea* and *A. chinensis*, *Planta Med.*, 62, 444–449.

Terasawa, K. (1990) The status of traditional Sino–Japanes (Kampon) medicine currently practised in Japan, In *Economic and Medicinal Plant Research*, (Wagner, H. and Farnsworth, N.R., Eds.), Academic Press, SanDiego, Vol. 4, 57–70.

Tomoda, M., Shimada, K., Konno, C., Sugiyama, K., and Hikino, H. (1984) Partial structure of panaxan A, a hypoglycemic glycan of *Panax ginseng* roots, *Planta Med.*, 50, 436–438.

Tomoda, M., Shimizu, N., Kanari, M., Gonda, R., Arai, S., and Okuda, Y. (1990) Characterization of two polysaccharides having activity on reticuloendotherial system from the root of *Glycyrrhiza uralensis*, *Chem. Pharm. Bull.*, 38, 1667–1671.

Tomoda, M., Ohara, N., Gonda, R., Shimizu, N., Takada, K., Satoh, Y., and Shiral, S. (1992) An acidic polysaccharide having immunological activities from the rhizome of *Cnidium officinale*, *Chem. Pharm. Bull.*, 40, 3025–3029.

Tomoda, M., Matsumoto, K., Shimizu, N., Gonda, R., and Ohara, N. (1993a) Characterization of a neutral and acidic polysaccharide having immunological activities from the root of *Paeonia lactiflora*, *Biol. Pharm. Bull.*, 16, 1207-1210.

Tomoda, M., Takeda, K., Shimizu, N., Gonda, R., Ohara, N., Takada, K., and Hirabayashi, K. (1993b) Characterization of two acidic polysaccharides having immunological activities from the root of *Panax ginseng*, *Biol. Pharm. Bull.*, 16, 22–25.

Tomoda, M., Hirabayashi, K., Shimizu, N., Gonda, R., Ohara, N., and Takada, K. (1993c) Characterization of two novel polysaccharides having immunological activities from the root of *Panax ginseng*, *Biol. Pharm. Bull.*, 16, 1087–1090.

Tomoda, M., Ohara, N., Shimizu, N., and Gonda, R. (1994a) Characterization of a novel glucan, which exhibits reticuloendotherial system-potentiating and anticomplementary activities, from the rhizome of *Cnidium officinale*, *Chem. Pharm. Bull.*, 42, 630–633.

Tomoda, M., Ohara, N., Shimizu, N., and Gonda, R. (1994b) Characterization of a novel heteroglucan from the rhizome of *Cnidium officinale* exhibiting high reticuloendotherial system-potentiating and anti-complementary activities, *Biol. Pharm. Bull.*, 17, 973–976.

Tomoda, M., Matsumoto, K., Shimizu, N., Gonda, R., Ohara, N., and Hirabayashi, K. (1994c) An acidic polysaccharide with immonological activities from the root of *Paeonia lactiflora*, *Biol. Pharm. Bull.*, 17, 1161–1164.

Tomoda, M., Miyamoto, H., Shimizu, N., Gonda, R., and Ohara, N. (1994d) Two acidic polysaccharides having reticuloendotherial system-potentiating activity from the raw root of *Rehmannia glutinosa*, *Biol. Pharm. Bull.*, 17, 1456–1459.

Tomoda, M., Miyamoto, H., Shimizu, N., Gonda, R., and Ohara, N. (1994e) Characterization of two polysaccharides having activity on the reticuloendotherial system from the root of *Rehmania glutinosa*, *Chem. Pharm. Bull.*, 42, 625–629.

Tomoda, M., Miyamoto, H. and Shimizu, N. (1994f) Structural features and anticomplementary activity of rehmannan SA, a polysaccharide from the root of *Rehmannia glutinosa*, *Chem. Pharm. Bull.*, 42, 1666–1668.

Tseng, J. and Chang, J.G. (1992) Suppression of tumor necrosis factor-alpha, interleukin-1 beta, interleukin 6 and granulocyte–monocyte colony stimulating factor secretion from human monocytes by extract of *Poria cocos*, *Zhonghua Min Guo Wei Sheng Wie Ji Mian Yi Xue Za Zhi*, 25, 1–11.

Wang, D-C. (1989) Influence of *Astragalus membranaceus* (AM) polysaccharide FB on immunologic function of human periphery blood lymphocyte, *Chung Hua Chung Liu Tsa Chih.*, 11, 180–183.

Wang, J., Ito, H., and Shimura, K. (1989) Enhancing effect of antitumor polysaccharide from *Astragalus* or *Radix hedysarum* on C3 cleavage production of macrophages in mice, *Jpn. J. Pharmacol.*, 51, 432–434.

Wang, L.-X. and Han, Z.-W. (1992) The effect of Astragalus polysaccharide on endotoxin-induced toxicity in mice, *Yao Hsuch Pao,* 27, 5–9.

Wang, Y. and Zhu, B. (1996) The effect of angelica polysaccharide on proliferation and differentiation of hematopoietic progenitor cell, *Chung Hua Hsueh Tsa Chih,* 76, 363–366.

Warsi, S.A. and Whelan, W.J. (1957) Structure of pachyman, the polysaccharide component of *Poria cocos, Chem. Ind.,* 1573.

Wei, X.-L. and Ru, X.-B. (1997) Effect of low-molecular-weight *Rehmannia glutinosa* polysaccharides on p 53 gene expression, *Chung Kuo Yao Li Hsueh Pao,* 18, 471–474.

WHO, Anonymous (1999) *WHO Monographs on Selected Medicinal Plants,* vol. 1, World Health Organization, Geneva, pp. 50–58 (Astragali Radix); pp. 95–104 (Cinnamomi Cortex); pp. 168–182 (Ginseng Radix); pp. 183–194 (Glycyrrhizae Radix); pp. 195–201 (Paeoniae Radix).

Wichtl, M. (1989), *Teedrogen,* Wissenshaftliche Verlagsgesellschaft mbH, Stuttgart, pp. 479–482 (Glycyrrhizae Radix).

Wu, Z.-Y. and Raven, P.H. (1998) *Flora of China,* vol. 18, Missouri Botanical Garden Press, St. Louis, p. 53 (Rehmanniae Radix).

Xie, F.-X. and Hu, T.S. (1994) *A Colour Atlas of Traditional Chinese Medicines with Text on Techniques of Their Cultivation,* Jin Dun Press, Beijing, p. 169 (Paeoniae Radix); p. 175 (Rehmanniae Radix); p. 177 (Angelicae Radix); p. 208 (Cnidii Rhizoma).

Yamada, H., Kiyohara, H., Cyong, J.-C., and Otsuka, Y. (1985) Studies on polysaccharides from *Angelica acutiloba*-IV. Characterization of an anticomplementary arabinogalactan from the roots of *Angelica acutiloba* Kitagawa, *Molec. Immunol.,* 22, 295–304.

Yamada, H., Komiyama, K., Kiyohara, H., Cyong, J.-C., Hirakawa, Y., and Otsuka, Y. (1990) Structural characterization of an antitumor pectic polysaccharide from the roots of *Angelica acutiloba, Planta Med.,* 56, 182–186.

Yasukawa, K., Kaminaga, T., Kitanaka, S., Tai, T., Nunoura, Y., Natori, S., and Takido, M. (1998) 3b-p-Hydroxybenzoyldehydrotumulosic acid from Poria Cocos and its anti-inflammatory effect, *Phytochemistry,* 48, 1357–1360.

Yoshikawa, M., Fukuda, Y., Taniyama, T., and Kitadawa, I. (1986) Absolute stereostructures of rehmaglutin C and glutinoside, a new iridoid lactone and a new chlorinated iridoid glucoside from Chinese Renmanial Radix, *Chem. Pharm. Bull.,* 34, 1403–2294.

Yu, K.-W., Kiyohara, H., Matsumoto, T., Yang, H.-C., and Yamada, H. (1998) Intestinal immune system modulating polysaccharides from rhizomes of *Atractylodes lancea, Planta Med.,* 64, 714–719.

Zhang, E.-Q. (1990) *The Chinese Materia Medica,* Publishing House of Shanghai University of Traditional Chinese Medicine, Shanghai, p. 24 (Cnidii Rhizoma); p. 70 (Angelicae Radix); p. 112 (Rehmanniae Radix); p. 120 (Paeoniae Radix); p. 230 (Cinnamomi Cortex); p. 290 (Cnidii Rhizoma); p. 406 (Chinese Angelica Root); p. 410 (Rehmanniae Radix); p. 416 (Paeoniae Radix).

Zhao, J.-F., Kiyohara, H., Sun, X.-B., Matsumoto, T., Cyong, J.-C., Yamada, H., Takemoto, N., and Kawamura, H. (1991a) *In vitro* immunostimulating polysaccharide fractions from roots of *Glycyrrhiza uralensis* Fisch. et DC., *Phytotherapy Res.,* 5, 206–210.

Zhao, J.-F., Kiyohara, H., Yamada, H., Takemoto, N., and Kawamura, H. (1991b) Heterogeneity and characterization of mitogenic and anticomplemenatry pectic polysaccharides from the roots of *Glycyrrhiza uralensis* Fisch et DC, *Carbohydr. Res.,* 219, 149–172.

3 Therapeutic Indications of Juzen-taiho-to in Modern Therapy

Toshihiko Hanawa

CONTENTS

3.1 INTRODUCTION

The origin and traditional use of Juzen-taiho-to (Shi-Quan-Da-Bu-Tang in Chinese) are found in a formula composed of ten crude drugs derived from the formula Shikunshi-to (Si-Jun-Zi-Tang in Chinese), which consists of four crude drugs—Ginseng Radix, Atractylodis Rhizoma, Poria, and Glycyrrhizae Radix—and another formula, Shimotsu-to (Si-Wu-Tang in Chinese), consisting of four crude drugs—Angelicae Radix, Paeoniae Radix, Cnidii Rhizoma, and Rehmanniae Radix (together with Hacchin-to [Pa-Chen-Tang in Chinese])—plus two additional drugs: Cinnamomi Cortex and Astragali Radix. The name *Juzen-taiho-to* means that the combination of the ten drugs brings out their full abilities to correct deficiency. The component crude drugs in Shikunshi-to supplement "Ki" (Qi in Chinese), which is vital energy and circulation of the vital energies within body, while those in Shimotsu-to supplement "Ketsu" (Xue in Chinese), which is a blood circulatory function. Therefore, the basic purpose of this prescription is to supplement Ki and Ketsu (Hanawa, 1994).

Shikunshi-to has been used for lack of vigor, fatigue, weak gastrointestinal function, and anorexia. It has also been used for anemia and symptoms called "Kekkyo" (Xue Xu in Chinese)—deficiency of blood circulatory function, such as dry sallow skin, rough lips, brittle nails, fuzzy head, dim sight, loss of hair, and bleeding. The additional drugs, Cinnamomi Cortex and Astragali Radix, are prescribed to potentiate the ability of both formulas.

Juzen-taiho-to is described in the Chinese *Pharmacopea* of the 12th century, *Wazai-kyokuho* (*He-ji-ju-fang* in Chinese), volume 5, Syokyomon, in a chapter stating inadequate resistance to disease (Chen, 1988). The indication for this formula is "Kyubyo kyoson" (impairment of various functions due to long-term disease), which may indicate chronic wasting diseases according to the modern interpretation. In the past, pulmonary tuberculosis was a representative disease for chronic wasting diseases. *Wazai-kyokuho* describes "withered yellow complexion" in the facial coloring of the patient, which indicates chronic anemia. This symptom is also treated with Juzen-taiho-to.

The Chinese *Pharmacopea* also contains the statement, "damaged ketsu and impaired function of Ki." Juzen-taiho-to is useful when Ki as action and Ketsu as a material basis supporting Ki (two major factors protecting human physiology) are damaged or impaired. This formula is also used for "impaired functions of the spleen (aspect of digestive function) and kidneys (urogenital function)," i.e., when the functions of the spleen (acquired energy source) and the kidneys (congenital energy source) are decreased. Juzen-taiho-to enhances gastrointestinal function and activates the antidisease responses. This formula has been used for physical exhaustion due to acquired diseases, as well as before and after giving birth and for various diseases associated with aging.

3.2 PRACTICAL USE IN MODERN MEDICINE

3.2.1 TREATMENT OF SYMPTOMS

Juzen-taiho-to is used for enhancement and recovery of physical strength and improvement of anemia after blood loss. It is effective in the acute exacerbation state of atopic dermatitis and dirty skin that is erosive and bleeds readily. Juzen-taiho-to is also used to overcome the rebound period after discontinuation of steroids, such as the markedly oozing eruptions of the face during the rebound period.

3.2.2 LONG-TERM USE

As well as being related to another Kampo formula, Hochu-ekki-to (Bu-Zhong-Yi-Qi-Tang in Chinese), Juzen-taiho-to is used for improving general conditions such as weak constitution and general malaise. This formula is particularly chosen for improving the constitution susceptible to immune-associated diseases. Skin mucosal erosion is a major indication. Ulcer, chronic purulent lesions, bleeding tendency, and anemia are also associated findings. Juzen-taiho-to has been used as adjuvant therapy for ulcerative colitis, stomatitis, anal fistulas, chronic sinusitis, chronic otitis media, pyorrhea, Bechet's disease, and poor general condition due to malignant tumors. Patients with dermatological conditions, malignant tumors, and other conditions have been treated with Juzen-taiho-to in one of the specialized Kampo clinics, Clinical Division of Oriental Medicine Research Center, in the Kitasato Institute, as shown in Table 3.1.

TABLE 3.1
Use of Juzen-Taiho-To in Disease Treatment at the Clinical Division of Oriental Medicine Research Center, The Kitasato Institute

Disease[a]	No. of Patients[a]
Dermatological diseases (atopic dermatitis, skin eruption, etc.)	63
Malignant tumors (postoperative, after radiotherapy, prevention of recurrence, etc.)	46
Gynecological diseases	4
Anemia	4
Autoimmune diseases	4
Ulcerative colitis	2
Others	9
Total	132

[a] Disease name and frequency of Juzen-taiho-to use in patients who underwent the first examination between January, 1998 and August, 2000.

3.3 TREATMENT OF ATOPIC DERMATITIS

Juzen-taiho-to is frequently used for atopic dermatitis at the Kampo Clinic of Oriental Medicine Research Center of the Kitasato Institute because many patients with adult type atopic dermatitis visit the clinic due to "rebound symptoms" occurring after they suddenly stop the excessive use of steroids for external use of their own choice method. Juzen-taiho-to restores the breakdown of the immune system resulting from sudden discontinuation of steroids, particularly repairing mucosal immunity of the skin. Hanawa (1995) presented the following representative case of a 29-year old male patient with atopic dermatitis:

The patient developed atopy when he was in the third year of high school and was treated with steroids for external use at a dermatological department. He had been using four to five tubes of hydrocortisone butyrate (Locoid® ointment) per month for 6 months. Subsequently, betamethasone valerate (Rinderon VG® ointment) was topically applied. However, because of no improvement in the eruptions, he visited another hospital.

On October 20, 1993, steroids for external use were discontinued according to the instruction of the physician, but the symptoms were gradually aggravated. The physician considered that the aggravation was not merely due to rebound. On December 3, 1993, the patient was referred to Kampo Clinic of Oriental Medicine Research Center to determine the presence or absence of secondary infection and undergo treatment. Kaposi varicelliform eruptions, erosion in the entire face, purulence, and crusts were observed. The trunk and four limbs also showed erythroderma and abrasion but were more normal compared with the face. The patient immediately entered the clinic's hospital, and the face was treated by disinfection with Isodine® and topical application of Gentacin® (antibiotic ointment) in the open state.

Cultures of local discharge revealed *Staphylococcus aureus*. Antiherpes IgM antibody titer was less than ×10. Thus, secondary infection with *S. aureus* was confirmed, but not the herpes virus. After treatment with Juzen-Taiho-To, the eruptions on the face rapidly improved, and the patient was discharged after 6 days.

When treatment experiences of acute exacerbation after rebound of atopic dermatitis and rosacea-like dermatitis at the research center are summarized, a good indication for Juzen-taiho-to is "an exhausted skin, with oozing erosion and desquamation without phase elevation after being peeled off once, that makes others feel like turning away."

3.4 CLINICAL USE FOR CANCER

Juzen-taiho-to is most frequently used for cancer, and many reports have suggested the following effects of this prescription:

- Improvement in the general condition before and after the operation
- Recovery from chronic exhaustion due to malignant tumors
- Prevention or reduction of the side effects of anticancer drugs
- Improvement in general condition after radiotherapy

Juzen-taiho-to is often prescribed to improve the general condition in cancer patients such as in the following case study (Watanabe et al., 2000):

A 52-year-old female patient with cancer of the left breast underwent an operation in November, 1987. In November, 1994, shadows were detected in her bilateral lung fields on x-ray films. Chemotherapy with 5-fluorouracil (5 FU) did not reduce the size of the shadows. In July, 1996, she visited the Kampo Outpatient Clinic of the author's research center and was treated with Hochu-ekki-to added to 5 g of coriolus (mycelium of *Coriolus versicolor*). On March 25, 1997, the left lung lesion was resected. On November 11, 1997, the right lung tumor was resected.

Except for certain periods before and after both operations, the patient was treated with Bukuryo-shigyaku-to (Fu-Ling-Si-Ni-Tang in Chinese).

In August, 1998, dyspnea developed and chest x-ray examination showed pleural effusion in the left lung. The patient was admitted and treated by systemic administration and intrathoracic infusion of adriamycin and administration of Kampo formulas Senkin-naitaku-sanryo (Chien-chin-Nei-Tuo-San-Liao in Chinese) and Ninjin-yoei-to (Ren-Shen-Yang-Rong-Tang in Chinese), but showed no improvement. She refused to receive chemotherapy because of exhaustion after chemotherapy performed on November 10 and was treated only with Kampo medicine thereafter. On December 1, treatment with Juzen-taiho-to mixed with 5 g coliolus was initiated again and added to 1 g of Aconti Tuber. Hemorrhagic pleural effusion from the drain was 300 ml per day but rapidly decreased after a change to Juzen-taiho-to mixed with 1 g Aconite Tuber and 5 g Lithospermi Radix on January 11, 1999; this was stopped on January 14. The thoracic drain was removed on January 15 and the patient was discharged on February 6. The subsequent course was favorable. Since December 10, 1999, she has been in a good general condition and is being treated on an outpatient basis.

Many studies have shown the effectiveness of Juzen-taiho-to based on an improvement in immunological parameters. Okamoto and Sairenji (1993) administered Juzen-taiho-to for 1 year to 23 patients who underwent gastrectomy and 16 who underwent colectomy and evaluated serial changes in cellular immune response and changes in blood lipids. The NK cell activity definitely increased at 1 and 3 months after the initiation of Juzen-taiho-to administration. Changes in NK cell activity were negatively correlated with changes in lipoprotein among blood lipids.

3.5 IMPROVEMENTS OF SIDE EFFECTS OF POSTOPERATIVE CHEMOTHERAPY AND EFFECTS ON IMMUNOLOGICAL PARAMETERS AND QUALITY OF LIFE (QOL)

To evaluate the usefulness of Juzen-taiho-to on advanced breast cancer, Adachi et al. (Adachi, 1989) allocated patients into two different treatment groups (Groups A and B) by the envelope method and compared the effects of treatment on survival, symptoms, and clinical examination values. Group A underwent endocrinological therapy and chemotherapy in combination with Juzen-taiho-to administration and Group B underwent only endocrinological therapy and chemotherapy. Of 130 patients entered into the study, 119 (58 in Group A and 61 in Group B) could be evaluated. Concerning the QOL of the patients, Juzen-taiho-to was useful for improving cold sensation in the hands and feet as well as anorexia, and had inhibitory effects on chemotherapy-induced leukopenia, especially lymphopenia.

Yamada et al. (1991) administered Juzen-taiho-to to patients who underwent curative gastric resection to improve the general condition and prevent the side effects of anticancer drugs; they compared the results with those in the nonadministration group. The hemoglobin level had increased 2 months after the operation in the administration group. The leukocyte count was markedly decreased 2 months after the operation in the nonadministration group, showing a significant difference between the two groups. CD8 (cluster of differentiation 8) was increased 2 months after the operation in the nonadministration group. Changes in the CD4/8 ratio did not significantly differ between the Juzen-taiho-to administration group and the nonadministration group, but was low 2 months or more after the operation in the nonadministration group. These results suggest the effects of Juzen-taiho-to on recovering from leukopenia and anemia associated with anticancer drugs.

Miura et al. (1985) evaluated the effects of Juzen-taiho-to in 59 patients with Stage II or III cancer who underwent curative resection and postoperative adjuvant multiple-drug chemotherapy (MFO chemotherapy) with mitomycin C (MMC), 5-FU, and OK-432 (Picibanil; biological response modifier derived from the weakly virulent Su strain of *Streptococcus pyogenes*). Juzen-taiho-to was used in combination with MFO chemotherapy in 29 of the 59 patients but not in the other 30. Juzen-taiho-to reduced the myelosuppressive action of MFO chemotherapy and promoted postoperative recovery of body weight. In addition, its preventive effects on transient liver dysfunction due to anticancer drugs were suggested.

Nagatomo (1992) randomly allocated 20 patients with unresectable hepatocellular cancer into two groups treated with cisplatin alone (10 patients) or in combination with Juzen-taiho-to (10 patients) and evaluated the effects of Juzen-taiho-to in reducing the side effects of cisplatin. Juzen-taiho-to significantly inhibited nausea and vomiting as side effects of cisplatin.

Kuboki et al. (1992) evaluated the effects of Juzen-taiho-to in 32 patients with unresectable gastric cancer or advanced cancer who underwent three-drug chemotherapy (EAP chemotherapy) with etoposide, adriamycin, and cisplatin. The mean duration showing a leukocyte count of less than $2000/mm^3$ was 5.5 days in the Juzen-taiho-to administration group and 10.7 days in the nonadministration group, thus showing a significant shortening of the duration in the administration group. The mean leukocyte count 4 weeks after initiation of EAP therapy was $5640/mm^3$ in the Juzen-taiho-to administration group and $2880/mm^3$ in the nonadministration group; this showed a significant recovery of the leukocyte count in the administration group. The degree of nausea, vomiting, and cancer pain before and after treatment decreased in the administration group.

Kosaka et al. (1994) treated 127 patients with cancer with tegafur (UFT) alone and 124 with UFT in combination with Juzen-taiho-to (total: 251 patients; 65 with gastric cancer, 103 with colorectal cancer, and 83 with breast cancer). The incidence of the side effects of the anticancer drug was lower in the group treated with UFT in combination with Juzen-taiho-to than in the group treated with UFT alone. For gastric cancer, improvement was observed in each QOL item 2 weeks after the initiation of administration in the drug combination group compared with the noncombination group. For colorectal cancer, the general feeling during the entire course improved in the drug combination group compared with the noncombination group, showing significant improvement in the latter after 1 month. For breast cancer, each item improved 2 weeks after the initiation of administration in the drug combination group; particularly, disease-associated symptoms showed a significant difference.

Fukagawa et al. (1986) administered Juzen-taiho-to to 44 patients with malignant tumors in the gynecological field to reduce complaints at the time of chemotherapy. Among the side effects of chemotherapy, digestive symptoms such as nausea, vomiting, and anorexia reduced in the administration group compared with the nonadministration group, showing decreases in the duration (days) and frequency of these symptoms and the frequency of the use of antiemetic drugs. Hematological examination showed a significantly lower incidence of leukopenia in the administration group than in the nonadministration group, suggesting the preventive effects of Juzen-taiho-to on leukopenia.

Go et al. (1985) evaluated the clinical effects of Juzen-taiho-to on symptoms and the results of examinations in the following patients with gynecological disease:

- Group A: 11 cancer patients treated with Juzen-taiho-to
- Group B: 12 cancer patients not treated with this prescription
- Group C: 10 surgically treated noncancer patients treated with Juzen-taiho-to
- Group D: 10 surgically treated noncancer patients not treated with this prescription

All patients in Groups A and B underwent cancer chemotherapy or immunotherapy and radiotherapy. In Groups A and C treated with Juzen-taiho-to, patients recovered physical strength rapidly after the operation. Group A showed a decrease in the incidence of side effects of chemotherapy and radiotherapy such as anorexia, general malaise, dizziness, stomatits, nausea, and vomiting. Even when these side effects developed, their degree was mild compared with the nonadministration groups. In addition, compared with the nonadministration group, marked weight gain was observed in Group A and a slight increase in cholesterol in Group C.

3.6 SIDE EFFECTS OF RADIOTHERAPY

Miyamoto et al. (1985) administered Juzen-taiho-to to 22 patients who complained of symptoms such as anorexia, general malaise, abdominal discomfort, diarrhea, nausea, radiation sickness, and constipation. They observed marked improvement in these symptoms in two patients (9.1%), moderate improvement in ten (45.4%), slight improvement in six (27.3%), and no change in four (18.2%).

Chiba et al. (1987) administered Juzen-taiho-to for more than 3 months to 18 of 27 patients with cervical cancer treated by radiotherapy, but not to another 9 patients, and compared the results between the two groups. In the administration group, anorexia and general malaise moderately to markedly improved in many patients, and diarrhea, abdominal discomfort, nausea, and constipation improved in most patients; radiotherapy was not discontinued due to leukopenia in any patient. Thus, administration of Juzen-taiho-to was useful for preventing leukopenia. In addition, a few patients showed improvement in radiation proctitis after long-term administration of Juzen-taiho-to.

3.7 ULCERATIVE COLITIS

Matsuo et al. (Matsuo, 2000) administered Juzen-taiho-to to eight patients with mild ulcerative colitis (UC) for 6 months: four with the proctitis type and four with left colitis type for 6 months. These researchers evaluated its effects in terms of the presence or absence of recrudescence. In five (62.5%) of the eight patients, no recrudescence was observed, and a reduction and withdrawal of steroids was possible. During the observation period, slight melena occurred in three patients, of whom two endoscopically showed recrudescence.

Decreased NK activity has been reported in UC patients treated with steroids for a relatively long period. We speculate that Juzen-taiho-to activates NK activity, enhancing decreased NK activity due to active UC colitis or steroid administration and that this immunomodulation maintains the remission of UC.

3.8 CAUTIONS FOR USE

As a component drug of Juzen-taiho-to, Rehmanniae Radix may cause gastrointestinal symptoms. When anorexia or diarrhea is observed, administration is discontinued. Because Glycyrrhizae Radix is part of the contents, attention should be paid to the possible development of edema and an increase in blood pressure.

REFERENCES

Adachi, I. The role of Kampo formula as supplementary therapy for progressive breast cancer (in Japanese). *Biotherapy,* 1989, 3: 782–788.

Chen, S-W. *Taihei-Keimin-Wazai-kyoku-ho (Tai-ping-hui-min-he-ji-ju-fang* in Chinese). Republished version (in Japanese). Kanseki-Isyo-Syuusei Enterprise Co. 1988, p. 96.

Chiba, T., Yokoyama, T., and Sejima, S. Clinical effect of Juzen-taiho-to and Hochu-ekki-to for irradiation therapy of cervical cancer of uterus (in Japanese). *J. Obstet. Gynecol. Chu-goku Shikoku Jpn.,* 1987, 35, 212–217.

Fukagawa, R., Matsumura, T., and Hamai, J. Improvement of symptoms induced by Juzen-taiho-to for patients with malignant disease (in Japanese). *Res. Obstet. Gynecol.,* 1986, 3, 20–27.

Go, M., Ichikawa, Y., and Osawa, M. Clinical experiments of Juzentaiho-to for patients of obstetrics and gynecology (in Japanese). *Obstet. Gynecol.,* 1985, 5, 539–544.

Hanawa, T. An outline of the use of Juzen-taiho-to in a traditional context, the knowledge of this Kampo pharmaceutical drug (in Japanese). *Yakuji-Nippo,* 1994, 11, 94–97.

Hanawa, T. Practical guidance to Kampo medicine (in Japanese). *Kanehara-Syuppan,* 1995, 198–200.

Kosaka, A., Kamiya T., and Sumiyama M. The efficacy and Influence on QOL of Juzen-taiho-to (TJ 48) -its reduction in the adverse effects of anti-cancer drugs (in Japanese). *Progressive Med.*, 1994, 14, 2259–2264.

Kuboki, S., Niizawa, H., and Takahashi, T. Evaluation of supplement therapy of Juzen-taiho-to for EPA therapy (in Japanese). *JAMA*, 1992, Suppl., 40–41.

Matsuo, T. Significance of Juzen-taiho-to for ulcerative colitis (in Japanese). *Jpn. J. Orient. Med.*, 2000; 50, 216.

Miura, F., Saito, T., and Nakamura, K. Postoperative combination therapy of chemotherapy with Juzen-taiho-to patients with digestive organ (especially gastric) cancer (in Japanese). *Surgical Ther.*, 1985, 27, 825–828.

Miyamoto, H., Shigematsu, N., and Yamashita, M. The protective effects of Juzen-taiho-to for irradiation therapy (in Japanese). *Diagn. Ther.*, 1985, 73, 153–159.

Nagatomo, H. Reduction effect of Juzen-taiho-to for adverse reactions induced by chemotherapy, especially reduced effects of nausea and vomiting (in Japanese). *JAMA*, 1992, Suppl., 38–39.

Okamoto, A. and Sairenji, I. The infuluence of Kampo medicines for improvement of QOL on chemotherapy of postoperative gastric cancer (in Japanese). *Igaku-no-Ayumi*, 1993, 167, 760–764.

Yamada, T., Nabeya, K., and Ri, S. Postoperative combination therapy of chemotherapy and Juzen-taiho-to patients with digestive organ (especially gastric) cancer (in Japanese), *Biotherapy*, 1991, 5, 1850–1856.

Watanabe, K., Suzuki, K., Ito, G., Muranushi, A., Ishino, S., and Hanawa, T. A case report of effective use of Juzen-taiho-to adding shikon (lithospermum root) for pleulitis with carcinoma (in Japanese). *Jpn. J. Orient. Med.*, 2000, 50, 136.

4 Immunological Properties of Juzen-taiho-to

Tsukasa Matsumoto and Haruki Yamada

CONTENTS

4.1 INTRODUCTION

Homeostasis of the human body, which is regulated by various physiological systems such as the immune, hemopoietic, endocrine, and central nervous systems, is one of the most important physiological functions. Its imbalance and breakdown have been considered as one of causes of several diseases, including atopic dermatitis, autoimmune diseases, and general malaise. Kampo medicine formulations consist of several component herbs, and their multiconstituents are thought to regulate the body systems and maintain homeostasis and thus overcome illness arising from complex diseases. Because it is expected that the immune system is involved in the maintenance of homeostasis, the immunomodulating activities of Kampo medicines are regarded as very important in contributing to their efficacy.

Juzen-taiho-to, which has traditionally been used for the treatment of a wide variety of diseases such as anemia, exhaustion, systemic fatigue, and general weakness (particularly after an operation), is considered an immunomodulating agent. This chapter focuses on the effects of Juzen-taiho-to on various aspects of the immune response. It will provide the scientific basis for establishing the clinical efficacy of Juzen-taiho-to. In most of the studies, except when otherwise specified, the Juzen-taiho-to used was in the form of granules, TJ-48, manufactured by Tsumura & Co. of Japan.

Two types of effector mechanisms mediate specific immune responses against foreign substances: those mediated by a cell product of the lymphoid tissues referred to as an antibody (humoral immunity) and those mediated by specifically sensitized lymphocytes or macrophages (cell-mediated immunity or cellular immunity). This chapter first describes the modulating activity of Juzen-taiho-to on humoral and cellular immunity and then shows its modulating activities on various aspects of immune systems.

4.2 MODULATION OF HUMORAL IMMUNITY

The antibodies or immunoglobulins are a group of glycoproteins present in the serum and tissue fluids of all mammals. Their production is induced when the host's lymphoid system comes into contact with immunogenic foreign molecules (antigen) and they bind specifically to the antigen. Antibodies are now recognized as one of most important humoral factors for host defense.

Enhancement of anti-sheep red blood cells (SRBC) response by Juzen-taiho-to was reported by Komatsu et al. (1986). When BDF1 mice were treated with 2.0 g/kg/day of Juzen-taiho-to for 7 consecutive days before and 4 consecutive days after the immunization of SRBC intravenously, the relative number of plaque-forming cells (PFC) in the spleen increased 2-fold as compared with the value for untreated mice (Figure 4.1). The maximum induction of anti-SRBC response was observed at 7 days' administration of Juzen-taiho-to and no further increase by longer treatment (Figure 4.2) was observed. After withdrawing Juzen-taiho-to, the augmentation of anti-SRBC response was maintained for 3 days and had disappeared by 7 days after the withdrawal. When Juzen-taiho-to at a dose of 1.0 g/kg was administered to 4-week-old lipopolysaccharide (LPS)-low responder C3H/HeJ mice for 4 weeks, Juzen-taiho-to also enhanced anti-SRBC antibody response by PFC assay (Ebisuno et al., 1990).

Immunodeficient states with low antibody response are often seen in the elderly. Effects of Juzen-taiho-to and its subfractions on anti-SRBC antibody response of mice were examined using older mice as a model (Kiyohara et al., 1995a). BALB/c mice (6 months old) produced a lower level of anti-SRBC-IgG in comparison with young BALB/c mice (8 weeks old); however, the anti-SRBC-IgG response of the older mice was stimulated significantly when Juzen-taiho-to (1.0 g/kg/day) was orally administered from 6 days before immunization (Figure 4.3). Among the subfraction of Juzen-taiho-to, polysaccharide fraction (F-5) also showed significant antibody response (Figure 4.3).

These results suggest that Juzen-taiho-to has potent immunomodulating activity, such as stimulation of antibody production, and that the polysaccharides are revealed as one group of important active ingredients responsible for the immunomodulating activity of Juzen-taiho-to. Kiyohara and Yamada describe details of the characterization of the active ingredient in Juzen-taiho-to in Chapter 7 of this book.

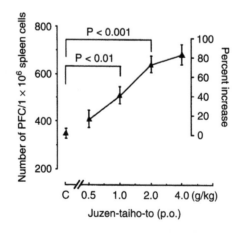

FIGURE 4.1 Effect of various concentrations of Juzen-taiho-to on the anti-SRBC response. BDF1 female mice were orally given Juzen-taiho-to for 7 days. Data were expressed as mean ± s.e.

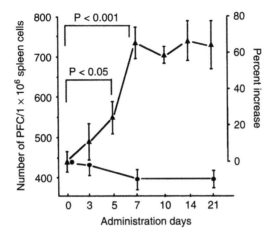

FIGURE 4.2 Effect of Juzen-taiho-to on the anti-SRBC response. BDF1 female mice were orally given Juzen-taiho-to (2.0 g/kg/day) for indicated days. Data were expressed as mean $\pm$ s.e. $\bullet$: control; $\blacktriangle$: Juzen-taiho-to.

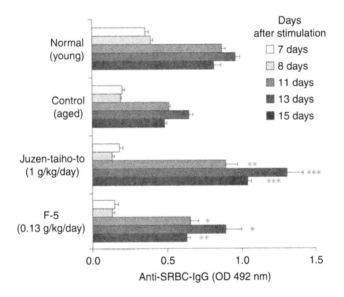

FIGURE 4.3 Effects of oral administration of Juzen-taiho-to and its polysaccharide fraction (F-5) on anti-SRBC antibody response of the aged BALB/c mice. Titers of anti-SRBC-IgG in plasma were measured at 7, 8, 11, 13, and 15 days after the stimulation. Data were expressed as mean $\pm$ s.e. $*p < 0.1$; $**p < 0.05$; $***p < 0.001$.

4.3 ANTICOMPLEMENTARY ACTIVITY

The complement system is composed of a group of structurally distinct proenzymes present in the blood plasma in an inactive form. This system is essential for the operation of the innate as well as the adaptive immune defense (Law and Reid, 1995). The complement proteins can be activated through three cascade pathways: the classical pathway; the alternative pathway; and the antibody-independent lectin pathway (Figure 4.4). The activation of the classical pathway is initiated by immune complexes containing immunoglobulin (Ig) M (IgM) and IgG antibodies. The alternative pathway is directly activated from complement component 3 (C3) by microorganisms or some

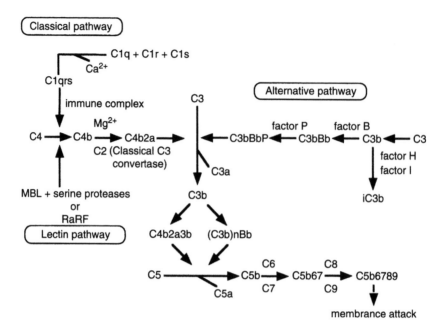

FIGURE 4.4 Activation steps of the complement system.

activators such as lipopolysaccharide (LPS) through an antibody-independent mechanism. Recently, an antibody-independent mannan-binding lectin (MBL; Ra-reactive factor, RaRF) pathway has also been established as the third activation pathway of complement systems (Thiel et al., 1997; Turner, 1996).

In general, the alternative and MBL pathways contribute to the early natural defense mechanism of the nonimmune host before production of the antibody. When normal human serum is incubated with some complement regulators and the remaining complement titer is measured by hemolysis of antibody sensitized sheep erythrocytes, the regulating substances for the hemolytic activity of complement are referred to as "anticomplementary" substances. Complement activators result in the decrease of hemolysis due to the reduced complement titer by the activation of the complement system; complement inhibitors result in the inhibition of hemolysis due to the inhibition of a certain step in the complement system by the coexistence of the inhibitors in the assay system. Therefore, the anticomplementary activity seen in the hemolytic assay includes activation and inhibition of the complement system.

Activation of the complement system generates a biological active complement fragment such as C3a and C5a. These released peptide mediators exert many biological activities such as the increment of local vascular permeability, chemotactic attraction of leukocytes, immune adherence, and modulation of antibody production. Therefore, complement activation contributes to inflammatory responses in addition to host defense reactions; inhibition of complement activation in inflammation would be a good therapeutic strategy for treating these diseases. In the activation of the complement system, a number of biological effects are induced (Table 4.1).

The effect of Juzen-taiho-to on the complement system has been examined (Yamada, 1989; Yamada et al., 1990). When Juzen-taiho-to was incubated with normal human serum as a source of complement, Juzen-taiho-to showed an anticomplementary activity in a dose-dependent manner. Among the subfractions of Juzen-taiho-to, anticomplementary activity was observed in water- and methanol-insoluble fractions (F-2) and the polysaccharide fraction (F-5), but not in low molecular weight fractions (F-1, F-3 and F-4) (Figure 4.5).

TABLE 4.1
**Immunomodulation Associated with Activation
of Complement System**

Thymus-dependent antibody response
Regulation of specific cyclical antibody production
Regulation of IgM-IgG switch
Modulation of T- and B-cell proliferation
Induction of suppressor or helper T-cells
Modulation of monokine or lymphokine release

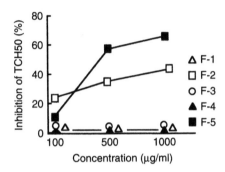

FIGURE 4.5 Anticomplementary activity of the subfractions of Juzen-taiho-to.

4.4 MODULATION OF CELLULAR IMMUNITY

The term "cell-mediated immunity" (cellular immunity) was originally used to refer to localized reactions to microorganisms, usually intracellular pathogens, mediated by lymphocytes and phagocytes. Certain subpopulations of lymphocytes, including natural killer (NK) cells and cytotoxic T-cells, and activated macrophages can attack target cells to which they are sufficiently closely bound.

The effects of Juzen-taiho-to on cellular immunity were investigated by using delayed type hypersensitivity (DTH) reaction in BALB/c mice sensitized with Type B influenza virus vaccine, mixed lymphocyte culture response (MLR), and cytotoxic T-lymphocyte (CTL) response (Takemoto et al., 1989a). The DTH reaction to the viral vaccine was induced by intradermal injection of 1000 HA units of the vaccine as the antigen into the backs of mice; 6 days later, 500 HA units of the vaccine were injected into the foot pad of the hind leg. After 24 h, the thickness of the foot pad was measured. When 2.0 g/kg/day of Juzen-taiho-to was orally administered for 13 consecutive days before sensitization, the DTH reaction was significantly enhanced in these mice (Figure 4.6).

The effect of Juzen-taiho-to on the response of mixed lymphocyte culture was also studied by the same research group (Takemoto et al., 1989a). Spleen cells subjected to oral Juzen-taiho-to were cultured for 4 days with irradiated BALB/c or C57BL/6 mouse spleen cells, and then the mixed culture was pulsed with ^{3}H-thymidine for 16 h before harvesting of the cells for measurement of the radioactivity. The spleen cells from BALB/c mice given 2.0 g/kg of Juzen-taiho-to orally for 7 consecutive days showed higher MLR (although not significantly) than those from spleen cells of untreated control mice.

The effect of Juzen-taiho-to on cytotoxic T-lymphocyte response was examined using two assay methods (Takemoto et al., 1989a). BALB/c mice were administered 2 g/kg/day of Juzen-taiho-to orally for 7 consecutive days, and their spleen cells, which were used as responder cells, were cultured with the mitomycin C (MMC)-treated spleen cells from C3H/He mice as stimulator cells

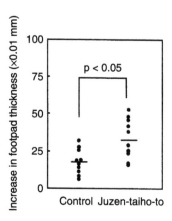

FIGURE 4.6 Enhancing effect of Juzen-taiho-to on DTH reaction. BALB/c mice were orally given Juzen-taiho-to (2.0 g/kg/day) for 13 days. The immunization and challenge with hemaglutinin were carried out on days 7 and 13, respectively. The thickness of foot pad was measured by dial thickness gauge. Results are expressed as an increase of foot pad thickness and bars are indicated for the mean of ten mice.

for 5 days. Then the cells were incubated with ^{51}Cr-loaded L929 cells as target cells. As shown in Table 4.2, the spleen cells from mice treated with Juzen-taiho-to showed significantly greater cytotoxic activity than those from spleen cells of untreated mice by ^{51}Cr-release assay. Virtually the same results were obtained when measured by the hemolytic plaque reduction technique using BALB/c mice-derived anti-SRBC hybridoma cells as a target.

Natural killer (NK) cells are found in normal animals that appear not to have been exposed to relevant antigens and are able to kill a variety of transformed, virus-infected, or embryo-derived cells *in vitro* in the absence of antibody. The effect of Juzen-taiho-to on NK activity in immuno-suppressed C3H/He mice, which were induced by *cis*-diaminedichloro-platinum (CDDP), was examined by means of ^{51}Cr-releasing cytotoxic assay using YAC-1 cell as a target (Ebisuno et al., 1990). An intraperitoneal injection of CDDP (5 mg/kg, i.p. twice; 1-week interval) attenuated NK activity in MBT-2 tumor-bearing mice; oral administration of Juzen-taiho-to significantly improved lowered NK activity. These results suggest that Juzen-taiho-to has an immunopotentiating effect on cellular immunity.

TABLE 4.2
Effect of Juzen-taiho-to on Induction of Cytotoxic T-Cells

Group	Stimulator	Percent of Cytotoxicity: Responder : Target	
		25 : 1	50 : 1
Control	−	1.7 ± 0.38	4.3 ± 0.58
	+	40.0 ± 3.32	66.6 ± 3.67
Juzen-taiho-to (2 g/kg, p.o.)	−	3.5 ± 0.66	3.8 ± 0.40
	+	50.1 ± 2.57	83.6 ± 1.64[a]

[a] $p < 0.01$.

Notes: BALB/c mice were orally given Juzen-taiho-to (2.0 g/kg) for 7 days. Responder cells (5×10^6 spleen cells) were cultured with stimulator cells (C3H/He spleen cells treated with MMC) for 5 days. Cytotoxic activity was measured by the methods of ^{51}Cr-release assay. Mean ± s.e. of four samples.

4.5 MODULATION OF MACROPHAGE FUNCTION

The engulfment and digestion of microorganisms is assigned to two major cell types recognized by Metchinikoff as microphages (mainly composed of neutrophils) and macrophages. As a rough generalization, it may be said that macrophages are at their best in combating bacteria, viruses, and protozoa, which are capable of living in the cells of a host, and that the neutrophils provide the major defense against pyrogenic bacteria. As well as T-cells, macrophages play a central role in cell-mediated immunity because they are involved in initiation of responses as antigen-presenting cells and in the effector phase as inflammatory, tumoricidal, and microbicidal cells, in addition to their regulatory functions.

Akagawa and colleagues (1996) studied the efficacy of orally administered Juzen-taiho-to on host-mediated antifungal actions in mice infected with *Candida albicans*. C3H/HeN and C3H/HeJ mice were infected intravenously with a lethal dose of *C. albicans* cells. They were orally administered with Juzen-taiho-to at a dose of 2.0 g/kg/day for 5 days starting on the day of infection. Juzen-taiho-to prolonged the survival period of infected mice of a C3H/HeJ strain, which is characteristic of functional deficiency of macrophages, but did not prolong that of infected mice of a C3H/HeN strain with normal macrophage function (Figure 4.7).

The possible immunopotentiating activity of Juzen-taiho-to on macrophages' function was also studied using *ex vivo* assay systems testing the phagocytic activity and generation of oxygen burst (Maruyama et al., 1988). BALB/c mice were treated with Juzen-taiho-to at various daily dosages for 7 consecutive days, and peritoneal exudate cells (PEC) and bone marrow cells were harvested from these treated and untreated control mice, respectively. The harvested cells were mixed with 5% fresh BALB/c mouse serum and viable *Candida parapsilosis* cells and then the mixtures were

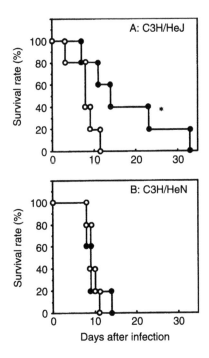

FIGURE 4.7 Efficacy of Juzen-taiho-to administered orally for elongation of survival periods of *Candida albicans*-infected C3H/HeJ or C3H/HeN mice. Normal C3H/HeJ (A) and C3H/HeN (B) mice were intravenously inoculated with 3×10^5 cells of *C. albicans* (day 0). They were orally administered Juzen-taiho-to at a dose of 2.0 g/kg (•) for 5 consecutive days from day 0. Control mice were orally administered 0.2 ml of water (○) in the same schedule as that of the Juzen-taiho-to group. *$p < 0.05$.

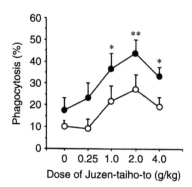

FIGURE 4.8 Effect of various doses of Juzen-taiho-to on phagocytic activity of peritoneal exudate cells (PEC) and bone marrow cells. Mice were orally administered various doses of Juzen-taiho-to for 7 days and the phagocytic activities of PEC (●) and bone marrow cells (○) were assayed. The effecter : target ratio was 10 : 1. Values represent the mean and s.d. of quadruplicate samples. $*p < 0.05$; $**p < 0.01$.

incubated for 3 h. Phagocytic activity was determined by measuring viable counts of *Candida* cells remaining in the supernatant of the incubation mixture.

The ability of PEC and bone marrow cells harvested from mice that had been treated with 1.0 g/kg/day or higher dosages of Juzen-taiho-to to engulf candida cells was significantly enhanced compared with corresponding cells from untreated mice (Figure 4.8). The enhanced phagocytosis was maintained for 5 days after the withdrawal of Juzen-taiho-to and returned to the basal level on day 7. The ability of PEC on phagocytosis of candida cells was also significantly enhanced in the presence of Juzen-taiho-to at a concentration of 3 μg/ml or above *in vitro* (Maruyama et al., 1988). The quantity of oxygen radicals produced during phagocytosis was measured using luminol excited to chemiluminescence by the oxygen radicals. The oxidative burst in response to zymosan was observed in PEC from Juzen-taiho-to treated mice (2.0 g/kg/day, p.o. for 7 days) by chemi-luminescence analysis (Maruyama et al., 1988).

These results suggest that the enhanced protective effect by Juzen-taiho-to is partly mediated by activation of macrophage function. Abe et al. (1998) also observed that Juzen-taiho-to has potent protective activity against lethal *Candida albicans* infection in immunosuppressed mice. Abe and colleagues describe details of Juzen-taiho-to's prevention of microbial infections in Chapter 6 of this book.

4.6 REDUCTION OF IMMUNOSUPPRESSION BY CHEMOTHERAPEUTIC ANTICANCER AGENT

In a clinical study, Kurokawa et al. (1989) reported that the action of the nonspecific immunosup-pression factor was lowered after administration of Juzen-taiho-to for postoperative gastric cancer patients. This observation may be partly explained by following basic research using mice. Komiyama and colleagues (1989) studied potentiation of therapeutic effects on hyperthermia and chemotherapy by Juzen-taiho-to using murine tumors. Tumor-bearing feet of mice treated with mitomycin C (MMC, 1.5 mg/kg, i.p. days 5, 8, and 12) and Juzen-taiho-to (days 1 to 40) were amputated on day 14; the sarcoma 180 tumor was reinoculated subcutaneously into the axillary region on day 19 and the subsequent tumor growth observed (Table 4.3). Results indicated that tumor growth and the incidence of tumors in the untreated group were suppressed. However, marked tumor growth was observed in the group administered MMC (Table 4.3). In contrast, the growth was reduced in the group given MMC plus Juzen-taiho-to when compared with that of those given MMC alone (Table 4.3).

TABLE 4.3
**Combined Effects of MMC and Juzen-taiho-to on S 180 Tumors
and Growth Rate of the Reinoculated Tumors**

Drug (mg/kg/day)			Mean Size of Reinoculated Tumor (mm^2)[b]		
Juzen-taiho-to	MMC	Tumor Weight[a]	Day 7	Day 14	Day 21
—	—	0.43 ± 0.02	– (0/7)[c]	28 ± 39.5 (2/7)	40 ± 57.2 (2/7)
—	1.5	0.42 ± 0.03	37 ± 6.9 (2/7)	133 ± 64.5 (6/7)	160 ± 93.2 (6/7)
2000	—	0.44 ± 0.05	– (0/7)	+ (1/7)	+ (1/7)
500	—	0.38 ± 0.05	– (0/7)	21 ± 29.6 (2/7)	60 ± 84.8 (2/7)
2000	1.5	0.38 ± 0.05	– (0/7)	100 ± 56.6 (3/7)	96 ± 65.3 (3/7)
500	1.5	0.35 ± 0.02	+ (1/7)	165 ± 132.6 (4/7)	249 ± 259.6 (4/7)

[a] Tumor-bearing feet were amputated on day 12 after tumor inoculation and weighed.

[b] S 180 tumor cells (5 × 10^5) were subcutaneously reinoculated in the axillary region on day 13 after the first tumor inoculation.

[c] Plus sign indicates small tumor nodules and minus signs indicate no tumor growth. Numbers in parentheses indicate number of tumor-bearing mice out of seven mice.

Notes: S 180 tumors cells (1 × 10^6) were injected subcutaneously into the foot pads of ICR mice. Mice were given Juzen-taiho-to (day 1 to 40) and MMC (days 5, 8, and 12).

These results suggest that Juzen-taiho-to not only potentiates the combined effects of MMC, but also reduces and/or eliminates the immunotoxicity of MMC in the host (Komiyama et al., 1989). Iijima et al. (1988) also reported protective effects of Juzen-taiho-to on the adverse effects, including immunotoxicity, of MMC. Juzen-taiho-to was administered orally at a daily dose of 2.0 g/kg for 7 days before the administration of MMC (8.4 mg/kg). Juzen-taiho-to reduced atrophy of the testis, thymus, and spleen caused by MMC and also recovered decreases of leucopenia, anemia, and body weight caused by MMC (Table 4.4 and Table 4.5).

Ebisuno et al. (1990) investigated the effects of Juzen-taiho-to on CDDP-induced immunosuppression by using PFC assay. When CDDP (5 mg/kg) was injected twice into aged C3H/He mice (13 to 15 months old) 1 week before and the day of immunization of SRBC, production of anti-SRBC antibody in the CDDP-treated mice was decreased in comparison with that of normal mice. However,

TABLE 4.4
Effects of Juzen-taiho-to on Organ Weights after MMC Administration

Treatment	n	Organ Weights (mg)				
		Liver	Kidney	Spleen	Testis	Thymus
Control	8	1403.1 ± 147.2	296.7 ± 35.3	64.4 ± 8.6	144.8 ± 12.3	52.8 ± 5.2
MMC (8.40 mg/kg)	10	1065.2 ± 228.4	224.1 ± 34.6	29.8 ± 11.7	79.9 ± 9.7	10.3 ± 7.0
+ Juzen-taiho-to (× 1)	10	992.0 ± 176.1	204.5 ± 25.4	27.8 ± 10.1	79.8 ± 10.5	8.5 ± 8.3
+ Juzen-taiho-to (× 7)	10	1131.6 ± 84.7	244.1 ± 30.7	48.3 ± 9.6[b]	99.7 ± 19.9[a]	22.2 ± 11.8[a]

[a] $p < 0.05$.

[b] $p < 0.01$.

Notes: Juzen-taiho-to was administered at a dose of 2 g/kg/day. Each value represents the mean ± s.d.

TABLE 4.5
Effects of Juzen-taiho-to on Hematological Values after Administration of MMC 8.40 mg/kg

Treatment	n	WBC ($\times 10^3$/mm3)	RBC ($\times 10^6$/mm3)	PCV (%)	Hb (g/dl)
Control	8	5.29 ± 1.84	9.19 ± 0.22	42.8 ± 1.2	13.9 ± 0.5
MMC	9	1.38 ± 0.70	7.84 ± 0.56	37.1 ± 2.9	12.3 ± 0.9
+ Juzen-taiho-to ($\times$ 1)	9	1.62 ± 0.66	8.05 ± 0.25	38.4 ± 1.1	12.4 ± 0.5
+ Juzen-taiho-to ($\times$ 7)	9	2.60 ± 0.75[a]	8.52 ± 0.11[a]	40.8 ± 0.8[a]	13.0 ± 0.2

[a] $p < 0.01$.

Notes: Juzen-taiho-to was administered at a dose of 2 g/kg/day. Each value represents the mean ± s.d.

oral administration of Juzen-taiho-to (1.0 g/kg/day) for 3 weeks before the injection of SRBC reduced the extent of immunosuppression.

Kiyohara et al. (1995a) also studied the effect of orally administered Juzen-taiho-to on mice whose immunocompromise was induced by CDDP. Intraperitoneal injection of CDDP (5 mg/kg/day) to BALB/c mice five times from 3 days before to 1 day after the immunization of SRBC resulted in suppression of anti-SRBC antibody response. However, oral administration of Juzen-taiho-to (1.0 g/ kg) from 4 days before the first injection of CDDP recovered anti-SRBC-IgM response to the level of normal mice. When polysaccharide fraction F-5 (0.13 g/kg/day) was orally administered, anti-SRBC-antibody response also recovered to the level of normal mice, as it did when Juzen-taiho-to was used.

CDDP also induces lowered NK activity. As described previously in this chapter, orally administered Juzen-taiho-to significantly improved lowered NK activity in tumor-bearing mice treated with CDDP (Ebisuno et al., 1990).

In addition to immunosuppression, CDDP shows severe renal toxicity, which is a dose-limiting factor in clinical use. Increment of blood urea nitrogen (BUN), an indicator of renal damage, by CDDP in mice was significantly suppressed by the oral administration of Juzen-taiho-to (Shibuya et al., 1987; Kiyohara et al., 1995b; Sugiyama et al., 1995). These results suggest that the combining treatment of Juzen-taiho-to with a chemotherapeutic agent is a new way to reduce adverse effects, including immunotoxity of chemotherapy.

4.7 REDUCTION OF MYELOSUPPRESSION BY CHEMOTHERAPEUTIC ANTICANCER AGENT AND RADIATION INJURY

Many immune system related cells arise from bone marrow hematopoietic cells. It has been widely known that chemotherapy and radiation therapy cause myelosuppression, which results in the induction of anemia and immunodeficiency. Kawamura et al. (1989) investigated the effects of Juzen-taiho-to on MMC-induced hematopoietic toxicity by means of assay of colony-forming units in spleen (CFU-S) and granulocytes–macrophage colony-forming cells (CFU-GM) in C57BL/6J mice. Intraperitoneal injection of MMC (1, 3, and 8 mg/kg) resulted in marked suppression of the numbers of CFU-S. Oral administration of Juzen-taiho-to (1.0 g/kg/day) did not protect the mice from the damage of hematopoietic function caused by MMC, but accelerated the recovery of CFU-S remarkably.

Effects of Juzen-taiho-to on the recovery of the hematopoietic system from radiation injury were also analyzed by using fibroblast colony-forming unit (CFU-f), CFU-S, CFU-GM, erythroid colony-forming unit (CFU-E), and erythroid burst-forming unit (BFU-E) assay systems (Ohnishi et al., 1989, 1990). The administration of Juzen-taiho-to (1.25% in drinking water) for 7 days enhanced recovery from hematopoietic injury by radiation due to the acceleration of the recovery of CFU-S.

These results suggest that Juzen-taiho-to has the ability to accelerate hematopoietic recovery from bone marrow injury, which closely relates to immunosuppression. Hisha and Ikehara describe details of the promotion of hemopoiesis by Juzen-taiho-to in Chapter 5 of this book.

4.8 MITOGENIC ACTIVITY

The activation of lymphocytes with various stimuli induces cell proliferation (mitogenic activity). The mitogenic activity of Juzen-taiho-to on murine lymphoid cells has been examined (Takemoto et al., 1989b). Spleen cells, lymph node cells, and thymocytes were taken from BALB/c mice and cultured for 72 h in RPMI-1640 medium with various concentrations of Juzen-taiho-to; the cultures were then pulsed with ^{3}H-thymidine for 4 h before harvesting for radioactive assay. Juzen-taiho-to at concentrations up to 200 μg/ml showed dose-dependent mitogenic activity on spleen cells and lymph node cells, but had no effect on thymocytes (Figure 4.9).

Juzen-taiho-to also responded against spleen cells taken from BALB/c (*nu/nu*) mice to the same extent as it did for spleen cells from BALB/c mice. The mitogenic activity caused by Juzen-taiho-to was abolished by pretreatment with anti-Ig antibody but was not affected by pretreatment with anti-Thy 1.2 antibody. Coculture of spleen cells with Juzen-taiho-to resulted in an increased number of sIgM-, sIgG-, or sIgD-positive cells according to fluorescent activated cell sorter (FACS) analysis; however, LPS increased sIgM- and sIgG-positive cells (Table 4.6). The increase of mitosis by Juzen-taiho-to was abolished in the absence of adherent cells, and readdition of adherent cells recovered the effect of Juzen-taiho-to. These data strongly suggest that Juzen-taiho-to is characterized as T-cell-independent and adherent cell-dependent B-cell mitogen. The mitogenic activity of Juzen-taiho-to may account for the augmentation of the antibody response.

To estimate components of Juzen-taiho-to responsible for its mitogenic activity, the active ingredients were studied, and a pectic polysaccharide fraction, named F-5-2, was obtained as the active fraction (Yamada, 1989; Yamada et al., 1990). When murine spleen cells were stimulated

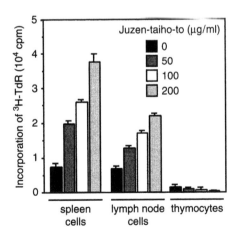

FIGURE 4.9 Mitogenic activity of Juzen-taiho-to on spleen cells, lymph node cells, and thymocytes. Cells were cultured with various concentrations of Juzen-taiho-to in 96 microculture plates for 3 days. ^{3}H-TdR incorporation was measured during the last 4 h of culture.

TABLE 4.6
FACS Analysis of Cells Proliferated with Juzen-taiho-to

Group	Percentage of Positive Cells			
	sIgM	sIgG	sIgD	Thy 1.2
Control	43.0 ± 2.18	46.7 ± 2.62	38.5 ± 2.30	46.5 ± 4.20
Juzen-taiho-to	59.0 ± 2.84	58.4 ± 4.20	50.9 ± 2.70	47.3 ± 4.95
LPS	69.7 ± 3.84	66.2 ± 4.99	30.7 ± 2.10	38.1 ± 3.95

Notes: Spleen cells were cultured with Juzen-taiho-to (200 µg/ml) or LPS (1.0 µg/ml) for 3 days. Cells were harvested and stained with each FITC-conjugated antibody. Mean ± s.e.

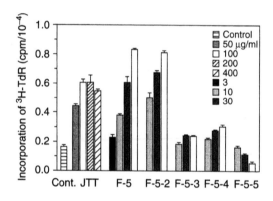

FIGURE 4.10 Mitogenic activity of polysaccharide subfractions of F-5 from Juzen-taiho-to (JTT). Note that F-5 is a crude polysaccharide fraction of Juzen-taiho-to (JTT) (see Chapter 7).

with F-5-2, significant B-cell mitogenic activity was observed; this was observed in Juzen-taiho-to as well, but at concentrations that were ten times lower (Figure 4.10). F-5-2 showed a different property from another B-cell mitogen, LPS, because it cannot fully differentiate into antibody secreting plasma cells (Takemoto et al., 1994).

B-cells that proliferated in response to F-5-2 were found not to undergo class switching from IgM- to IgG- producing cells. However, the B-cells induced by F-5-2 produced IgM antibody in response to interleukin 6 (IL-6) and an antigen (SRBC), but not IgG antibody. F-5-2 induced the expression of IL-6R not only on IgM$^+$ and IgG$^+$ B-cells but also on IgD$^+$ B-cells (Takemoto et al., 1994). These results suggest that F-5-2 is a new type of B-cell mitogen and differentiates B-cells to antibody secreting plasma cells in the presence of adherent cells (macrophages) that produce IL-6. Details of the active ingredient in Juzen-taiho-to are described by Kiyohara and Yamada in Chapter 7 of this book.

4.9 MODULATION OF CYTOKINE PRODUCTION

Released by activation of various immune cells, cytokines play an important role in the regulation and potentiation of specific- and nonspecific immune reactions. The effect of Juzen-taiho-to on the production of interferon-γ (IFN-γ) and IL-2 from the phytohemagglutinin (PHA)-stimulated human peripheral blood mononuclear cells has been investigated *in vitro* (Sakagami et al., 1988). Production of IFN-γ and IL-2 were significantly enhanced in the presence of Juzen-taiho-to in the culture media.

TABLE 4.7
Effects of Juzen-taiho-to on Cytokine Production of Macrophages and Spleen Cells

Juzen-taiho-to Concentration (μg/ml)	IL-1 (u/ml)[a]	IL-2 (u/ml)[b]	IL-3 (u/ml)[c]	IL-6 (ng/ml)[d]
0	2.0 ± 0.10	39.0 ± 1.50	7.3 ± 0.44	0.69 ± 0.07
10	2.9 ± 0.47	45.0 ± 3.00	8.0 ± 0.41	Not done
30	3.7 ± 0.05[e]	54.8 ± 0.15[e]	8.9 ± 0.23	Not done
100	5.3 ± 0.73[f]	70.5 ± 0.01[e]	7.9 ± 0.15	0.97 ± 0.08[e]
300	5.4 ± 0.60[e]	69.8 ± 8.25[f]	7.2 ± 0.08	0.99 ± 0.01[f]

[a] Measured by bioassay using PUA and spleen cells from C3H/HeJ mouse.
[b] Measured by IL-2 dependent CTLL-2 cells.
[c] Measured by IL-3 dependent FDC-D2 cells.
[b] Measured by FLISA.
[e] $p < 0.01$.
[f] $p < 0.05$.
Notes: Proteose–peptone-induced peritoneal macrophages were used for IL-1 production. Spleen cells were used for IL-2, IL-3, and IL-6 production. Mean ± s.e.

Enhanced cytokine production by Juzen-taiho-to has also been reported by Takemoto (1996). Proteose–peptone-induced murine peritoneal macrohages or spleen cells were cultured with Juzen-taiho-to in the presence of LPS or concanavalin A (ConA). IL-3 production was not changed in comparison with the control. However, the production of IL-1, IL-2, and IL-6 was enhanced by the stimulation of Juzen-taiho-to (Table 4.7). Immunomodulating activity of Juzen-taiho-to may be exerted partly by this modulation of cytokine production.

4.10 MODULATION OF GASTRIC MUCOSAL IMMUNE SYSTEM

The inner surface of the intestinal tract possesses a large area of mucosal membranes that are continuously exposed to various substances in the intestinal lumen. The gut-associated lymphoreticular tissues (GALT) exist on the intestinal mucosal sites and play an important role in host defense, including IgA response in the mucosal immune system (McNabb and Tomasi, 1981). Because Kampo medicines are taken orally, the gastric mucosal immune system may act as one of the major targets for the expression of pharmacological activity. Some active substances in the Kampo medicines are absorbed or affected in the intestine and may modulate the gastric mucosal immune system.

Because Juzen-taiho-to affects the hematopoietic system, the effects of Juzen-taiho-to on Peyer's patch cells mediated hematopoietic response were investigated (Hong et al., 1998). When the conditioned medium of Peyer's patch cells obtained from C3H/HeJ mice orally administered Juzen-taiho-to for 7 consecutive days was added to a culture medium of bone marrow cells, proliferation of bone marrow cells was significantly enhanced compared with that of mice receiving water alone (Table 4.8). When Peyer's patch cells from untreated C3H/HeJ mice were stimulated with various concentrations of Juzen-taiho-to *in vitro* and the resulting conditioned medium was used for the stimulation of bone marrow cells, the cells were also proliferated in a dose-dependent manner.

The contents of granulocyte–macrophage colony-stimulating factor (GM-CSF) and IL-6 in the conditioned medium from Peyer's patch cells stimulated by Juzen-taiho-to were increased compared with those from control (Figure 4.11). The population of Gr-1 antigen positive cells in the bone marrow cell culture also increased as a marker of activation/differentiation of bone marrow cells. This observation suggests that Juzen-taiho-to modulates the intestinal immune system, including the cytokine network.

TABLE 4.8
**Effect of Orally Administered Juzen-taiho-to on Peyer's Patch
Cell-Mediated Hematopoietic Response of Bone Marrow Cells**

Treatment	Dose (g/kg/day, p.o.)	Fluorescence Intensity
Exp. 1		
Control (vehicle)		1679 ± 96
Juzen-taiho-to	1.0	1907 ± 142[a]
Juzen-taiho-to	2.0	2281 ± 264[b]
Exp. 2		
Control (vehicle)		1470 ± 212
Juzen-taiho-to	0.5	1975 ± 252[a]
Juzen-taiho-to	1.0	1934 ± 357[c]

[a] $p < 0.05$.
[b] $p < 0.01$.
[c] $p < 0.1$.

Notes: Peyer's patch cells obtained from C3H/HeJ mice ($n = 3$ to 4) were orally administered Juzen-taiho-to for 7 consecutive days and were pooled and cultured for 5 days. The bone marrow cells from untreated mice at a density of 2.5×10^4 cells/well were cultured for 6 days in the presence of the conditioned medium of Peyer's patch cells. The proliferation of bone marrow cells was measured by Alamar Blue assay. Water alone was administered for normal control mice. Mean $\pm$ s.d. of quadruplicate cultures.

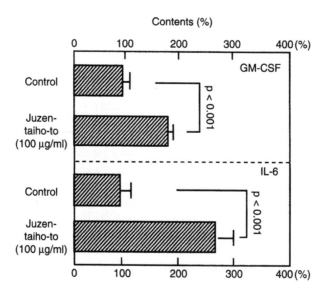

FIGURE 4.11 Effect of Juzen-taiho-to on GM-CSF and IL-6 productions of Peyer's patch cells from untreated mice. Cells were pooled and cultured at a density of 2×10^5 cells/well with or without the addition of Juzen-taiho-to for 5 days *in vitro*, and the resulting cell-free supernatants were subjected to ELISA. Data were expressed as percent of control of mean $\pm$ s.d. of quadruplicate assays.

4.11 MODULATION OF TH1–TH2 BALANCE

Murine helper T-cells are classified into Th1 and Th2 subsets according to their cytokine secretion pattern (Mosmann et al., 1986; Mosmann and Coffman, 1989). The ratio of the two subsets (Th1–Th2 balance) has important effects on the balance of cellular immunity and humoral immunity (Liblau et al., 1995; Charlton and Laffertym, 1995; Druet et al., 1996). IFN-γ and IL-4 are key cytokines with antagonistic actions towards each other; they provide distinct help for inflammatory or humoral immune responses and are expressed by Th1 and Th2 cells, respectively.

The effect on cytokine production of oral administration of Juzen-taiho-to to mice has been investigated (Matsumoto et al., 2000a). Lymphocytes from spleen (SPN), mesenteric lymph node (MLN), and Peyer's patches (PP) from the mice orally administered Juzen-taiho-to (1.0 g/kg) for 2 weeks were stimulated with concanavalin A (Con A), and the resulting conditioned medium was tested for cytokine production by enzyme-linked immunosolvent assay (ELISA). As shown in Figure 4.12, the IFN-γ secretions from MLN and PP of mice administered Juzen-taiho-to increased over twofold compared to those of the cells from the control mice. The ratios of IFN-γ/IL-4 (Th1–Th2 balance) from MLN and PP were increased 9.4- and 2.4-fold, respectively. These results suggest that Juzen-taiho-to regulates differentiation of the lymphocytes in gut-associated lymphoid tissue (GALT) toward the Th1 type cell. Oral administration of Juzen-taiho-to resulted in the enhancement of IL-5 secretion from MLN and PP in comparison with the control (Figure 4.12).

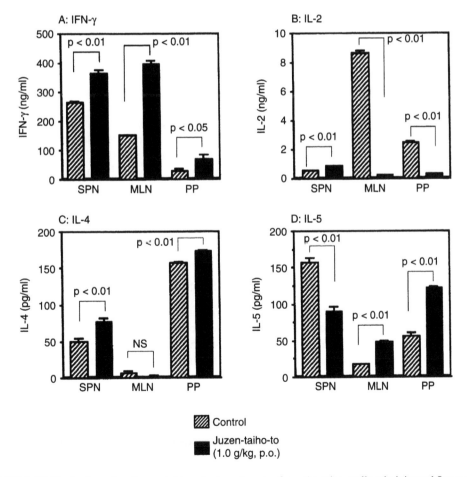

FIGURE 4.12 Production of various cytokines by lymphocytes from the mice orally administered Juzen-taiho-to in response to stimulation with ConA. The lymphocytes (2.5×10^6 cells/ml) were cultured for 3 days in the presence of ConA (5 μg/ml). Culture supernatants were harvested and tested for concentration of cytokines.

IL-5 synthesized from Th2 type cells has been shown to selectively induce sIgA$^+$ B cells that differentiate to IgA-secreting cells and to enhance IgA synthesis (McGhee et al., 1989; Coffman et al., 1987; Harriman et al., 1988; Beagley et al., 1988). It has also been reported that IFN-γ is important for the production of the secretory component of epithelial cells in mucosa (Sollid et al., 1987; Phillips et al., 1990). These findings suggest that the enhancement of IFN-γ and IL-5 secretions in GALT by administration of Juzen-taiho-to may stimulate IgA response in gastric mucosa.

4.12 MODULATION OF HEPATIC LYMPHOCYTE FUNCTION

The liver is now recognized as one of the important immune organs with its abundance in lymphocytes in addition to Kupffer cells (Knook et al., 1977; Zahlten et al., 1978; Itoh et al., 1988). All blood streams from the digestive tract run into the liver via the portal vein. Consequently, active substances in Kampo medicines absorbed in the intestine are expected to be directly brought into the liver. Thus, the liver seems to be one of most important organs on which to exert the immunopharmacological effects of Kampo medicines.

The effects of orally administered decoction of Juzen-taiho-to on cytokine production in hepatic lymphocytes were studied in mice (Matsumoto et al., 2000a). The number of IFN-γ- and IL-4-producing cells was not changed by administration of Juzen-taiho-to (1.0 g/kg) for 2 weeks. However, Juzen-taiho-to was found to increase IFN-γ, as well as IL-4, IL-5, and IL-6 secretions from stimulated hepatic lymphocytes, whereas IL-2 secretion was reduced (Figure 4.13). CD4/CD8 ratio and $\alpha\beta$/$\gamma\delta$ T-cell receptor (TCR) ratio in hepatic lymphocytes were not changed. However, flow cytometric analysis revealed that the population of CD3-positive intermediate cells in NK-positive cells (NKT cells) was increased after oral administration of Juzen-taiho-to (Figure 4.14).

It has been reported that NKT cells can produce a large amount of cytokines such as IFN-γ and IL-4 (Arase et al., 1993, 1996; Yoshimoto and Paul, 1994) and have potent antitumor effects against leukemia and solid tumors. From these results, it may be concluded that a rise in the NKT cell population contributes, at least partially, to the modulating effect of Juzen-taiho-to on cytokine production in liver lymphocytes. Juzen-taiho-to enhanced transcription of IL-12 mRNA in liver, and the induction of NKT cells by Juzen-taiho-to was reduced by injection of 2-chloroadenosine, which can eliminate macrophages (Saito and Yamaguchi, 1985), suggesting that macrophages and the production of IL-12 in the liver may contribute to this NKT induction.

4.13 PREVENTION OF AUTOIMMUNE DISEASES

Autoimmune disease is the result of several immunological abnormalities or disorders arising from unknown etiology and is commonly characterized by excessive proliferation of T- and B-lymphocytes and production of autoantibodies. The effect of Juzen-taiho-to on autoimmune diseases was studied using autoimmune-prone NZB/NZW F1 mice (Jeng et al., 1994). When Juzen-taiho-to was administered orally to the mice at a dose of 0.2 g/kg for 7 months, it had the beneficial effect of causing prolonged survival and reduced urine protein concentration. This is a one of the markers of renal damage by immune complexes, compared with those of the control mice administered Juzen-taiho-to (Table 4.9).

It has been reported that NKT cells have potent protective activity against autoimmunity (Sharif et al., 2002). The induction of NKT cells by Juzen-taiho-to (Matsumoto et al., 2000a) described previously in this chapter may contribute to this preventive effect on autoimmune disease in NZB/BZW F1 mice. These results suggest that Juzen-taiho-to is a potential therapeutic drug for prevention of the development of autoimmune diseases in humans.

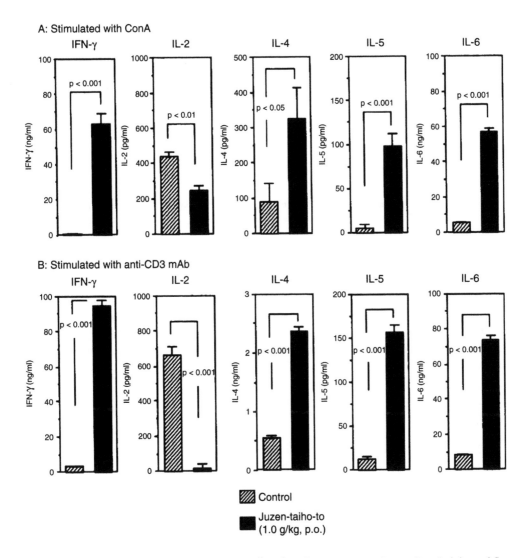

FIGURE 4.13 Production of various cytokines by liver lymphocytes from mice orally administered Juzen-taiho-to in response to stimulation with ConA or anti-CD3 mAb. Liver lymphocytes (2.5×10^6 cells/ml) were cultured for 2 days in the presence of ConA or anti-CD3 mAb. Culture supernatants were harvested and tested for concentration of cytokines. Values shown are the mean ± s.d. from triplicate experiments.

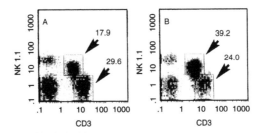

FIGURE 4.14 Analysis of phenotype of liver lymphocytes from mice orally administered Juzen-taiho-to by flow cytometric analysis. At 18 h after last administration of Juzen-taiho-to, hepatic lymphocytes were obtained and stained with specific antibody. Hepatic lymphocytes from control mice (A) and from mice administered Juzen-taiho-to (B). The numbers show the percentage of each population and are indicated by arrows.

TABLE 4.9
Effects of Herbal Medicine on Mortality Rate and Proteinuria
of Autoimmune-Prone NZB/NZW F1 Mice

Exp. No. (Sex)	Treatment Group	Mortality Rate	Urine Protein Concentration (mg/dl)
1 (Male)	Control	5/10	220 ± 78
	Juzen-taiho-to	2/10	75 ± 49[a]
2 (Female)	Control	5/10	235 ± 81
	Juzen-taiho-to	3/10	175 ± 64

[a] $p < 0.05$.

Note: Juzen-taiho-to was administered orally for 6 or 7 months in experiments 1 and 2, respectively.

4.14 CONCLUSION

All of the previously mentioned data strongly suggest that Juzen-taiho-to has potent biological response modifying activities. In addition to humoral and cellular immunity, Juzen-taiho-to modulates functions of the gastric mucosal immune system. It also stimulates the hemopoietic system to potentiate the therapeutic effects of chemotherapy and radiotherapy and to prevent and ameliorate from immunosuppression due to anticancer drugs such as mitomycin C and *cis*-diaminedichloroplatinum.

REFERENCES

Abe, S., Tansho, S., Ishibashi, H., Inagaki, N., Komatsu, Y., and Yamaguchi, H. (1998) Protective effect of oral administration of a traditional medicine, Juzen-Taiho-To, and its components on lethal *Candida albicans* infection in immunosuppressed mice. *Immunopharmacology,* 20, 421–431.

Akagawa, G., Abe, S., Tansho, S., Uchida, K., and Yamaguchi, H. (1996) Protection of C3H/He J mice from development of *Candida albicans* infection by oral administration of Juzen-taiho-to and its component, Ginseng Radix: possible roles of macrophages in the host defense mechanisms. *Immunopharmacol. Immunotoxicol.,* 18, 73–89.

Arase, H., Arase, N., Nakagawa,K., Good, R.A., and Onoe, K., 1993. NK1.1 + CD4 + CD8- thymocytes with specific lymphokine secretion. *Eur. J. Immunol.,* 23, 307–310.

Arase, H., Arase, N., and Saito, T., 1996. Interferon γ production by natural killer (NK) cells and NK1.1 + T-cells upon NKR-P cross-linking. *J. Exp. Med.,* 183, 2391–2396.

Beagley, K.W., Eldridge, J.H., Kiyono, H., Everson, M.P., Koopman, W.J., Honjo, T., and McGhee, J.R. (1988) Recombinant murine IL-5 induces high rate IgA synthesis in cycling IgA-positive Peyer's patch B-cells. *J. Immunol.,* 141, 2035–2042.

Charlton, B. and Laffertym, K.J. (1995) The Th1/Th2 balance in autoimmunity. *Curr. Opin. Immunol.,* 7, 793–798.

Coffman, R.L., Shrader, B., Carty, J., Mosmann, T.R., and Bond, M.W. (1987) A mouse T-cell product that preferentially enhances IgA production. 1. Biologic characterization. *J. Immunol.,* 139, 3685–3690.

Druet, P., Sheela, R., and Pelletier, L. (1996) Th1 and Th2 cells in autoimmunity. *Chem. Immunol.,* 63, 138–170.

Ebisuno, S., Hirano, A., Ohkawa, T., Maruyama, H., Kawamura, H., and Hosoya, E. (1990) Immunomodulating effects of Juzen-Taiho-To in immunosupressive mice. *Biotherapy,* 4, 112–116.

Harriman, G.R., Kunimoto, D.Y., Elliott, J.F., Paetkau, V., and Strober, W. (1988) The role of IL-5 in IgA B cell differentiation. *J. Immunol.,* 141, 3033–3039.

Hong, T., Matsumoto, T., Kiyohara, H., and Yamada, H. (1988) Enhanced production of hematopoietic growth factors through T cell activation in Peyer's patches by oral administration of Kampo (Japanese herbal) medicine, "Juzen-taiho-to." *Phytomedicine,* 5, 353–360.

Iijima, O., Fujii, Y., Kobayashi, Y., Kuboniwa, H., Murakami, C., Sudo, K., Aburada, M., and Hosoya, E. (1988) Protective effects of Juzen-taiho-to on the adverse effects of Mitomicin C. *J. Jpn. Soc. Cancer Ther.,* 23, 1277–1282.

Itoh, H., Abo, T., Sugawara, S., Kanno, A., and Kumagai, K. (1988) Age-related variation in the proportion and activity of murine liver natural killer cells and their cytotoxicity against regenerating hepatocytes. *J. Immunol.,* 141, 315–323.

Jeng, K.-C., Hsiao, S.-H., Chen, S.-F., Lan, J.-L., Yang, L.-L., and Lin, T.-D. (1994) Immunological effect of extract of *Tripterygium wilfordii* and Shih-chuan-da-bo-tang in NZB/NZW F1 mice. *J. Chin. Med.,* 5, 51–60.

Kawamura, H., Maruyama, H., Takemoto, N., Komatsu, Y., Aburada, M., Ikehara, S., and Hosoya, E. (1989) Accelerating effect of a Japanese Kampo medicine on recovery of murine hematopoietic stem cells after administration of mitomycin C. *Int. J. Immunother.,* 5, 35–42.

Kiyohara, H., Matsumoto, T., Takemoto, N., Kawamura, H., Komatsu, Y., and Yamada, H. (1995a) Effect of oral administration of a pectic polysaccharide fraction from a Kampo (Japanese herbal) medicine "Juzen-taiho-to" on antibody response of mice. *Planta Med.,* 61, 429–434.

Kiyohara, H., Matsumoto, T., Komatsu, Y., and Yamada, H. (1995b) Protective effect of oral administration of a pectic polysaccharide fraction from a Kampo (Japanese herbal) medicine "Juzen-taiho-to" on adverse effects of *cis*-diaminedichloroplatinum. *Planta Med.,* 61, 531–534.

Knook, D.L., Blansjaar, N., and Sleyster, E.C. (1977) Isolation and characterization of Kupffer and endothelial cells from the rat liver. *Exp. Cell Res.,* 109, 317–329.

Komatsu, Y., Takemoto, N., Maruyama, H., Tsuchiya, H., Aburada, M., Hosoya, E., Shinohara, S., and Hamada, H. (1986) Effect of Juzen-taihoto on the anti-SRBC response in mice. *Jpn. J. Inflam.,* 6, 405–408.

Komiyama, K., Zhibo, Y., and Umezawa, I. (1989) Potentiation of the theraputic effects of chemotherapy and hypothermia on experimental tumor and reduction of immunotoxicity of mitomycin C by Juzentaihoto, a Chinese herbal medicine. *Jpn. J. Cancer Chemother. (Gan To Kagaku Ryoho),* 16, 251–257.

Kurokawa, T., Tamakuma, S., Imai, J., Yamamoto, T., and Taguchi, J. (1989) Immunologic examination of Juzentaiho-to (TJ-48) for postoperative gastric cancer. *Jpn. J. Cancer Chemother. (Gan To Kagaku Ryoho),* 16 (4 Pt2-2), 1506–1510.

Law, S.K.A. and Reid, K.B.M. (Eds.) (1995) *Complement,* 2nd ed. IRL Press, New York, Oxford University Press.

Liblau, R.S., Singer, S.M., and McDevitt, H.O. (1995) Th1 and Th2 CD4+ T-cells in the pathogenesis of organ-specific autoimmune diseases. *Immunol. Today,* 16, 34–38.

Maruyama, H., Kawamura, H., Takemoto, N., Komatsu, Aburada, M., and Hosoya, E. (1988) Effect of Juzentaihoto on phagocytes. *Jap. J. Inflam.,* 8, 461–465.

Matsumoto, T., Sakurai, M.H., Kiyohara, H., and Yamada, H. (2000a) Orally administered decoction of Kampo (Japanese herbal) medicine, "Juzen-taiho-to," modulates cytokine secretion and induces NKT cells in mouse liver. *Immunopharmacology,* 46, 149–161.

Matsumoto, T. and Yamada, H. (2000b) Orally administered Kampo (Japanese herbal) medicine, "Juzen-taiho-to," modulates cytokine secretion in gut associated lymphoreticular tissues in mice. *Phytomedicine,* 6, 425–430.

McGhee, J.R., Mestecky, J., Elson, C.O., and Kiyono, H. (1989) Regulation of IgA synthesis and immune response by T-cells and interleukins. *J. Clin. Immunol.,* 9, 175–199.

McNabb, P.C. and Tomasi, T.B. (1981) Host defense mechanisms at mucosal surfaces. *Annu. Rev. Microbiol.,* 35, 477–496.

Mosmann, T.R., Cherwinski, H., Bond, M.W., Giedlin, M.A., and Coffman, R.L. (1986) Two types of murine helper T-cell clone. I. Definition according to profiles of lymphokine activities and secreted proteins. *J. Immunol.,* 136, 2348–2357.

Mosmann, T.R. and Coffman, R.L. (1989) TH1 and TH2 cells: different patterns of lymphokine secretion lead to different functional properties. *Annu. Rev. Immunol.,* 7, 145–173.

Ohnishi, Y., Yasumizu, R., and Ikehara, S. (1989) Preventive effect of TJ-48 on recovery from radiation injury. *Jpn. J. Cancer Chemother. (Gan To Kagaku Ryoho),* 16 (4 Pt. 2-2) 1494–1499.

Ohnishi, Y., Yasumizu, R., Fan, H.X., Liu, J., Takao-Liu, F., Komatsu, Y., Hosoya, E., Good, R.A., and Ikehara, S. (1990) Effects of Juzen-taiho-to (TJ-48), a traditional Oriental medicine, on hematopoietic recovery from radiation injury in mice. *Exp. Hematol.,* 18, 18–22.

Phillips, J.O., Everson, M.O., Moldoveanu, Z., Lue, C., and Mestecky, J. (1990) Synergistic effect of IL-4 and IFN-γ on the expression of polymeric Ig receptor (secretory component) and IgA binding by human epitherial cells. *J. Immunol.*, 145, 1740–1744.

Saito, T. and Yamaguchi, J. (1985) 2-Chloroadenosine: a selective lethal effect to mouse macrophages and its mechanism. *J. Immunol.*, 134, 1815–1822.

Sakagami, Y., Mizoguchi, Y., Miyajima, K., Kuboi, H., Kobayashi, K., Kioka, K., Takeda, H., Shin, T., Morisawa, S., and Yamamoto, S. (1988) Antitumor activity of Shi-quan-da-bu-tang and its effects on interferon-γ and interleukin 2 production. *Allergy (Tokyo)*, 37, 57–60.

Sharif, S., Arreaza, G.A., Zucker, P., Mi, Q.S., and Delovitch, T.L. (2002) Regulation of autoimmune disease by natural killer T-cells. *J Mol Med.*, 80, 290–300.

Shibuya, K., Satoh, M., Hasegawa, T., Naganuma, A., and Imura, N. (1987) Protective effect of Chinese medicines on toxic side effects of *cis*-diaminedichloroplatinum in mice. *Yakugaku Zasshi*, 107, 511–516.

Sollid, L.M., Kvale, D., Brandtzaeg, P., Markussen, G., and Thorsby, E. (1987) Interferon-γ enhances expression of secretory component, the epitherial receptor for polymeric immunoglobulins. *J. Immunol.*, 138, 4303–4306.

Sugiyama, K., Ueda, H., Ichio, Y., and Yokota, M. (1995) Improvement of cisplatin toxicity and lethality by Juzen-taiho-to in mice. *Biol. Pharm. Bull.*, 18, 53–58.

Takemoto, N. (1996) The study on the effect of the Kampo medicines "Juzen-taiho-to" on immune systems. Doctoral dissertation (Hokuriku University).

Takemoto, N., Kiyohara, H., Maruyama, H., Komatsu, Y., Yamada, H., and Kawamura, H. (1994) A novel type of B-cell mitogen isolated from Juzen-taiho-to (TJ-48), a Japanese traditional medicine. *Int. J. Immunopharmacol.*, 16, 919–929.

Takemoto, N., Kawamura, H., Maruyama, H., Komatsu, Y., Aburada, M., and Hosoya, E. (1989a) Effect of TJ-48 on murine cellular immunity. *Jpn. J. Inflam.*, 9, 49–52.

Takemoto, N., Maruyama, H., Kawamura, H., Komatsu, Y., Aburada, M., and Hosoya, E. (1989b) Mitogenic activity of Juzentaihoto (TJ-48) on murine lymphoid cells. *Jpn. J. Inflam.*, 9, 137–140.

Thiel, S., Vorup–Jensen, T., Stover, C.M., Schwaeble, W., Laursen, S.B., Paulsen, K., Willis A.C., Eggleton, P., Hamsen, S., Holmskov, U. et al. (1997) A second serine protease associated with mannan-binding lectin that activates complement. *Nature*, 386, 506–510.

Turner, W.T. (1996) Mannose-binding lectin: the pluripotent molecule of the innate immune system. *Immunol. Today*, 17, 532–540.

Yamada, H. (1989) Chemical characterization and biological activity of the immunologically active substances in Juzen-taiho-to. *Jpn. J. Cancer Chemother. (Gan To Kagaku Ryoho)*, 16 (4 Pt 2-2), 1500–1505

Yamada, H., Kiyohara, H., Cyong, J.C., Takemoto, N., Komatsu, Y., Kawamura, H., Aburada, M., and Hosoya, E. (1990) Fractionation and characterization of mitogenic and anticomplementary active fractions from Kampo (Japanese herbal) medicine "Juzen-taiho-to." *Planta Med.*, 56. 386–391.

Yoshimoto, T. and Paul, W.E. (1994) CD4 pos. NK1.1 pos T-cells promptly produce interleukin 4 in response to *in vivo* challenge with anti-CD3. *J. Exp. Med.*, 179, 1285–1295.

Zahlten, R.N., Hagler, H.K., Nejtek, M.E., and Day, C.J. (1978) Morphological characterization of Kupffer and endothelial cells of rat liver isolated by counterflow elutriation. *Gastroenterology*, 75, 80–87.

5 Hemopoiesis-Stimulatory Effects of Juzen-taiho-to

Hiroko Hisha and Susumu Ikehara

CONTENTS

5.1 INTRODUCTION

In recent decades, the hemopoietic system has been well elucidated by the development of culture systems and the discovery of hemopoiesis-stimulatory cytokines. Figure 5.1 shows the differentiation pathway of blood cells from pluripotent hemopoietic stem cells (P-HSCs). Hemopoiesis is maintained by a few P-HSCs with extensive developmental and self-renewal potential. The P-HSCs have cross contact with stromal cells in the bone marrow and are controlled by cytokines produced by stromal cells and adhesion molecules on these cells. These P-HSCs differentiate into cytokine-reactive HSCs and then lineage-specific progenitor cells under the influence of various cytokines produced from bone marrow stromal cells, macrophages, fibloblasts, and endothelial cells, and finally into mature blood cells.

Traditionally, Juzen-taiho-to has been administered to patients with anemia, anorexia, or fatigue. Several reports have demonstrated that Juzen-taiho-to enhances peripheral blood counts in cancer patients who have been administered phase-specific drugs and/or have received radiation therapy (Nabeya and Ri, 1983). This effect was also observed in animal models, where it was shown to

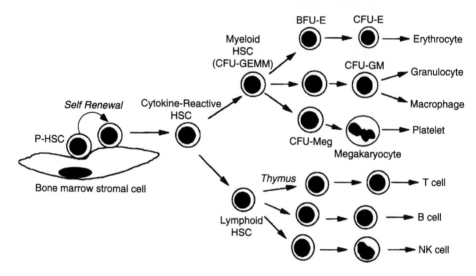

FIGURE 5.1 Hemopoietic system.

prolong the survival of irradiated mice (Yonezawa, 1976; Hosokawa, 1985). The oral administration of Juzen-taiho-to prolongs the survival of tumor-bearing mice injected with mitomycin C (MMC, an antitumor drug) (Aburada et al., 1983; Kawamura et al., 1989). Our previous study showed that Juzen-taiho-to enhances day-14 CFU-S (colony forming unit in spleen on day 14) but not day-9 CFU-S, when administered orally to mice before radiation and syngeneic bone marrow transplantation (BMT) (Ohnishi et al., 1990; Ikehara et al., 1992).

CFU-S are hemopoietic colonies formed in the spleen of recipient mice that have been lethally irradiated and injected with donor bone marrow cells. Day-14 CFU-S represent primitive HSCs and day-9 CFU-S represent more mature HSCs. This result indicates that Juzen-taiho-to contains active substances that stimulate the proliferation of HSCs. Another report indicates that the radioprotective effects of extracts of *Acanthopanax senticosus* Harms (Shigoka), which belongs to the ginseng family, were caused by the stimulation of proliferation and self-renewal of HSCs (Miyamomae and Frindel, 1988). In spite of the definite effect of Juzen-taiho-to on hemopoiesis, to our knowledge, no reports on the purification and identification of the active compounds in this prescription have been issued.

Recently, we have attempted to isolate the HSC-stimulating substances in Juzen-taiho-to (using the pharmaceutical preparation TJ-48 manufactured by Tsumura & Co.) and to identify their chemical structure (Hisha et al., 1997). We have found that at least some of the active substances in Juzen-taiho-to are fatty acids, which are contained at a concentration of 0.1%. These standard fatty acids, obtained commercially, can also stimulate the proliferation of HSCs *in vitro* and increase day-8 and -14 CFU-S counts in MMC-treated mice *in vivo*. We discuss here the mechanisms behind the action of these active compounds. The effect of these fatty acids on bone marrow cells (BMCs) from normal donors and patients with bone marrow dysfunction are also demonstrated.

5.2 PURIFICATION AND IDENTIFICATION OF HEMOPOIESIS-STIMULATORY SUBSTANCES FROM JUZEN-TAIHO-TO

5.2.1 FRACTIONATION OF JUZEN-TAIHO-TO AND IDENTIFICATION OF ACTIVE SUBSTANCES

Juzen-taiho-to was fractionated as shown in Figure 5.2 (Yamada et al., 1990). In order to examine the stimulatory activity of these fractions for primitive HSCs, we performed the HSC proliferation assay using HSC-enriched fractions (low density: LD cells) (Ogata et al., 1995) and an irradiated stromal

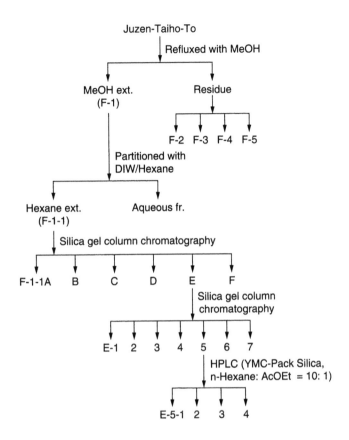

FIGURE 5.2 Purification of Juzen-taiho-to. (From Hisha, H. et al., 1997, *Blood*, 90, 1022–1030. With permission.)

cell line (MS-5, C3H mice bone marrow derived) (Itoh et al., 1989). These fractions were added to the culture system and, 2 weeks later, HSC proliferation was measured by ^{3}H-thymidine uptake.

The hexane-soluble fraction (F-1-1) stimulated the proliferation of HSCs at various concentrations (0.1 to 10 μg/ml) (Table 5.1, Exp. 1). Activity was found in F-1-1 E or F-1-1 F but not F-1-1 B and F-1-1 D. F-1-1 E was further fractionated using silica gel column chromatography to obtain E-1 to E-7. E-4 to E-7 showed stimulatory activity (Table 5.1, Exp. 2), but not E-1 to E-3 (data not shown). Elution patterns of silica gel column chromatography and thin layer chromatography revealed that each fraction was a mixture of several compounds, some of which were similar in each fraction (data not shown). E-5 was further purified to obtain the active substances using HPLC, as shown in Figure 5.3. Four main peaks (E-5-1 to E-5-4) were found. The minor peaks were mixed together and collected as E-5-5. As shown in Table 5.1, Exp. 3, all fractions had the capacity to stimulate the proliferation of HSCs, although the optimal doses differed.

The ^{1}H-NMR, GC, and GC-EI-MS analyses revealed that E-5-2 and E-5-3 contained at least four kinds of fatty acids, respectively: palmitic acid, stearic acid, oleic acid, and linolenic acid (Figure 5.4). E-5-1 also contained several fatty acids. Pure, commercially available substances were tested for their ability to stimulate the HSCs (Figure 5.5). Oleic acid and linolenic acid increased the proliferation of HSCs to about twice the level of the control, whereas palmitic and stearic acid induced a slight increase. E-5-4 was mainly composed of β-sitosterol, but low contamination also occurred with fatty acids. When all the fatty acids were removed from E-5-4, its stimulatory activity disappeared. Pure, commercially available β-sitosterol had no stimulatory effect on the HSCs (data not shown).

Finally, six fatty acids (stearic, palmitic, oleic, linolenic, vaccenic, and linoleic acids) were found in Juzen-taiho-to (Figure 5.4) when Juzen-taiho-to was directly extracted with chloroform/MeOH

TABLE 5.1
Stimulation Index of Various Fractions Obtained from Juzen-taiho-to for HSC Proliferation

	Exp. 1				
	F-1-1				
	F-1-1	B	D	E	F
0.1 µg/ml	2.09[a]	1.17	1.28	1.80	0.87
0.5 µg/ml	1.96	0.90	1.11	1.95	1.50
1 µg/ml	1.77	1.10	1.16	2.71	1.68
2.5 µg/ml	1.55	0.89	0.14	0.22	0.32
5 µg/ml	1.38	N.D.	N.D.	N.D.	N.D.
10 µg/ml	1.46	N.D.	N.D.	N.D.	N.D.

	Exp. 2						
	F-1-1						WEHI-3
	E	F	E-4	E-5	E-6	E-7	Cond. Med. (5%)[b]
0.1 µg/ml	N.D.	N.D.	1.01	1.21	1.17	0.97	1.60
0.5 µg/ml	N.D.	N.D.	1.09	1.43	1.28	1.06	
1 µg/ml	1.98	1.42	1.30	1.49	1.95	1.68	

	Exp. 3						
							WEHI-3
	E-5	E-5-1	E-5-2	E-5-3	E-5-4	E-5-5	Cond. Med. (5%)
0.05 µg/ml	N.D.	1.41	1.16	1.51	1.37	1.48	1.22
0.1 µg/ml	0.97	1.60	1.45	1.56	1.67	1.41	
0.2 µg/ml	N.D.	1.49	1.29	1.51	1.49	1.45	
0.5 µg/ml	1.37	N.D.	N.D.	N.D.	N.D.	N.D.	
1 µg/m	1.40	N.D.	N.D.	N.D.	N.D.	N.D.	

[a] Stimulation index was calculated using the following formula: stimulation index = ^{3}H-thymidine uptake on sample well/^{3}H-thymidine uptake on control well. Values more than 1.3 are in bold face.
[b] As a positive control, 5% of WEHI-3 conditioned medium was added to the culture medium instead of the sample.

(2:1). These standard fatty acids were tested for their stimulatory activity. Vaccenic and linoleic acids showed poor stimulatory activity. These findings indicated that fatty acids of C_{16} to C_{18}, such as stearic, palmitic, oleic, and linolenic acids, are the substances in Juzen-taiho-to capable of stimulating HSC proliferation (particularly linolenic and oleic acids, as shown in Figure 5.5).

5.2.2 HSC-Stimulatory Activity of Other Standard Fatty Acids

We next examined various standard fatty acids of C_4 to C_{24} for their capacity to stimulate the proliferation of HSCs. The stimulatory activity was found in petroselinic, petroselaidic, elaidic, and γ-linolenic acids of C_{18}. Behenic, erucic, and docosahexaenoic acids of C_{22} also exhibited such activity (Table 5.2). The chemical structure of elaidic acid resembles that of oleic acid: they are *cis* or *trans* form, respectively (Figure 5.4). Further studies are required to clarify the relationship between their chemical structure and biological activity.

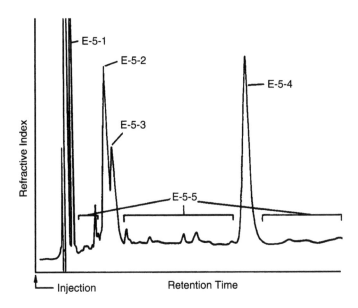

FIGURE 5.3 HPLC pattern of subfraction (E-5) of Juzen-taiho-to. E-5 was further fractionated by HPLC using n-hexane-AcOEt (10:1) as eluent and the four main peaks collected as E-5-1~4. The minor peaks were mixed together as E-5-5. (From Hisha, H. et al., 1997, *Blood*, 90, 1022–1030. With permission.)

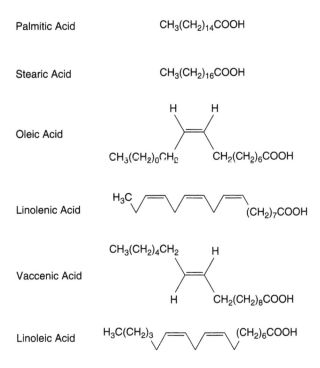

FIGURE 5.4 Chemical structures of fatty acids.

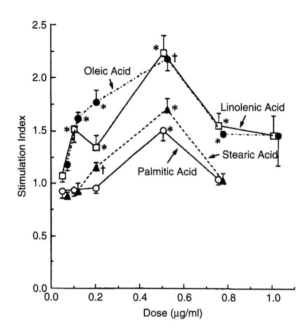

FIGURE 5.5 Stimulation of HSC proliferation by standard fatty acids. LD cells were cultured on MS-5 with or without various concentrations of standard commercially available fatty acids. 3H-thymidine incorporation of wells was measured and compared. The data are mean ± SD. *: Different from control (culture without fatty acids) (P < 0.05); †: (P < 0.005). (From Hisha, H. et al., 1997, *Blood*, 90, 1022–1030. With permission.)

Oleic and linolenic acids seem to be compounds commonly found in various Kampo medicines; both are found in Ninjin-yoei-to (Ren-Shen-Yang-Rong-Tang in Chinese), and Sho-seiryu-to (Xia-Qing-Long-Tang in Chinese), although their proportions are different from those found in Juzen-taiho-to (data not shown). The former Kampo medicine has been reported to stimulate hemopoiesis, but the latter does not. Indeed, we have found that the Ninjin-yoei-to extracts had a similar stimulatory effect on HSCs to Juzen-taiho-to, whereas Sho-seiryu-to had no effect (data not shown). It is therefore conceivable that the combinations and contents of the fatty acids in prescriptions are important for their clinical efficacy.

It is known that stearic acid is metabolized in the body and is converted to oleic acid. Accordingly, it is likely that the metabolism has an additional and definite effect on the efficacy of herbal medicines.

5.2.3 Effects of Oleic Acid and Linolenic Acid on Proliferation of Highly Purified HSCs

In the preceding experiments, we used partially purified HSCs (LD cells). This fraction contained HSCs and progenitor cells. Recently, we have established a new method of purifying P-HSCs from 5-FU-injected mice; a small number of the highly purified HSCs (lineage⁻, CD71⁻, H-2^high HSCs) were found to restore long-term hemopoiesis in irradiated recipients (Ogata et al., 1995). Therefore, we examined the effects of these fatty acids on the P-HSCs.

The P-HSCs were cultured with irradiated (20 Gy) MS-5 cells or syngeneic bone marrow stromal cells with or without fatty acids. A higher stimulatory activity of fatty acids was observed in the P-HSCs than in the LD cells, as shown in Table 5.3. This *in vitro* data also indicate that the fatty acids in Juzen-taiho-to stimulate the proliferation of P-HSCs rather than progenitor cells.

TABLE 5.2
Stimulation Index of C_4 to C_{24} Standard Fatty Acids for HSC Proliferation

	Fatty Acid Standard	0.2 µg/ml	0.5 µg/ml	1 µg/ml
4:0	Butyric acid		0.97	1.03
8:0	Caprylic acid		1.02	0.98
10:0	Decanoic acid		1.00	1.00
11:0	→Undecanoic acid	0.84	1.36	1.36
14:1(*cis*-9)	Myristoleic acid		1.25	1.24
16:0	**Palmitic acid**	0.96	1.54	
16:1(*cis*-9)	Palmitoleic acid		1.11	1.27
18:0	**Stearic acid**[a]	1.14	1.65	
18:1(*cis*-6)	→Petroselinic acid[b]	1.39	1.28	1.63
18:1(*trans*-6)	→Petroselaidic acid	0.80	1.75	1.62
18:1(*cis*-9)	**Oleic acid**	1.45	1.89	1.56
18:1(*trans*-9)	→Elaidic acid	1.68	1.70	1.82
18:1(*cis*-11)	*cis*-Vaccenic acid	1.29	0.84	0.92
18:1	11-*trans*-Octadecenoic acid	0.83	0.90	1.02
18:2(*cis*-9,-12)	**Linoleic acid**	1.35	1.14	1.24
18:2(*trans*-9,-12)	Linolelaidic acid	1.05	1.07	1.22
18:3(*cis*-6,-9,-12)	→γ-Linolenic acid	1.47	0.93	1.05
18:3(*cis*-9,-12,-15)	**Linolenic acid**	1.78	1.87	1.51
19:1	*cis*-10-Nonadecenoic acid	0.95	1.24	1.24
20:1	5-Eicosenoic acid	1.05	1.21	1.28
20:4(*cis*-5,-8,-11,-16)	Arachidonic acid		1.08	0.89
20:5(*cis*,-8,-11,-14,-17)	Eicosapentaenoic acid		1.03	0.79
22:0	→Behenic acid		1.59	1.37
22:1(*cis*-13)	→Erucic acid		1.23	1.35
22:6(*cis*-4,7,10,13,16,19)	→Docosahexaenoic acid		1.24	1.60
24:0	Lignoceric acid	0.94	1.11	1.24
24:1(*cis*-15)	Nervonic acid	0.98	0.93	1.12

[a] Fatty acids in bold face are found in Juzen-taiho-to.

[b] Fatty acids indicated by arrows are not found in Juzen-taiho-to, but show a stimulatory activity of more than 1.3.

TABLE 5.3
Comparison of Stimulatory Activity of Oleic and Linolenic Acids on Proliferation of LD Cells and P-HSCs

	LD Cells		P-HSCs		
	Linolenic Acid	Oleic Acid	Linolenic Acid		Oleic Acid
Fatty Acid	MS-5	MS-5	MS-5	BM Ad Cells	BM Ad Cells
0.5 µg/ml	1.98 ± 0.29[a]	1.59 ± 0.13[a]	2.46 ± 0.36[a]	3.02 ± 0.14b	1.71 ± 0.04[a]
1 µg/ml	1.47 ± 0.29[a]	1.65 ± 0.18[b]	2.98 ± 0.29[b]	4.90 ± 0.04[b]	1.83 ± 0.39[a]
2 µg/ml	N.D.	N.D.	4.44 ± 0.55[b]	N.D.	N.D.

[a] Different from control ($P < 0.05$).

[b] ($P < 0.005$).

Notes: P-HSCs were cultured on 20-Gy irradiated MS-5 cells or syngenic bone marrow stromal cells in the presence of fatty acids, and cell proliferation was measured 10 to 14 days later. LD cells were also cultured on MS-5 cells in the presence of oleic acid or linolenic acid for 4 days. Data are mean ± se.

5.3 MECHANISM OF ACTION OF FATTY ACIDS

To clarify the mechanisms behind the action of these fatty acids, we examined whether they act directly on the P-HSCs or if they first act on the stromal cells, followed by the proliferation of P-HSCs. LD cells or MS-5 cells were preincubated with oleic or linolenic acid for 2 to 4 hours and, after washing, cultured with nontreated MS-5 cells or LD cells, respectively. Table 5.4 indicates that short preincubation of MS-5 cells with linolenic acid can stimulate the proliferation of LD cells (Exp. 1). The same finding was observed when P-HSCs (instead of LD cells) were used (Exp. 2). Oleic acid also showed similar stimulatory activity (Exp. 3).

Figure 5.6 indicates the staining pattern of MS-5 cells preincubated with linolenic acid and then stained with anti-MHC class I and II antibodies as well as adhesion molecule antibodies. It is well known that IFNγ stimulates the expression of MHC class I antigen. The increase in MHC class I expression was observed in the preincubation with linolenic acid (particularly a high concentration of 5 µg/ml), although the intensity was weak in comparison with IFNγ. Expressions of MHC class II antigens and adhesion molecules (CD54 and CD106e) were also enhanced slightly at the concentration of 5 µg/ml. A similar enhancement was observed when oleic acid was preincubated with

TABLE 5.4
Effects of Preincubation of HSCs or MS-5 with Oleic or Linolenic Acids on HSC Proliferation

	Exp. 1			
	Preincubation of Low-Density Cells		**Preincubation of MS-5**	
Linolenic Acid	**2 h**		**2 h**	**4 h**
1 µg/ml	0.39 ± 0.06		1.99 ± 0.30^a	2.22 ± 0.33^a
	Exp. 2			
	Preincubation of P-HSCs		**Preincubation of MS-5**	
Linolenic Acid	**2 h**		**2 h**	
0.5 µg/ml	N.D.		2.36 ± 0.34^a	
1 µg/ml	0.86 ± 0.18		1.98 ± 0.30^a	
	Exp. 3			
	Preincubation of MS-5			
Oleic Acid	**4 h**			
1 µg/ml	1.86 ± 0.23^a			
5 µg/ml	2.33 ± 0.16^b			
10 µg/ml	2.56 ± 0.25^b			
1 µg/ml	2.44 ± 0.37^a			
10 µg/ml	1.76 ± 0.02^b			

[a] Different from control ($P < 0.05$).
[b] ($P < 0.005$).

Notes: LD cells or MS-5 were preincubated with linolenic acid for 2 or 4 hours, then washed two times. These cells were cultured with nontreated MS-5 cells or with nontreated LD cells, respectively. The incorporation of ³H-thymidine was measured 5 days later (Exp.1). P-HSCs (instead of LD cells) were used in Exp. 2 and Exp. 3. The incorporation of ³H-thymidine was assessed 10 days later. Stimulation indices: sample well/control well, in which nontreated LD cells or P-HSCs were cultured on nontreated MS-5 cells. Mean ± se.

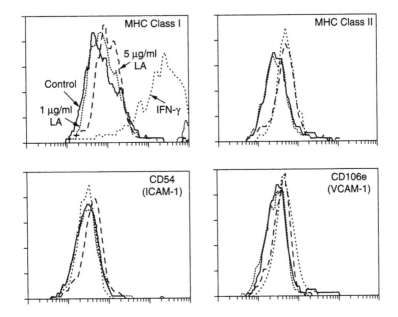

FIGURE 5.6 Enhancement of surface molecule expression on MS-5 by linolenic acid. (From Hisha, H. et al., 1997, *Blood*, 90, 1022–1030. With permission.)

MS-5 (data not shown). No increase in the expression of CD44 molecules, however, was observed, even if MS-5 had been preincubated with oleic or linolenic acid at the concentration of 5 μg/ml (data not shown). When butyric acid was incubated with MS-5, no increase in the expression of the five antigens was observed (data not shown); this is compatible with the finding in Table 5.2.

When oleic acid, linolenic acid, or a mixture of both was added to the stromal cell line at the concentrations of 0.05 to 10 μg/ml, these fatty acids did not show any stimulatory effects on the proliferation of MS-5 cells (data not shown). A similar observation was also obtained when a hybridoma-producing actibody was cultured with these fatty acids at the doses of 0.01 to 5 μg/ml (data not shown). Moreover, the fatty acids could not stimulate the proliferation of P-HSCs in the absence of stromal cells (data not shown). These observations indicate that the fatty acids in Juzen-taiho-to act directly on the MS-5 cells, resulting in the proliferation of P-HSCs.

Although it is not yet clear exactly how these fatty acids induce the proliferation of HSCs, it seems likely that the signal for HSC proliferation is mediated by stromal cells stimulated by the fatty acids because the preincubation of MS-5 cells with oleic acid and linolenic acid resulted in the proliferation of HSCs (Table 5.4). In addition, incubation of MS-5 with oleic and linolenic acids increased the expression of adhesion molecules (CD54 and CD106e) and MHC class I as well as class II antigens on the stromal cells. One of a number of other possible mechanisms might give rise to the signal for HSC proliferation. For example, linolenic acid stimulates the production of early acting cytokines from MS-5 cells. Moreover, it is conceivable that this acid is utilized for the biosynthesis of the glycolipids and/or phospholipids important for cell growth.

5.4 EFFECT OF LINOLENIC ACID ON LONG-TERM CULTURE OF P-HSCs

The effects of linolenic acid on long-term culture (Dexter et al., 1977; Issaad et al., 1993) were examined. When the P-HSCs were cultured for 6 weeks on the MS-5 preincubated with 1 or 5 μg/ml of linolenic acid, the numbers of cobblestone colonies and nonadherent cells per flask significantly increased (Table 5.5). Figure 5.7 is a photograph of cobblestone colonies (hemopoietic colonies formed

TABLE 5.5
Stimulatory Effect of Linolenic Acid on Proliferation
of P-HSCs in Long-Term Culture

	Per Flask	
	No. of Cobblestone Colonies	No. of Non-ad. Cells ($\times 10^4$)
Control	4.3 ± 0.9	1.7 ± 0.2
Preincubation:		
1 μg/ml	33.7 ± 5.2[a]	9.4 ± 1.4[a]
5 μg/ml	49.3 ± 9.3[a]	12.2 ± 2.1[a]
Coculture:		
1 μg/ml	43.7 ± 5.8[a]	12.1 ± 1.5[a]
5 μg/ml	6.3 ± 2.8[b]	1.7 ± 0.2[b]

[a] Different from control ($P < 0.05$).
[b] Not significant.

Notes: MS-5 were preincubated with linolenic acid (1 or 5 μg/ml) for 1 day and washed two times. P-HSC was cultured on the MS-5 and the culture medium was replaced weekly with fresh medium. The number of cobblestone colonies and nonadherent cells per flask were counted 6 weeks later. Mean ± SE.

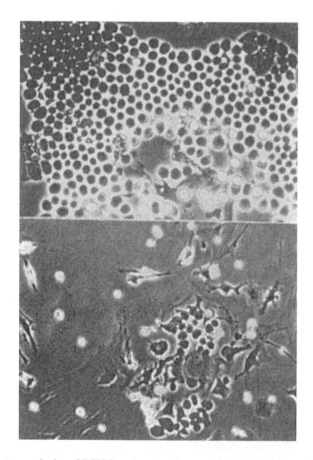

FIGURE 5.7 Cobblestone colonies of P-HSCs cultured with nontreated MS-5 (left) or MS-5 preincubated with 5 μg/ml of linolenic acid (right) for 6 weeks. (From Hisha, H. et al., 1997, *Blood*, 90, 1022–1030. With permission.)

among MS-5). Similar findings were obtained when P-HSCs and MS-5 were cocultured with 1 µg/ml of linolenic acid. A high concentration (5 µg/ml) of linolenic acid, however, did not show any stimulatory effect.

5.5 *IN VIVO* EFFECTS OF FATTY ACIDS ON RECOVERY OF HEMOPOIESIS IN MMC-TREATED MICE

Whether the fatty acids also stimulate the proliferation of HSCs *in vivo* was investigated. The recovery of hemopoiesis in MMC-treated mice administered various doses of oleic acid was assessed using CFU-s assay. The experimental protocol is shown in Figure 5.8. The administration of MMC significantly decreased day-8 and day-14 CFU-S counts (Figure 5.9). An increase in CFU-S counts, however, was observed when oleic acid was given to the mice at concentrations of 1 to 1000 mg/Kg. The most significant effect was observed at a dose of 1 mg/Kg on day 8 and 10 mg/Kg on day 14. Similar (but slightly lower) stimulation was also observed in mice administered linolenic acid (data not shown).

5.6 HEMOPOIESIS-STIMULATORY EFFECT OF FATTY ACIDS ON HUMAN BONE MARROW CELLS

5.6.1 STIMULATORY EFFECT OF FATTY ACIDS ON HEMOPOIESIS OF NORMAL HUMAN BMCs

The stimulatory activity of fatty acids on human hemopoietic cells was also investigated. There was no stimulatory effect of fatty acids on human whole bone marrow cells, lineage-negative cells (HSC-enriched fraction), or CD34$^+$ cells (CD34: a marker for human HSCs) (Krause et al., 1996), when cultured in a semisolid culture system, a methylcellulose assay containing human cytokines (Figure 5.10). However, these fatty acids showed a stimulatory effect on a long-term culture system of human BMCs, where HSCs can proliferate and differentiate on a murine stromal cell

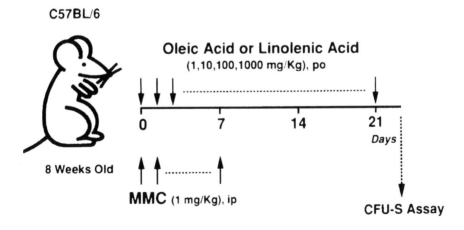

FIGURE 5.8 Experimental protocol of oleic acid administration into MMC-treated mice. Various concentrations of fatty acids were given orally to MMC-treated mice for 3 weeks and their bone marrow cells then injected into a syngeneic recipient for CFU-S assay. The number of hemopoietic colonies in the recipient spleen was counted 8 or 14 days later.

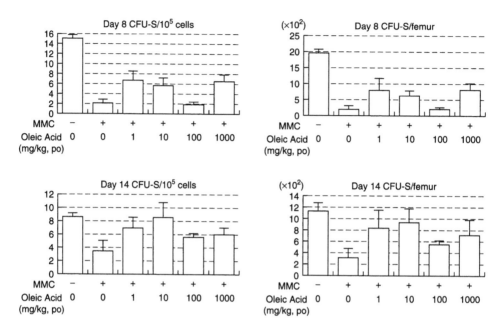

FIGURE 5.9 Effect of oleic acid on CFU-S recovery after MMC injection. The data represent mean ± SD of four mice. (From Hisha, H. et al., 1997, *Blood*, 90, 1022–1030. With permission).

line (MS-5) in the presence of a low concentration of human cytokines (SCF, Flt-3 ligand, IL-3, IL-6, LIF, G-CSF) (Figure 5.11). Increases in the number of cobblestone colonies, nonadherent cells, and CFU-C (progenitor cells were detected in a methylcellulose assay) were observed when BMCs were cultured on MS-5 pretreated with linolenic or oleic acid. These observations agree with the finding in the murine system that the fatty acids act on the proliferation of P-HSCs through the stromal cells.

5.6.2 Marked Stimulatory Effect of Fatty Acids on BMCs from Patients with Shwachman Syndrome

Shwachman et al. (1964) reported a unique disease: the patient showed not only diarrhea due to exocrine pancreatic insufficiency but also neutropenia and anemia due to bone marrow hypoplasia. This syndrome is also characterized by short stature, skeletal abnormalities, and various other less common findings; dysfunctions in the pancreas and bone marrow are most commonly observed (Mack et al., 1996). Up to now, it has not been well elucidated why pancreatic insufficiency and bone marrow hypoplasia concur in this syndrome.

Several hypotheses for this puzzling phenomenon have been put forth. First, several genes controlling the functions of the pancreas and hemopoiesis share each other or exist near each other. Therefore, a mutation or deletion of one locus, caused by chromosome breakage (Tada et al., 1987; Fraccaro et al., 1988) or reciprocal translocation (Masuno et al., 1995), is thought to elicit both dysfunctions. Second, copper deficiency in the newborn is thought to elicit this symptom because a similar symptom can be induced in rats by feeding them food lacking copper (Paterson and Wormsley, 1988). To date, 45 reports concern this syndrome and the general treatment, which is the administration of enzymes from the pancreas such as pancreatin.

In some cases, G-CSF is administered for severe neutropenia and infections, although its effect is short lived (Grill et al., 1993; Ventura et al., 1995). However, a patient in our hospital apparently benefited from Juzen-taiho-to. The patient was administered Juzen-taiho-to (2.5 g/day) orally when

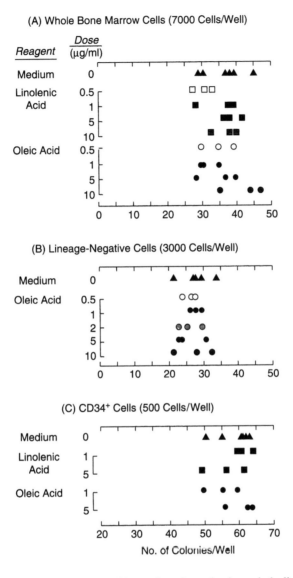

FIGURE 5.10 No effect of linolenic or oleic acid on colony formation in methylcellulose assay of (A) whole BMCs; (B) lineage-negative cells; and (C) CD34+ cells (HSCs) from a normal donor. BMCs were cultured in methyl cellulose containing various cytokines (Epo, SCF, GM-CSF, IL-3, and G-CSF), with or without linolenic or oleic acid. The colonies (composed of >50 cells) were counted 20 days later.

she showed pancytopenia, after which she recovered from the neutropenia and anemia (Kohdera et al., manuscript in preparation) (Figure 5.12). Thereafter, a recovery of the hemopoietic function was observed with an increase in the dose (2.5 to 7.5 g/day).

In this report, we studied two children with Shwachman syndrome: the patient in our hospital (No. 1) and a patient at another hospital (No. 2). When linolenic acid was added to the methylcellulose assay of BMCs from patient No. 1, an increase in the cluster counts was observed (Figure 5.13, Exp. 1). However, the patient with aplastic anemia had no stimulatory effect on BMCs. A slight increase in colony formulation was observed in patient No. 2 (Figure 5.13, Exp. 2).

When BMCs from patient No. 2 were cultured with MS-5 pretreated with the fatty acids in long-term culture, the numbers of cobblestone colonies, nonadherent cells, and CFU-C significantly

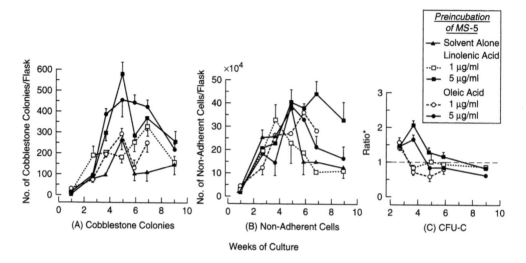

FIGURE 5.11 Long-term culture of normal BMCs on MS-5 preincubated with fatty acids. *: Ratio = no. of CFU-C per 10^4 cells collected from sample well/no. of CFU-C per 10^4 cells collected from control well. (From Hisha, H. et al., 2002, *Stem Cells*, 20, 311–319. With permission.)

increased (Figure 5.14). Because the concentration of fatty acids in the blood is lower than normal levels in these patients, it is conceivable that bone marrow hypoplasia is induced or made worse by low levels of fatty acids in the blood, due to pancreatic insufficiency. This case may give us important clues to the relationship between fatty acids and hemopoiesis.

Very recently, we have observed that CFU-C counts increase up to twice the control when BMCs from a patient with MDS (myelodysplastic syndrome) are cocultured with extracts of

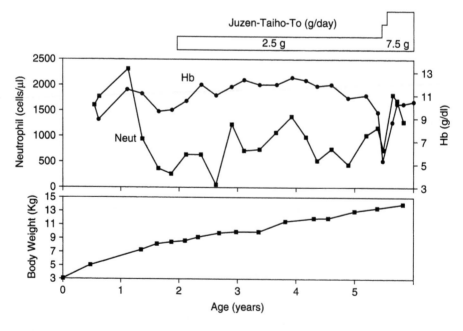

FIGURE 5.12 Hemopoietic recovery by administration of Juzen-taiho-to to a patient (No.1) with Shwachman syndrome. (From Hisha, H. et al., 2002, *Stem Cells*, 20, 311–319. With permission.)

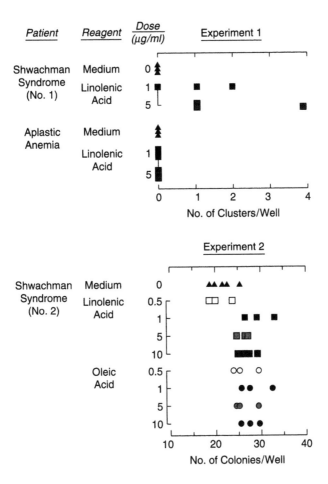

FIGURE 5.13 Stimulatory effect of fatty acid on colony formation in methylcellulose assay of BMCs from patients with Shwachman syndrome. BMCs were cultured in methylcellulose containing various cytokines (Epo, SCF, GM-CSF, IL-3, and G-CSF), with or without linolenic or oleic acid (3×10^4/well). The clusters (composed of 10 to 50 cells) were counted 20 days later.

Juzen-taiho-to (data not shown). This indicates the possibility that Juzen-taiho-to has a marked stimulatory effect for other bone marrow diseases.

5.7 DISCUSSION

Several lines of evidence suggest that Juzen-taiho-to has immunological activities in addition to stimulating hemopoiesis. In this chapter, we have focused on its stimulatory activities for HSCs. It has been shown that P-HSCs generate multilineage colonies on the bone marrow stromal feeder layer, but not in methylcellulose supplemented with various combinations of early- to late-stage-acting cytokines (Ogata et al., 1992; Berardi et al., 1995). These findings indicate that the direct interaction of P-HSCs with stromal cells is essential to the proliferation of P-HSCs. We have therefore established an HSC proliferation assay system using a cell line (MS-5). In the present study, we have demonstrated, using this assay and also long-term culture of P-HSCs, that two fatty acids (oleic and linolenic acids) are potent substances promoting the proliferation of P-HSCs.

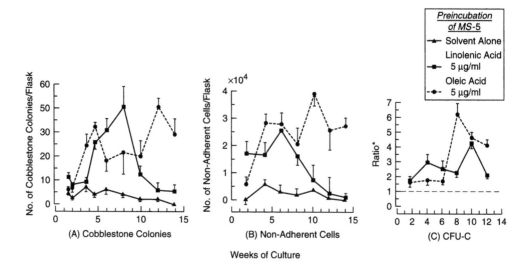

FIGURE 5.14 Long-term culture of BMCs from a patient (No.2) with Shwachman syndrome on MS-5 preincubated with fatty acids. *: Ratio = no. of CFU-C per 10^4 cells collected from sample well/no. of CFU-C per 10^4 cells collected from control well. (From Hisha, H. et al., 2002, *Stem Cells*, 20, 311– 319. With permission.)

It is well known that oleic acid is an unsaturated fatty acid commonly found in animals as well as in plants and that linolenic acid is an essential fatty acid that animals cannot biosynthesize. Oleic acid appears to be an essential substance for cell growth because it is added to the culture medium at a concentration of 5.6 µg/ml together with dipalmitoyl lecithin and cholesterol as a substitute for FCS. Recently, Inaoka et al. have isolated the fatty acids (palmitic, stearic, oleic, linolenic, and eicosanoic acids) from *Vitex rotundifolia* Linne. fil; these fatty acids stimulate hair regrowth (manuscript in preparation). This suggests that the fatty acids (at least oleic acid) are common cell growth factors. In the present study, however, oleic and linolenic acids did not appear to have a significant stimulatory effect on MS-5 and hybridoma cells (data not shown). Stimulation indices over 4.0 were obtained when highly purified P-HSCs were cocultured with stromal cells in the presence of these fatty acids (Table 5.3). Accordingly, it seems that these fatty acids are essential for the proliferation and/or differentiation of P-HSCs.

In the present study, we have used P-HSCs purified by the method described previously (Ogata et al., 1995). We have recently established a method for further purifying P-HSCs (lineage⁻/CD71⁻/H-2^{high}/c-kit$^{<low}$) (Doi et al., 1997). Using these further purified P-HSCs, we have confirmed that oleic acid enhances the proliferation of P-HSCs in the presence of stromal cells (data not shown). To our knowledge, this is the first report indicating that oleic and linolenic acids have HSC-stimulatory activity. The data presented here might provide new insights into the mechanism underlying HSC proliferation.

It is possible that other substances with the capacity to stimulate the proliferation of P-HSCs are contained in Juzen-taiho-to together with the fatty acids. Indeed, we find a fraction of acidic polysaccharides shows such activity (data not shown). Experiments to purify and identify such substances are now in progress.

5.8 SUMMARY

- It is well-known that Juzen-taiho-to has the capacity to stimulate hemopoiesis. Here we show that at least some of the active substances in Juzen-taiho-to are unsaturated fatty acids (oleic and linolenic acids).

- These standard fatty acids show a significant stimulatory effect on HSCs in an HSC proliferation assay and a long-term bone marrow culture. The administration of oleic acid into MMC-treated mice enhances CFU-S counts on day 14 to twice that of the control group.
- The stimulatory effect of these fatty acids is mediated by stromal cells, rather than by direct action on HSCs.
- A patient with Shwachman syndrome, showing exocrine pancreatic insufficiency and severe bone marrow hypoplasia, recovered from pancytopenia by the administration of Juzen-taiho-to.

This observation and the preceding experimental results indicate that fatty acids are essential substances for hemopoiesis and that their low levels in the blood, due to pancreatic insufficiency, cause or make worse pancytopenia.

ACKNOWLEDGMENT

The authors thank Drs. H. Yamada, M. Sakurai-H., and H. Kiyohara of Kitasato Institute (Tokyo, Japan) for the purification and identification of the active substances in Juzen-taiho-to. We are also grateful to Drs. U. Kohdera and Y. Kobayashi in the Department of Pediatrics, Kansai Medical University, to Dr. M. Hirayama in the Department of Pediatrics, Mie University, for donating clinical data and bone marrow samples, and to Ms. K. Ando for manuscript preparation.

GLOSSARY

BMCs: bone marrow cells
BMT: bone marrow transplantation
CD34: cell surface molecule expressed on human HSCs
CD44: adhesion molecule phagocytic glycoprotein-1
CD54: adhesion molecule ICAM-1
CD71: transferrin receptor
CD106e: adhesion molecule VCAM-1
CFU-C: colony forming unit in culture
CFU-S: colony forming unit in spleen
c-kit: tyrosine kinase-type receptor, receptor of SCF
Epo: erythropoietin
Flt-3 ligand: fms-like tyrosine kinase 3
5-FU: 5-fluorouracil
G-CSF: granulocyte colony stimulating factor
H-2: mouse major histocompatibility complex
HSCs: hemopoietic stem cells
IFNγ: interferon-γ
IL-3: interleukin-3
IL-6: interleukin-6
LIF: leukemia inhibitory factor
MMC: mitomycin C
MS-5: murine stromal cell line that can support hemopoiesis
MHC: major histocompatibility complex
SCF: stem cell factor
WEHI-3 Cond. Med.: culture supernatant of WEHI-3 (a mouse lymphocyte-like cell line) that secretes growth factors for HSCs

REFERENCES

Aburada, M., Takeda, S., Ito, E., Nakamura, M., and Hosoya, E. (1983) Protective effects of Juzentaihoto, dried decoctum of 10 Chinese herbs mixture, upon the adverse effect of mitomycin C in mice. *J. Pharm. Dyn.* 6, 1000–1004.

Berardi, A.C., Wang, A., Levine, J.D., Lopez, P., and Scadden, D.T. (1995) Functional isolation and characterization of human hematopoietic stem cells. *Science,* 267, 104–108.

Dexter, T.M., Allen, T.D., and Lajtha, L.G. (1977) Conditions controlling the proliferation of hemopoietic stem cells *in vitro. J. Cell. Physiol.,* 91, 335–344.

Doi, H., Inaba, M., Yamamoto, Y., Taketani, T., Mori, S., Sugihara, A., Ogata, H., Toki, J., Hisha, H., Inaba, K., Sogo, S., Adachi, S., Matsuda, T., Good, R.A., and Ikehara, S. (1997) Pluripotent hemopoietic stem cells are c-kit$^{<low}$. *Proc. Natl. Acad. Sci. USA.,* 94, 2513–2517.

Fraccaro, M., Scappaticci, S., and Arico, M. (1998) Shwachman syndrome breakage. *Hum. Genet.,* 79, 194.

Grill, J., Bernaudin, F., Dresch, C., Lemerle, S., and Reinert, P. (1993) Treatment of neutropenia in Shwachman's syndrome with granulocyte growth factor (G-CSF). *Arch. Fr. Pediatr.,* 50, 331–333.

Hisha, H., Yamada, H., Sakurai-H., M., Kiyohara, H., Li, Y., Yu, C., Takemoto, N., Kawamura, H., Yamaura, K., Shinohara, S., Komatsu, Y., Aburada, M., and Ikehara, S. (1997) Isolation and identification of hematopoietic stem cell-stimulating substances from Japanese herbal medicine, Juzen-Taiho-to (TJ-48). *Blood,* 90, 1022–1030.

Hisha, H., Kohdera, U., Hirayama, M., Yamada, H., Iguchi-Uehira, T., Fan, T., Cui, Y., Yang, G., Li Y., Sugiura, K., Inabe, M., Kobayashi, Y., and Ikehara, S. (2002) Treatment of Shwachman syndrome by Japanese herbal medicine (Juzen-taiho-to): stimulatory effects of its fatty acids on hemopoiesis in patients. *Stem Cell,* 20, 311–319.

Hosokawa, Y. (1985) Radioprotective effects of oriental drugs on mice. *Ther. Res.,* 2, 1029–1034.

Ikehara, S., Kawamura, H., Komatsu, Y., Yamada, H., Hisha, H., Yasumizu, R., Ohnishi-inoue, Y., Kiyohara, H., Hirano, M., Aburada, M., and Good, R.A. (1992) Effects of medical plants on hemopoietic cells. In H. Friedman (Ed.), *Microbial Infections,* Plenum Press, New York, 319–330.

Inaoka, Y., Kuroda, H., and Fukushima, M. (manuscript in preparation).

Issaad, C., Croisille, L., Katz, A., Vainchenker, W., and Coulombel, L. (1993) A murine stromal cell line allows the proliferation of very primitive human CD34^{++}/CD38^{-} progenitor cells in long-term cultures and semisolid assays. *Blood,* 81, 2916–2924.

Itoh, K., Tezuka, H., Sakoda, H., Konno, M., Nagata, K., Uchiyama, T., Uchino, H., and Mori, K.J. (1989) Reproducible establishment of hemopoietic supportive stromal cell lines from murine bone marrow. *Exp. Hematol.* 17, 145–153.

Kawamura, H., Maruyama, H., Takemoto, N., Komatsu, Y., Aburada, M., Ikehara, S., and Hosoya, E. (1989) Accelerating effect of Japanese Kampo medicine on recovery of murine hematopoietic stem cells after administration of mitomycin C. *Int. J. Immunotherapy,* 5, 35–42.

Krause, D.S., Fackel, M.J, Clvin C.I., and Way, W.S. (1996) CD34: structure, biology and clinical utility. *Blood,* 87, 1–13.

Mack, D.R., Forstner, G.G., Wilschanski, M., Freedman, M.H., and Durie, P.R. (1996) Shwachman syndrome: exocrine pancreatic dysfunction and variable phenotypic expression. *Gastroenterology,* 111, 1593–1602.

Masuno, M., Imaizumi, K., Nishimura, G., Nakamura, M., Saito, I., Akagi, K., and Kuroki, Y. (1995) Shwachman syndrome associated with *de novo* reciprocal translocation (t96;12)(q16.2;q21.2). *J. Med. Genet.* 32, 894–895.

Miyamomae, T. and Frindel, E. (1988) Radioprotection of hemopoiesis conferred by *Acanthopanax senticosus* Harms (Shigoka) administered before or after irradiation. *Exp. Hematol.* 16, 801–806.

Nabeya, K. and Ri, S. (1983) Effect of oriental medical herbs on the restoration of the human body before and after operation. *Proc. Symp. Wakan–Yaku,* 16, 201–206.

Ogata, H., Bradley, G.W., Inaba, M., Ogata, N., Ikehara, S., and Good, R.A. (1995) Long-term repopulation of hematolymphoid cells with only a few hemopoietic stem cells in mice. *Proc. Natl. Acad. Sci. USA,* 92, 5945–5949.

Ogata, H., Taniguchi, S., Inaba, M., Sugawara, M., Ohta, Y., Inaba, K., Mori, K.J., and Ikehara, S. (1992) Separation of hematopoietic stem cells into two populations and their characterization. *Blood.* 80, 91–95.

Ohnishi, Y., Yasumizu, R., Fan, H., Liu, J., Takao-Liu, F., Komatsu, Y., Hosoya, E., Good, R. A., and Ikehara, S. (1990) Effect of Juzen-taiho-toh (TJ-48), a traditional Oriental medicine, on hematopoietic recovery from radiation injury in mice. *Exp. Hematol.,* 18, 18–22.

Paterson, C.R. and Wormsley, K.G. (1988) Hyposesis: Shwachman's syndrome of exocrine pancreatic insufficiency may be caused by neonatal copper deficiency. *Ann. Nutr. Metab.*, 32, 127–132.

Shwachman, H., Diamond, L.K., Oski, F.A., and Khaw, K.-T. (1964) The syndrome of pancreatic insufficiency and bone marrow dysfunction. *J. Pediatr.* 65, 645–663.

Tada, H., Ri, T., Yoshida, H., Ishimoto, K., Kaneko, M., Yamashiro, Y., and Shinohara, T. (1987) A case of Shwachman syndrome with increased spontaneous chromosome breakage. *Hum. Genet.*, 77, 289–291.

Ventura, A., Dragovich, D, Luxardo, P., and Zanazzo, G. (1995) Human granulocyte colony-stimulating factor (rHuG-CSF) for treatment of neutropenia in Shwachman syndrome. *Hematologica*, 80, 227–229.

Yamada, H., Kiyohara, H., Cyong, J.-C., Takemoto, N., Komatsu, Y., Kawamura, H., Aburada, M., and Hosoya, E. (1990) Function and characterization of mitogenic and anti-complementary active fractions from kampo (Japanese herbal) medicine "Juzen-Taiho-To." *Plant Med.*, 56, 386–391.

Yonezawa, M. (1976) Restration of radiation injury by intra-peritoneal injection of ginseng extract in mice. *Radiat. Res.*, 17, 111–113.

6 Preventive Effects of Juzen-taiho-to on Infectious Diseases

Shigeru Abe, Nobuo Yamaguchi, Shigeru Tansho, and Hideyo Yamaguchi

CONTENTS

6.1 INTRODUCTION

Kampo medicines have immunostimulating activity for ill patients. Our idea is that these medicines can be applied to prophylactic and therapeutic strategies for opportunistic infection, such as that offered by Nomoto (1996) and Yamaguchi (1996). Juzen-taiho-to has been used clinically for patients with anemia, weakness, or debility. Experimental studies have revealed that orally administered Juzen-taiho-to shows multiple immunopharmacological effects, particularly those enhancing macrophage functions (Maruyama et al., 1988), humoral immunity (Komatsu et al., 1986), and cell-mediated immunity (Takemoto et al., 1989), as reviewed previously (Yamaguchi, 1992; Tei, 1996).

This chapter reviews experimental studies on the protective action of Juzen-taiho-to towards various microbial diseases as well as those of related immunopharmacological activities. It focuses especially on our recent studies of *in vivo*, *ex vivo*, and *in vitro* antifungal effects of Juzen-taiho-to.

6.2 ANTIBACTERIAL EFFECTS

Satomi et al. (1989a) reported that the treatments of ddy mice with Juzen-taiho-to (pharmaceutical preparation TJ-48 manufactured by Tsumura & Co.) or several other Kampo prescriptions were effective in protecting them from lethal infection by *Pseudomonas aeruginosa*, in which excessive production of tumor necrosis factor (TNF-α) might be responsible for the morbidity. They also reported that Juzen-taiho-to protected mice from acute lethality caused by the administration of murine TNF-α and suppressed the production of leukotrienes (Satomi et al., 1989b). Experimental infection of *P. aeruginosa* in leukopenic mice was also prevented by Ninjin-yoei-to (pharmaceutical preparation TJ-108 by Tsumura & Co.), a similar prescription to Juzen-taiho-to (see Appendix 1 and Appendix 2) (Miura et al., 1992). They also showed that this protective activity related to the acceleration of hematopoiesis on leukopenic mice.

Hamada et al. (1988) investigated the effect of Juzen-taiho-to on the survival time of *Salmonella enteritidis*-infected mice transplanted with EL-4 leukemia. They found that oral administration of Juzen-taiho-to at a dose of 1 g/kg/day clearly restored the macrophage functions depressed by leukemia and prolonged the life span of the infected mice. They provided several pieces of evidence suggesting that the protective activity of Juzen-taiho-to against salmonella infection may be due to the augmentation of antimicrobial activity of macrophages, through chemotaxis and phagocytosis.

Shimizu et al. (1997) found that Hochu-ekki-to (pharmaceutical preparation TJ-41 by Tsumura & Co.), a prescription similar to Juzen-taiho-to (see Appendix 1 and Appendix 2), clearly prevented experimental infection of *S. enteritidis* in mitomycin C-induced immunosuppressed mice. Recently, Shimizu et al. (1998) and Yamaguchi et al. (1999) compared the efficacy of Hochu-ekki-to and Juzen-taiho-to against infection of salmonella and methicillin-resistant *Staphylococcus aureus* (MRSA) in mice. They suggested that the protective action of Juzen-taiho-to is more potent against MRSA, an extracellular parasite, than against salmonella, an intracellular one.

6.3 ANTIVIRAL AND ANTIMALARIAL EFFECTS

Yamaura et al. (1996) investigated the effects of Juzen-taiho-to on the course of nonlethal *Plasmodium berghei* XAT infection in CBA mice. Administration of Juzen-taiho-to at a dose of 2 g/kg orally once a day during the observation period suppressed the development of parasitaemia of the infected mice. Mice treated with Juzen-taiho-to showed higher levels of specific antibody titers, especially IgG2a isotype of immunoglobulins, in the serum than did the control mice. Production of interferon-γ by spleen cells of mice treated with Juzen-taiho-to was also higher than that of the control mice.

These results suggest that Juzen-taiho-to is effective in suppressing the course of parasitaemia through stimulation of cellular immunity. This finding seems reasonable because Juzen-taiho-to was reported to modulate the differentiation of T-helper cell subsets so that TH1 dominated over TH2 (Takemoto et al., 1996; Matsumoto and Yamada, 1999–2000; Iijima et al., 1999).

We have seen no report about preventive effects of Juzen-taiho-to against experimental viral infections, but some of the crude drugs used in Juzen-taiho-to, such as Ginseng Radix and Glycyrrhizae Radix, are known to protect against viral infections. Mori et al. (2000) recently examined the effects of Hochu-ekki-to on viral infection and reported that its oral administration increased the survival rate of influenza virus-infected mice. Hochu-ekki-to did not show any inhibitory effect on the replication of influenza virus *in vitro*. These researchers observed lowered virus titers in bronchoalveolar lavage fluid (BALF) of mice and suggested that the efficacy might be related to the rapid elevation of interferon α in BALF of mice treated with Hochu-ekki-to. The antiviral activity of Hochu-ekki-to may be explained by its augmentation action of NK cell function as reported by Ohno et al. (1988). Hochu-ekki-to contains Zizyphi Fructus, which strongly augments NK cell activity (Yamaoka et al., 1996), but Juzen-taiho-to does not, as shown in Appendix 1 and Appendix 2.

6.4 ANTIFUNGAL EFFECTS

Fungal infections are the major cause of mortality in patients with severely impaired host defenses, including neutropenic patients with acute leukemia and other patients receiving immunosuppressive therapy. In view of the unacceptably high mortality of systemic fungal infections in immunosuppressed patients, immunostimulating agents with protective activity against fungal infections applicable to these individuals must be developed. Juzen-taiho-to is considered to be one of the promising candidates to meet this need.

6.4.1 Protective Effect in Normal and Immunosuppressed Mice

In a series of studies on the prevention of fungal infection by Juzen-taiho-to (TJ-48), we evaluated *in vivo* efficacy of orally administered Juzen-taiho-to against systemic candidasis in mice. Uchida et al. (1995) investigated the effects of oral administration of Juzen-taiho-to on candida infection in mice of an outbred mouse strain, ICR. Normal mice were challenged intravenously with *Candida albicans* and given Juzen-taiho-to orally once daily for 5 consecutive days, beginning 3 h after candida inoculation.

The treatment with Juzen-taiho-to inhibited the candida growth in kidneys of the infected mice and prolonged their life span in a dose-dependent manner with an optimal daily dose of 2 g/kg. Juzen-taiho-to was also shown to protect against candida infection in cyclophosphamide (CY)-treated immunosuppressed mice, similar to neutropenic patients intensively treated with anticancer chemotherapeutic agents (Satonaka et al., 1996). This efficacy was significantly increased when combined with an antifungal agent, fluconazole.

Juzen-taiho-to was found to be similarly efficacious against intravenous infection by *Aspergillus fumigatus* in mice, but not by *Cryptococcus neoformans*. It is well known that host defense against infection of *A. fumigatus* depends on phagocytes such as macrophages and neutrophils. In contrast, that of *Cr. neoformans* is dependent on specific cellular immunity, as reviewed previously (Abe et al., 1994). Therefore, therapeutic failure of Juzen-taiho-to against cryptococcus infection may be a characteristic of its immunomodulating action, especially in a relatively short term.

The protective effect of Juzen-taiho-to against candida infection was compared with those of other Kampo formulas (prescription)—for example, Hochu-ekki-to or Ninjin-yoei-to in neutropenic mice (Abe et al., 2000). CY-induced neutropenic mice were orally given Juzen-taiho-to or another prescription for 4 consecutive days (day 4 to day 1). They were then challenged intravenously with a lethal dose of *Candida albicans* (day 0). Table 6.1 shows that all three prescriptions appeared to prolong the life span of candida-infected mice.

An oral dose of 1 g/kg/day of Juzen-taiho-to prolonged life remarkably. This suggests that pretreatment with Juzen-taiho-to strongly protected the mice from candida infection and that the protective activity against *Candida* infection is present not only in Juzen-taiho-to but also in other Kampo formulas. However, no significant effect was observed with the administration

TABLE 6.1
Prophylactic Effect of Kampo Medicine against *C. albicans* Infection in Leukopenic Mice

Dose g/kg/day	Juzen-taiho-to		Ninjin-yoei-to		Hochu-ekki-to	
	S/T	(%)	S/T	(%)	S/T	(%)
0	3/18	17				
0.5	1/6	17	0/6	0		
1	8/12	67[b]	6/12	50[a]	4/12	33
2	4/12	33	6/12	50[a]	2/6	33

[a] $p < 0.05$
[b] $p < 0.01$

Notes: Mice were pretreated intraperitoneally with cyclophosphamide on day 4 and inoculated intravenously with 6×10^4 CFU of *C. albicans* on day 0. Each group of mice was given orally distilled water or Kampo medicine from day 4 to day 1. S/T: Mice surviving/mice tested.

of Juzen-taiho-to at a dose of 2 g/kg/day, suggesting that there is an optimal dose rate for the prophylactic effect. As described earlier, when Juzen-taiho-to was administered after candida infection, the optimal dose was 2 g/kg/day. This difference in an effective dose is not inconceivable because the prophylactic efficacy of Juzen-taiho-to depends on the immunological condition of the host animal (Abe et al., 1997b).

Polak–Wyss (1991) reported that opportunistic fungal infections were difficult to cure by immunomodulators in glucocorticoid-induced immunosuppressed hosts. Glucocorticoids induce many types of immunosuppression: depressed cytokine production, functional inhibition of macrophages (Schaffner, A. and Schaffner, T., 1987), and neutrophils (Nohmi et al., 1994; Abe et al., 1996). The protective effect of Juzen-taiho-to on experimental candidiasis was examined in mice immunosuppressed by injection of prednisolone (Abe et al., 2000). Treatments with a daily dose of 1 g/kg/day of Juzen-taiho-to or Hochu-ekki-to (Abe et al., 1999) for 4 consecutive days from day 0, but not from day 4, significantly prolonged the life span of thus pretreated candida-infected mice.

The reasons for this characteristic effective administration timing in prednisolone-treated mice cannot be clearly explained in relation to the activities of Juzen-taiho-to or Hochu-ekki-to. However, Morrissey and Charrier (1991) reported that interleukin-1 administration to C3H/HeJ mice after, but not prior to, infection increased resistance to salmonella, which may result from an insufficient production of cytokines in the host. We believe that this is one of the first findings of the effectiveness of a clinical herbal medicine against an infectious disease in glucocorticoid-treated mice (Abe et al., 1999).

6.4.2 IMMUNOPHARMACOLOGICAL ACTION TOWARD THE ANTIFUNGAL HOST DEFENSE MECHANISM

The *in vivo* antifungal activity of Juzen-taiho-to has been suggested to be associated with its immunological activity because Juzen-taiho-to has no direct fungicidal activity *in vitro* (Uchida et al., 1995). To study immunological mechanisms by which the anti-candida action of Juzen-taiho-to is exerted, Akagawa et al. (1996) first investigated the protective activity against lethal infection with *C. albicans* in two different inbred strains of mice: C3H/He N and C3H/He J. Macrophages from these mice strains have been proved to have high and low responsiveness to bacterial lipopolysaccharide (LPS) (Vogel, 1992).

Both groups were infected intravenously with a lethal dose of 3×10^5 *C. albicans* cells. In the C3H/He J mice, Juzen-taiho-to treatment at a dosage of 2 g/kg/day for 5 consecutive days starting on the day of infection was effective in prolonging the animal life span; however, no such protective effect was observed in candida-infected C3H/He N mice. Thus, an *in vivo* anti-candida effect of oral Juzen-taiho-to is obtained in hosts with the relatively suppressed activity of macrophages.

Abe et al. (1997a) investigated the effects of Juzen-taiho-to on an *in vivo* macrophage function, the capacity to produce TNF-α, in candida-infected mice. ICR mice were inoculated intravenously with *C. albicans*. Candida infection severely depressed the capacity of the mice to produce TNF-α after stimulation by an intravenous injection of a TNF-α -inducer, OK432, a *streptococcal* preparation. Oral administration of Juzen-taiho-to at a dose of 2 g/kg at 6 h after the infection, however, significantly prevented this repression of the capacity to produce TNF-α in *C. albicans*-infected mice (see Figure 6.1).

The effects of Juzen-taiho-to on macrophage function were studied with macrophages collected from C3H/He J mice by measuring their *C. albicans* growth-inhibitory activity in the presence and absence of Kampo medicine (Akagawa et al., 1996). Thioglycolate-induced peritoneal macrophages from these mice partially inhibited the mycelial growth of *C. albicans* as estimated by ^{3}H-glucose uptake, and the anti-candida activity was strengthened when the macrophages were incubated with Juzen-taiho-to.

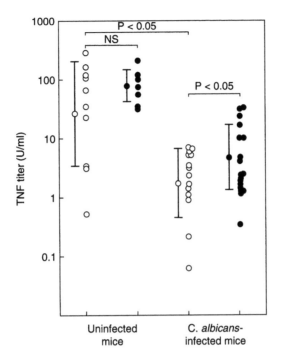

FIGURE 6.1 Effect of Juzen-taiho-to (TJ-48) on TNF-α production of individual OK432-triggered mice with or without *C. albicans* infection. Mice were intravenously inoculated with 3×10^6 cells of *C. albicans* or saline and 3 h later received an oral dose of 2 g/kg of Juzen-taiho-to (●) or distilled water (○). All mice tested were intravenously injected with OK432 20 h after *Candida* infection and, 2 h later, serum specimens were obtained for TNF assay. Small symbols indicate individual values and large circles and bars represent their means and standard deviation, respectively.

6.4.3 ACTIVE COMPONENTS OF JUZEN-TAIHO-TO INVOLVED IN ITS ANTIFUNGAL EFFECT

Abe et al. (1998) sought to determine which component herbs (crude drugs) in Juzen-taiho-to were effective in preventing this lethal infection of *C. albicans*. Table 6.2 and other results show that Ginseng Radix, Glycyrrhizae Radix, Atractylodis Lanceae Rhizoma, and Cnidii Rhizome have a life-prolonging effect even when used alone. Therefore, we can speculate that Juzen-taiho-to employs the total sum of the activities of these effective components.

Akagawa et al. (1996) reported the effects of these ten crude drugs of Juzen-taiho-to on anti-candida activity of macrophages of C3H/HeJ mice. None of the herbal extracts used alone without murine macrophages at concentrations up to 100 µg/ml had significant effect on candida growth (data not shown). As shown in Figure 6.2, supplementation of an extract from Atractylodis Lanceae Rhizoma, Ginseng Radix, Cnidii Rhizoma, or Angelica Radix at a concentration of 20 or 100 µg/ml clearly augmented the anti-candida activity of macrophages. The effect of a Glycyrrhizae Radix extract appeared more complex; it facilitated anti-candida activity of macrophages at a concentration of 20 µg/ml but was suppressive at 100 µg/ml. Therefore, the extracts with protective activity against candida infection (those of Ginseng Radix, Atractylodis Lanceae Rhizoma, and Cnidii Rhizoma) can strengthen the ability of macrophages to inhibit the growth of candida *in vitro*.

Furthermore, in Chinese and Japanese traditional medicine, Ginseng Radix and Atractylodis Lanceae Rhizoma are classified as typical medicines that make up for a deficiency of "Ki" (Qi in Chinese, meaning vital energy) (Raku, 1988), and their stimulating effects on immunity are

TABLE 6.2
Protective Effects of Component Herbs in Juzen-taiho-to against Lethal
C. albicans Infection in Cyclophosphamide-Treated Mice

Pretreatment with	Surviving/Tested[a]	Mean Survival Period (days)	Tested/Control (%)
Control	0/8	4.88 ± 3.89	100
Glycyrrhizae Radix	0/6	9.83 ± 4.38	202
Atractylodis Lancae rhizoma	0/6	8.17 ± 8.26	168
Ginseng Radix	0/6	10.33 ± 6.73	212
Cnidii Rhizoma	1/6	8.67 ± 9.38	178
Angelicae Radix	0/6	3.33 ± 1.97	68
Poria	1/6	8.00 ± 9.54	164
Rehmanniae Radix	0/6	4.33 ± 2.87	89
Astragali Radix	0/6	2.50 ± 1.18	51

[a] Survival ratio on day 28.

Notes: Mice were treated with CY at day 4 and given distilled water orally (control group) or extracts of each component herb (crude drug) at a daily dose of 350 mg/kg (except for 140 mg/kg for Glycyrrhizae Radix) for 4 concecutive days starting at day 4. They were intravenously challenged with _C. albicans_ on day 0 and were then followed for 28 days.

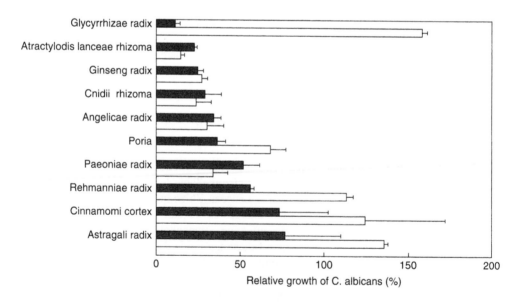

FIGURE 6.2 Effect of ten herbal component extracts of Juzen-taiho-to (TJ-48) on anti-_Candida_ activity of murine macrophages. Macrophages prepared from TGL-peritoneal cells (2×10^5 cells/well) of C3H/He J mice were cultured for 14 h with _C. albicans_ in the medium containing 20 (■) and 100 (□) μg/ml of an herbal extract. _Candida_ growth was estimated by ^{3}H-glucose(d.p.m.). Effects of ten herbal extracts on the inhibition of _C. albicans_ growth by macrophages were tested and are represented in terms of relative growth of _C. albicans_. Relative growth of _C. albicans_ (%) was calculated by the following formula: (^{3}H-glucose incorporation by _C. albicans_ in the presence of macrophages with an herbal extract)/(^{3}H-glucose incorporation by _C. albicans_ in the presence of macrophages alone) × 100.

well recognized. Actually, Miyazawa et al. (1990) found that Ginseng Radix, Glycyrrhizae Radix, Atractylodis Lanceae Rhizoma, and Rehmanniae Radix in Juzen-taiho-to show the antitumor activity in CY-pretreated mice.

Additionally, Haranaka et al. (1985) reported that Cnidii Rhizoma had immunostimulating and antitumor activities in a murine model when administered orally. Jin and Kurashige (1996) also reported that Cnidii Rhizoma and Astragali Radix augment production of TNF-α in carcinogen-induced immunosuppressed rats. Putting all these findings together, these components of Juzen-taiho-to are considered responsible for the immunostimulating activities offered by the prescription. This is also consistent with our recently acquired knowledge that the defense mechanism of infected mice against candida infection is related to strengthening the macrophage function *in vivo* (Abe et al., 1997b).

No clear conclusion can yet be made on the effects of the other six crude drugs (Cinnamomi Cortex, Astragali Radix, Poria, Angelicae Radix, Rehmanniae Radix, and Paeoniae Radix); further study changing the conditions of administration, etc. is required to identify their efficacy. Sugiyama et al. (1988), however, have shown the protective effect of Juzen-taiho-to on inflammatory nephrotoxicity of mice manifested by the administration of cisplatin (see Sugiyama, Chapter 9 in this book). They reported that the effective components are Cinnamomi Cortex, Astragali Radix, Poria, and Angelicae Radix—those without prophylactic activity to candida infection, as described earlier. The crude drugs composing Juzen-taiho-to are therefore believed to fall into at least two groups with different activities.

Takemoto (1996) reported the chemical composition and immunological activities of hot water extracts of these ten crude drugs of Juzen-taiho-to. As shown in Table 6.3, we noticed the interesting correlation that the crude drugs with the prophylactic activity against candida infection contain a relatively large quantity of extractable polysaccharides.

TABLE 6.3
Protective Activities for Candida Infection and Polysaccharide Yields of Ten of Juzen-taiho-to Crude Drugs

Crude Drug	Protection of Candida-Infected Mice[a]	Augmentation of Anti-Candida Activity of Macrophages[b]	Crude Polysaccharides by Hot-Water Extraction[c] (%)	Mitogenic Activity of Each Polysaccharide for Spleen Cells[d]
Glycyrrhizae Radix	+	+	3.4	+
Atractylodis Lancae Rhizoma	+	+	5.3	+
Ginseng Radix	+	+	6.9	+
Cnidii Rhizoma	+	+	5.0	−
Angelicae Radix	−	+	5.5	−
Poria	−	+	0.3	−
Paeoniae Radix	−	±	1.9	−
Rehmanniae Radix	−	−	2.6	−
Cinnamomi Cortex	−	−	1.0	−
Astragali Radix	−	−	3.4	+

[a] See Abe, S. et al., 1998, *Immunopharm. Immunotoxicol.*, 20, 421–431.
[b] See Akagawa, G. et al., 1996, *Immunopharm. Immunotoxicol.*, 18, 73–89.
[c] See Takemoto, N., 1996, doctoral thesis. Hokuriku University (in Japanese).
[d] See Yamada, H. et al., 1992,

Inagaki et al. (2001) found that active principles in Atractylodis Lanceae Rhizoma were extracted by hot water and precipitated with ethanol. A polysaccharide fraction containing galacturonic acid was obtained and the fractions clearly lengthened the life span of candida-infected mice. Its effective dose by oral administration is about 10 to 20 mg/kg body weight, which is estimated to be 1/50 that of Juzen-taiho-to. These results suggest that one active principle of Atractylodis Lanceae Rhizoma is a pectic polysaccharide.

Yamada et al. (1990) and Yu et al. (1998) extracted a polysaccharide fraction containing galacturonic acid from all the herbs of Juzen-taiho-to and showed it to have immunostimulating activity. We can therefore assume that some of these polysaccharides must have major roles in preventive effects against fungal infections. However, we must also recognize the possibility that some part of the preventive effects of Juzen-taiho-to (TJ-48) may be attributable to other components, including the saponins, ginsenosides, and glycyrrhizin (Kumazawa et al., 1989), which have macrophage-stimulating activity, and some fatty acids with stimulating activity for hematopoietic stem cells (Hisha et al., 1997; see Hisha and Ikehara, Chapter 5 in this book).

6.5 CONCLUSION

All of the previously mentioned data strongly suggest that Juzen-taiho-to has potential usefulness in the adjunctive immunotherapy for several types of opportunistic microbial infections. One of the most important findings here is that Juzen-taiho-to and its components can effectively prevent deep-seated mycoses such as candida and aspergillus infection through activation of macrophages. This efficacy was suggested to be achievable in leukopenic and glucocorticoid-treated hosts. Such an approach to the evaluation of Juzen-taiho-to on the basis of medical science will facilitate its clinical application for the preventive control of infectious disease in various types of immunosuppressed patients.

REFERENCES

Abe, S., Ishibashi, H., Tansho, S., Hanazawa, R., Komatsu, Y., and Yamaguchi, H. (2000) Protective effect of oral administration of several traditional Kampo medicines on lethal *Candida* infection in immunosuppressed mice. *Jpn. J. Med. Mycol., Nippon Ishinkin Gakkai Zasshi,* 41, 115–119 (in Japanese)

Abe, S., Tansho, S., Ishibashi, H., Akagawa, G., Komatsu, Y., and Yamaguchi, H. (1999) Protection of immunosuppressed mice from lethal *Candida* infection by oral administration of a Kampo medicine, Hochu-ekki-to. *Immunopharm. Immunotoxicol.,* 21, 331–342.

Abe, S., Tansho, S., Ishibashi, H., Inagaki, N., Komatsu, Y., and Yamaguchi, H. (1998) Protective effect of oral administration of a traditional medicine, Juzen-taiho-to, and its components on lethal *Candida albicans* infection in immunosuppressed mice. *Immunopharm. Immunotoxicol.,* 20, 421–431.

Abe, S., Akagawa, G., Tansho, S., Ochi, H., Osumi, M., Komatsu, Y., Uchida, K., and Yamaguchi, H. (1997a) Repression of the capacity to produce tumor necrosis factor alpha in *Candida albicans*-infected mice and reversal of the repression by oral Juzen-taiho-to treatment. *Jpn. J. Med. Mycol.,* 38, 183–187 (in Japanese).

Abe, S., Tansho, S., Uchida, K., and Yamaguchi, H. (1997b) Roles of activated macrophages in host defense mechanisms against *Candida* infection. *Jpn. J. Med. Mycol.,* 38, 223–227 (in Japanese).

Abe, S., Ohnishi, M., Nohmi, T., Katoh, M., Tansho, S., and Yamaguchi, H. (1996) A glucocorticoid antagonist, mifepristone affects anti-*Candida* activity of murine neutrophils in the presence of prednisolone *in vitro* and experimental candidasis of prednisolone-treated mice *in vivo*. *FEMS Immunol. Med. Microbiol.,* 13, 311–316.

Abe, S. and Yamaguchi, H. (1994) Host defense mechanisms and immunotherapy for fungal infections. *Teikyo Med. J.,* 17, 291–301 (in Japanese).

Akagawa, G., Abe, S., Tansho, S., Uchida, K., and Yamaguchi, H. (1996) Protection of C3H/HeJ mice from development of *Candida albicans* infection by oral administration of Juzen-taiho-to and its component, Ginseng Radix: possible roles of macrophages in the host defense mechanisms. *Immunopharm. Immunotoxicol.,* 18, 73–89.

Hamada, M., Fujii, Y., Yamamoto, H., Miyazawa, Y., Shui, S.M., Tung, Y.C., and Yamaguchi, N. (1988) Effect of a Kampo medicine, zyuzentaihoto, on the immune reactivity of tumor-bearing mice. *J. Ethnopharm.*, 24, 311–320.

Haranaka, K., Satomi, N., Sakurai, A., Haranaka, R., Okada, N., and Kobayashi, M. (1985). Antitumor activities and tumor necrosis factor producibility of traditional Chinese medicines and crude drugs. *Cancer Immunol. Immunother.*, 20, 1–5.

Hisha, H., Yamada, H., Sakurai, M.H., Kiyohara, H., Li, Y., Yu, C., Takemoto, N., Kawamura, H., Yamaura, K., Shinohara, S., Komatsu, Y., Aburada, M., and Ikehara, S. (1997). Isolation and identification of hematopoietic stem cell-stimulating substances from Kampo (Japanese herbal) medicine, Juzen-taiho-to. *Blood*, 90, 1022–1030.

Iijima, K., Sun, S., Cyong, J-C., and Jyonouchi, H. (1999) Juzen-taiho-to, a Japanese herbal medicine, modulates type 1 and type 2 T-cell responses in old BALB/c mice. *Am. J. Chin. Med.*, 27, 191–203.

Inagaki, N., Komatsu, Y., Sasaki, H., Kiyohara, H., Yamada, H., Ishibashi, H., Tansho, S., Yamaguchi, H. and Abe, S. (2001). Suggestive evidence showing that an active principle in rhizomes of *Atractylodis lancea* for protective activity in *Candida*-infected mice is polysaccharides. *Planta Medica*, 67, 428–431.

Jin, R. and Kurashige, S. (1996). Effects of Chinese herbs on macrophage functions in N-butyl-N-butanolnitrosoamine-treated mice. *Immunopharmacol. Immunotoxicol.*, 18, 105–114.

Komatsu, Y., Takemoto, N., Maruyama, H., Tsuchiya, H., Aburada, M., Hosoya, E., Shinohara, S., and Hamada, H. (1986) Effect of Juzen-taiho-to on the anti-SRBC response in mice. *Jpn. J. Inflam.*, 6, 405–408 (in Japanese).

Kumazawa, Y., Takimoto, H., Nishimura, C., Kawakita, T., and Nomoto, K. (1989). Activation of murine peritoneal macrophages by Saikosaponin A, Saikosaponin D, and Saikogenin D. *Int. J. Immunopharm.*, 11, 21–28.

Maruyama, H., Kawamura, H., Takemoto, N., Komatsu, Y., Aburada, M., and Hosoya, E. (1988) Effect of Kampo medicines on phagocytes. *Jpn. J. Inflam.*, 8, 65–66 (in Japanese).

Matsumoto, T. and Yamada, H. (1999–2000) Orally administered Kampo (Japanese herbal) medicine, "Juzen-Taiho-To" modulates cytokine secretion in gut-associated lymphoreticular tissues in mice, *Phytomedicine*, 6, 425–430.

Miura, S., Takimoto, H., Yoshikai, Y., Kumazawa, Y., Yamada, A., and Nomoto, K. (1992). Protective effect of Ren-shen-yang-rong-tang (Ninjin-yoei-to) in mice with drug-induced leukopenia against *Pseudomonas aeruginosa* infection. *Int. J. Immunopharm.*, 14, 1249–1257.

Miyazawa, Y., Fujii, Y., Ichikawa, M., and Yamaguchi, N. (1990) Effective components of herbal medicine (Juzen-taiho-to) for antitumor activity and nonspecific immune responses in tumor-bearing mice. *Proc. 49th. Annu. Meet. Jpn. Soc. Cancer Res.*, Sapporo, Japan, 314 (in Japanese).

Mori, K., Kido, T., Daikuhara, H., et al. (1999) Effect of Hochu-ekki-to (TJ-41), a Japanese herbal medicine, on the survival of mice infected with influenza virus. *Antiviral. Res.*, 44, 103–111.

Morisawa, S. (1989) Progress in fundamental and clinical research In *Ginseng '89* (Ohura, H., Kumagaya, A., Shibata, S., and Takagi, K., Eds.), 187–196. Kyoritsu Press, Tokyo (in Japanese).

Morrissey, P.J. and Charrier, K. (1991) Interleukin-1 administration to C3H/He J mice after but not prior to infection increases resistance to *Salmonella typhimurium*. *Infect. Immun.*, 59, 4729–4731.

Nohmi, T., Abe, S., Tansho, S., and Yamaguchi, H. (1994) Suppression of anti-*Candida* activity of murine neutrophils by glucocorticoids. *Microbiol. Immunol.*, 38, 977–982.

Nomoto, K. (1996) Kampo and host defense. *J. New Remedies Clinics*. 45, 1272–1279 (in Japanese).

Ohno, S. (1988) Effects of a traditional Chinese medicine "Hochu-ekki-to" on natural killer activity. *Allergy*, 37, 107–114 (in Japanese).

Polak–Wyss, A. (1991) Protective effect of human granulocyte colony stimulating factor (hG-CSF) on *Candida* infections in normal and immunosuppressed mice. *Mycoses*, 34, 109–115.

Raku, W. (1988) *Immunology and Kampo Medicine*, Taniguchi Press, Tokyo (in Japanese).

Satomi, N., Sakurai, A., Haranaka, R., and Haranaka, K. (1989a) Traditional Chinese medicine and drugs in relation to the host-defense mechanism. In *The Influence of Antibiotics on the Host–Parasite Relationship* (G. Gillissen, W. Opferkuch, G. Peters, and G. Pulverer, Eds.), 77–86, Springer–Verlag, Berlin/Heidelberg.

Satomi, N., Sakurai, A., Iimura, F., Haranaka, R., and Haranaka, K. (1989b) Japanese modified traditional Chinese medicines as preventive drugs of the side effects induced by tumor necrosis factor and lipopolysaccharide. *Mol. Biother.*, 1, 155–162.

Satonaka, K., Ohashi, K., Nohmi, T., Yamamoto, T., Abe, S., Uchida, K., and Yamaguchi, H. (1996) Prophylactic effect of *Enterococcus faecalis* FK-23 preparation on experimental candidiasis in mice. *Microbiol. Immunol.,* 40, 217–222.

Schaffner, A. and Schaffner, T. (1987) Glucocorticoid-induced impairment of macrophage antimicrobial activity: mechanisms and dependence on the state of activation. *Rev. Infect. Dis.,* 9, S620.

Shimizu, S., Furuno, H., Horiguchi, A., et al. (1997) Establishment of mouse model for salmonella infection and trial of immunomodulating therapy using Hochu-ekki-to. *Jpn. Toyo-Igaku Zasshi,* 48, 369–379 (in Japanese).

Shimizu, S., Wang, X-X., Matsuno, H., and Yamaguchi, N. (1998) Effect of the herbal medicines, Juzen-taiho-to, Hotyu-ekki-to and Tohki-rikuoh-to, on infectious immunity to MRSA in MMC-treated mice. *Prog. Med.,* 18, 753–758 (in Japanese).

Sugiyama, K., Yokota, M., Ueda, H., and Ichio, Y. (1988) Protective effect of Chinese medicines on renal toxicity in cisplatin-treated mice (NO.1). *J. Med. Pharm. Soc. WAKAN-YAKU.,* 5, 290–291 (in Japanese).

Takemoto, N., Kawamura, H., Maruyama, H., Komatsu, Y., Aburada, M., and Hosoya, E. (1989) Effect of TJ-48 on murine cellular immunity. *Jpn. J. Inflam.,* 9, 49–52 (in Japanese).

Takemoto, N. (1996) Study on activities of a Kampo prescription, Juzen-taiho-to, to immune systems. Doctoral thesis. Hokuriku University (in Japanese).

Tei, M. (1996) (Cyong, J-C) Juzen-taiho-to (2). *Kampo Igaku,* 20, 57–63 (in Japanese).

Uchida, K., Abe, S., Komatsu, Y., Akagawa, G., and Yamaguchi, H. (1995) Therapeutic efficacy of oral treatment with a traditional medicine, Juzen-taiho-to, against experimental deep-seated mycoses in mice. *Jpn. J. Med. Mycol.,* 36, 47–51 (in Japanese).

Vogel, S.N. (1992) The LPS gene; insights into the genetic and molecular basis of LPS responsiveness and macrophage differentiation. In *Tumor Necrosis Factors: the Molecules and Their Emerging Role in Medicine* (Beutler, B., Ed.), 485–513. Raven Press, New York.

Yamada, H., Kiyohara, H., Cyong, J-C., Takemoto, N., Komatsu, Y., Kawamura, H., Aburada, M., and Hosoya, E. (1990) Fractions and characterization of mitogenic and anticomplementary active fractions from Kampo (Japanese herbal) medicine Juzen-taiho-to. *Planta Med.,* 56, 386–391.

Yamaguchi, H. (1992) Immunomodulation by medicinal plants. In *Microbial Infections: Role of Biological Response Modifiers* (Friedman, H., Klein, T.W., and Yamaguchi, H., Eds.), 287–297. Plenum Press, New York.

Yamaguchi, H. (1996) Deep-seated mycoses and Kampo medicine. *Prog. Med.,* 16, 1723–1737 (in Japanese).

Yamaguchi, N., Wang, X-X., Ogata, Y., and Shimizau, M. (1999) Preventive effects of Kampo "Hozai" against experimental animal infections with extracellular and intracellular microbial parasites. *Prog. Med.,* 19, 986–996 (in Japanese).

Yamaoka, Y., Kawakita, T., Kaneko, M., and Nomoto, K. (1996) A polysaccharide fraction of Zizyphi fructus in augmenting natural killer activity by oral administration. *Biol. Pharm. Bull.,* 19, 936–939.

Yamaura, H., Kobayashi, F., Komatsu, Y., Tsuji, M., Sato, H., Shirota, F., Shirasaka, R., and Waki, S. (1996) A traditional Oriental herbal medicine, Juzen-taiho-to has suppressive effect on nonlethal rodent malaria by means of stimulation of host immunity. *Jpn. J. Parasitol.,* 45, 6–11.

Yu, K.W., Kiyohara, H., Matsumoto, T., Yang, H.C., and Yamada, H. (1998) Intestinal immune system modulating polysaccharides from rhizomes of *Atractylodes lancea. Planta Med.,* 64, 714–719.

7 The Search for Active Ingredients of Juzen-taiho-to

Hiroaki Kiyohara and Haruki Yamada

CONTENTS

7.1   INTRODUCTION

Juzen-taiho-to has been used to promote improvement of the debilitated general conditions of patients recovering from surgery or suffering from chronic diseases. The prescription has been applied to the treatment of anemia, the prevention of adverse effects of chemotherapy and radiation therapy on cancer patients, and the prevention of rebound reactions of steroids for chronic atopic skin disease. From these clinical effects of Juzen-taiho-to, it is expected that the formula expresses the clinical effects through immune and hemopoietic systems. By three-dimensional (3-D) HPLC analysis, 17 kinds of low molecular weight compounds can be detected by photodiode array detector in the decoction of Juzen-taiho-to; these substances are also isolated from the component crude drugs of Juzen-taiho-to (Figure 7.1, Table 7.1). Among the ingredients, xanthotoxin, glycyrrhizin, cinnamaldehyde, and paeoniflorin have been reported to have biological actions related to the immune system (Table 7.1).

Juzen-taiho-to consists of ten kinds of component herbs. In Kampo medicines, some combination effect between ingredients originated from the same or different component herbs have been

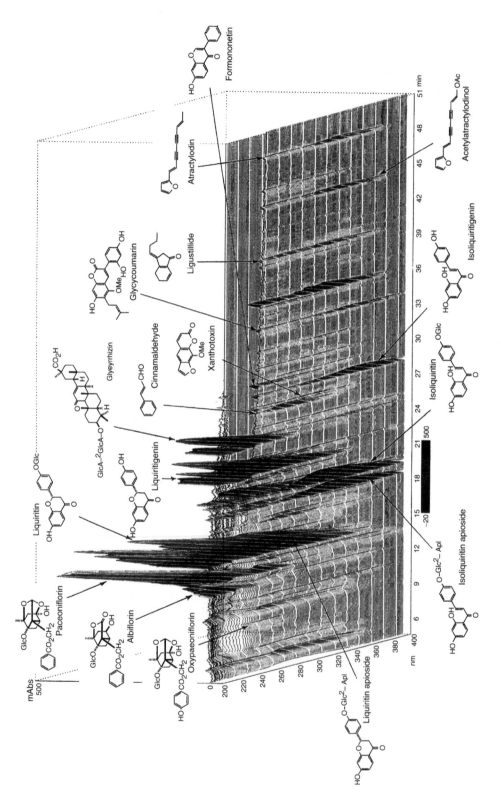

FIGURE 7.1 Chemical profile of decoction of Juzen-taiho-to (TJ-48) analyzed by three-dimensional HPLC.

TABLE 7.1
Low Molecular Weight Ingredients Found in Decoction of Juzen-taiho-to by 3-D HPLC

Ingredient	Original Component Herb	Pharmacological Activity	Ref.
Paeoniflorin	Paeoniae Radix	Inhibition of binding of sex hormones to their binding proteins	Tamaya et al., 1986
		Inhibition of androgen production	Takeuchi et al., 1991
		Hypoglycemic activity	Hsu et al., 1997
		Protection of scopolamine-induced amnesia	Abdel–Hafez et al., 1998
Albiflorin	Paeoniae Radix	Inhibition of pentylene tetrazol-induced EEG power spectrum changes and extracellular calcium concentration charges (*in vivo, po*)	Sugaya et al., 1991
Oxypaeoniflorin	Paeoniae Radix	No activity reported	
Isoliquiritin	Glycyrrhizae Radix	Inhibition of castor oil-induced diarrhea based on inhibition of cyclooxygenase-2 (cox-2)(*in vivo, po*)	Kase et al., 1999
		Inhibition on angiogenesis	Kobayashi et al., 1995
Liquiritin	Glycyrrhizae Radix	Weakly suppressive activity on development of colonic aberrant crypt foci induced by azoxymethane (*in vivo, po*)	Kawamori et al., 1995
Liquiritin apioside	Glycyrrhizae Radix	No activity reported	
Liquiritigenin	Glycyrrhizae Radix	No activity reported	
Isoliquiritigenin	Glycyrrhizae Radix	Antioxidative and superoxide scavenging activity	Haraguchi et al., 1998
		Accumulative effect of Ca^{2+} ion in venticular heart muscle	Wegener and Nawrath, 1997a
		Relaxation of aortic smooth muscle	Wegener and Nawrath, 1997b
		Inhibition of aldose reductase	Aida et al., 1990
		Antiplatelete activity	Tawata et al., 1992
		Inhibition of c-AMP phosphodiesterase	Kusano et al., 1991
		Antitumor promoting activity	Yamamoto et al., 1991
Glycycoumarin	Glycyrrhizae Radix	No activity reported	
Glycyrrhizin	Glycyrrhizae Radix	Restorative effect on hepatic function	Yamamura et al., 1997
		Hepatoprotective activity	Hase et al., 1997
		Inhibition of Phlebovirus replication	Nagai et al., 1991
		Enhancing activity on drug absorption in intestine	Crance et al., 1997
		Antidiarrheal effect	Imai et al., 1999
		Inhibition of prostaglandin E_2 production	Kase et al., 1999
			Ohuchi et al., 1981

(*continued*)

TABLE 7.1
(Continued)

Ingredient	Original Component Herb	Pharmacological Activity	Ref.
		Stabilization of lysosome	Shiki et al., 1986
		Antiviral activity against varicella-zoster virus (VZV)	Baba and Shigeta, 1987
		Antiviral activity against HIV (inhibition of cytopathic activity)	Ito et al., 1987
		Antiviral activity against encephalitis virus	Badam, 1997
		Antimutagenic activity	Tanaka et al., 1987
		Inhibition of Na^+-K^+-ATPase	Itoh et al., 1989
		Inhibition of testosterone synthesis	Takeuchi et al., 1991
		Mineral corticoid activity through alternation in cortisol metabolism	Kageyama et al., 1992
		Induction of contrasuppressor T-cell	Kobayashi et al., 1993a
		Stimulation of host resistance against tumors	Suzuki et al., 1992
		Modulation of phosphorylation of lipocortin I by A kinase	Ohtsuki et al., 1992
		Inhibition of cytotoxicity of T-cells	Yoshikawa et al., 1997
		Antithrombin activity	Francischetti et al., 1997
		Inhibition of hyaluronidase (GL binding protein)	Furuya et al., 1997
Cinnamaldehyde	Cinnamomi Cortex	Antimutagenic activity	Ohta et al., 1983
		Antiinflammatory activity	Claeson et al., 1993
		Inhibition of lymphocyte proliferation and modulation of T-cell differentiation	Koh et al., 1998
Ligustilide	Angelicae Radix	Tumoricidal activity	Kwon et al., 1998
		Suppressive effect on development of pentobarbital-induced sleep in isolated mice (effect on central nonadrenergic and/or GABA (A) system)	Matsumoto et al., 1998
Xanthotoxin	Angelicae Radix	Weak inhibition on proliferation of primary culture of mouse aorta smooth muscle cell	Kobayashi et al., 1993b
Formononetin	Glycyrrhizae Radix	Anti-inflammatory activity	Chen et al., 1995
		Hypolipidemic activity	Sharma, 1979
		Antimicrobial activity	Mitscher et al., 1980
		Estrogenic activity	Juniewicz et al., 1988
		Inhibitory activity against alcohol dehydrogenase	Keung, 1993
		Antioxidative activity	Ruiz–Larrea et al., 1997
Atractylodin	Atractylodis Lanceae Rhizoma	No activity reported	

TABLE 7.2
Pharmacologically Active Ingredients Found in Juzen-taiho-to

Biological System	Ingredient	Pharmacological Activity	Ref.
Systemic immune system	Pectins	Mitogenic activity against B cells	Kiyohara et al., 1991
		Complement activating activity	Kiyohara et al., 1991
	Pectic arabinogalactans	Complement activating activity	Kiyohara et al., 1993
	Pectic heteroglycans	IL-2 production enhancing activity	Kiyohara et al., 1993
	Glycoprotein (?)	Anticomplementary activity	Yamada et al., 1990
Intestinal immune system	Lignin-carbohydrate complexes	Stimulation of growth factor production for bone marrow cells from Peyer's patch cells	Kiyohara et al., 2000
	Polysaccharides with/without arabino-3,6-galactan moiety		Kiyohara et al., 1997
Hemopoietic system	Unsaturated C_{18} fatty acids	Bone marrow stem cell proliferation enhancing activity	Hisha et al., 1997
	Pectic polysaccharides		Sakurai et al., 1996
Other	Sodium malate	Protective activity against lethal and renal toxicity of anticancer drug (*cis*-diaminedichloroplatinum)	Sugiyama et al., 1994 Ueda et al., 1997
		Protective activity against immunosuppression by anticancer drug (*cis*-diamminedichloroplatinum)	Kiyohara et al., 1995a
	Polysaccharide	Protective activity against lethal and renal toxicity of anticancer drug (*cis*-diaminedichloroplatinum)	Kiyohara et al., 1995b

observed during decoction in the pharmacological aspect (Hosoya, 1988). It also has been shown that the glycyrrhizin content in the decoction is changed by the coexistence of other component herbs during decoction (Tomimori and Yoshimoto, 1980). The coexistence of Ephedra Herba or Coptis Rhizoma with Glycyrrhizae Radix decreased the content of glycyrrhizin in the extract, but that of Magnoliae Cortex increased it (Tomimori and Yoshimoto, 1980).

Many substances with primary amine groups such as amino acid, peptides, or proteins are known to interact with the glycosidic hydroxy group (Maillard reaction). It has been reported that some amino acids react with sugars to form 4-(2-formyl-5-hydroxymethylpyrrol-1-yl)-butyric acid during heat processing of Aconiti Tuber (Matsui et al., 1998). This compound increases peripheral blood flow. It is recognized that active ingredients related to the clinical effects of Kampo formulas must be assessed in relation to those found in decoctions of the prescription. During the past 10 years, several active ingredients in a pharmaceutical preparation of dried extract (TJ-48 by Tsumura & Co) of Juzen-taiho-to have been clarified, as follows (Table 7.2).

7.2 ACTIVE INGREDIENTS RELATED TO THE EFFECT OF JUZEN-TAIHO-TO ON THE SYSTEMIC IMMUNE SYSTEM

It has been clarified that Juzen-taiho-to (TJ-48) expresses *in vitro* and/or *in vivo* antibody production-enhancing activity, complement-activating activity, mitogenic activity against B-cells, and IL-2 and IL-6 production-enhancing activity as the immunopharmacological activity on systemic immune system (described in Chapter 4). TJ-48 was fractionated by methanol extraction, water extraction, dialysis, and ethanol precipitation in order to divide it into the fractions containing low molecular weight ingredients, intermediately sized ingredients, and high molecular weight ingredients. As shown in Figure 7.2, five fractions (F-1 to F-5) were obtained and tested for the immunopharmacological activities.

The methanol-soluble fraction (F-1), which contained low molecular weight substances, showed no activity related to the systemic immune system; most of the activities were observed in the crude polysaccharide fraction (F-5). The water and methanol-insoluble fraction (F-2) expressed only complement-activating activity (Yamada et al., 1990; Kiyohara et al., 1993). F-2 comprised 91.4% carbohydrate, which consisted mainly of glucose, in addition to small amounts of protein (3.9%).

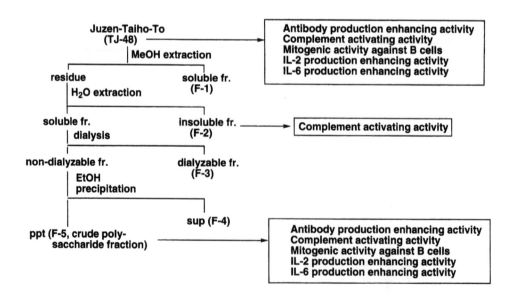

FIGURE 7.2 Immunomodulating activity on systemic immune system of fractions from Juzen-taiho-to.

F-5 was composed of about 80% carbohydrate, which mainly comprised rhamnose, arabinose, galactose, glucose, and galacturonic acid, in addition to 16.8% protein.

In order to clarify which molecules in F-2 or F-5 contribute to the expression of the immunopharmacological activities, these fractions were treated with proteinase (enzymic degradation of protein moiety) or periodate (chemical degradation of carbohydrate moiety), and the complement-activating activities of the products were compared with original F-2 and F-5. Although the activity of F-5 was decreased only by periodate oxidation, the activity of F-2 was reduced by both treatments, thus suggesting that only polysaccharide molecules in F-5 contribute to the activity and that polysaccharide and protein molecules are responsible for the activity of F-2 (Yamada et al., 1990).

Because the polysaccharide fraction (F-5) was suggested to be a major active fraction for the expression of *in vitro* immunopharmacological activity of Juzen-taiho-to on the systemic immune system, the role of polysaccharides in F-5 was estimated on the expression of *in vivo* antibody production-enhancing activity of Juzen-taiho-to. When the fractions (F-1 to F-5) from Juzen-taiho-to were injected intraperitoneally into mice immunized with sheep red blood cells (SRBC), Juzen-taiho-to and F-5 increased antibody titer against SRBC in the blood stream. However, the other fractions had no effect (Kiyohara et al., 1995a). Oral administration of F-5 also stimulated antigen-specific antibody production not only in normal and aged mice but also in mice compromised immunologically by treatment with the anticancer drug cisplatin (Kiyohara et al., 1995a). These results indicate that polysaccharide molecules in Juzen-taiho-to play important roles for the expression of immunopharmacological activities on the systemic immune system.

The immunopharmacologically active polysaccharides were further purified from F-5 in relation to complement-activating activity, mitogenic activity, and IL-2 production-enhancing activity. Figure 7.3 shows the 23 kinds of polysaccharides purified from F-5 (Kiyohara et al., 1991, 1993). Among these polysaccharides, nine showed only complement-activating activity and three expressed only IL-2 production-enhancing activity. Complement-activating and mitogenic activities were found in 11 polysaccharides. Structural analysis showed that some complement-activating polysaccharides and complement-activating and mitogenic polysaccharides comprised a large proportion of the galacturonan region (polymerized α-(1→4)-linked galacturonic acid). These polysaccharides also consisted of small amounts of the region termed "ramified," in which several kinds of olligosaccharide chains rich in arabinose and galactose are attached to the acidic core (rhamnogalacturonan core) comprising (1→4)-linked galacturonic acid and (1→2,4)-branched rhamnose (Figure 7.3).

The results of the structural characterization suggest that these active polysaccharides are classified into pectins in the pectic polysaccharide group of plant cell wall polysaccharides. The active sites of the mitogenic pectins were further examined. Mitogenic activities of the pectins were increased with increasing molecular weights of the active pectins (Figure 7.4), but molecular weights of the ramified regions in the pectins were similar. These observations assumed that the size of galacturonan regions in the active pectins related to the degree of the activity (Kiyohara et al., 1991). However, the polygalacturonic acid with large molecular weight from orange did not show the activity (Table 7.3). The ramified region of the active polysaccharide still showed a significant mitogenic activity. The result suggests that ramified regions are active sites for the expression of mitogenic activity and that galacturonan regions may modify the activity.

The other complement-activating polysaccharides and IL-2 production-enhancing polysaccharides mainly comprise the ramified region and small amounts of the galacturonan region. The side chains in the ramified region of these polysaccharides are longer than those in the complement-activating and mitogenic pectins. The complement-activating polysaccharides had arabinogalactan side chains, whereas IL-2 production enhancing polysaccharides consisted of side chains containing more complex component sugars. From these characteristics, the complement-activating polysaccharides are classified as pectic arabinogalactans and the IL-2 production-enhancing polysaccharides as pectic heteroglycans in the pectic polysaccharide group.

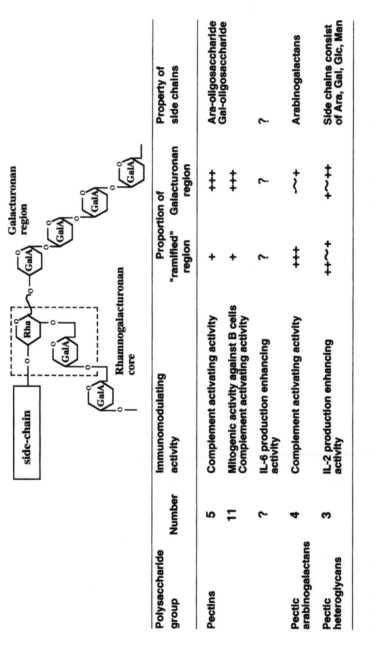

Polysaccharide group	Number	Immunomodulating activity	Proportion of "ramified" region	Galacturonan region	Property of side chains
Pectins	5	Complement activating activity	+	+++	Ara-oligosaccharide Gal-oligosaccharide
	11	Mitogenic activity against B cells Complement activating activity	+	+++	
	?	IL-6 production enhancing activity	?	?	?
Pectic arabinogalactans	4	Complement activating activity	+++	~+	Arabinogalactans
Pectic heteroglycans	3	IL-2 production enhancing activity	++~+	+~++	Side chains consist of Ara, Gal, Glc, Man

FIGURE 7.3 Immunomodulating pectic polysaccharides on systemic immune system found in Juzen-taiho-to.

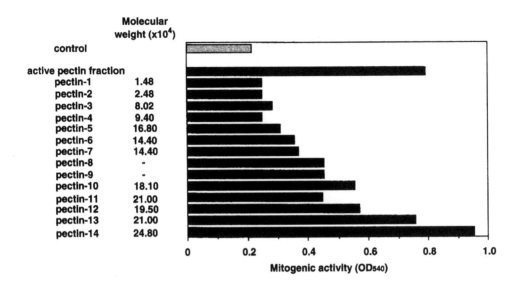

FIGURE 7.4 Mitogenic activity of pectic polysaccharides purified from Juzen-taiho-to.

Juzen-taiho-to has ten crude drugs, among which the polysaccharide fraction obtained from Glycyrrhizae Radix showed the most potent complement-activating and mitogenic activities (Yamada et al., 1992). When Glycyrrhizae Radix was depleted from the prescription of Juzen-taiho-to, the hot-water extract of Glycyrrhizae Radix-depleted prescription significantly reduced both activities compared with that of Juzen-taiho-to. Periodate oxidation or pronase digestion indicated that carbohydrate molecules in the polysaccharide fraction from Glycyrrhizae Radix contributed to the expression of both activities. Three kinds of polysaccharides (GR-2IIa, IIb, and IIc) from Glycyrrhizae Radix were purified from the polysaccharide fraction; GR-2IIa and IIb had only complement-activating activity, whereas GR-2IIc showed complement-activating and mitogenic activities (Zhao et al., 1991a, b).

TABLE 7.3
Mitogenic Activity of Polygalacturonic Acid and Enzymic Digestion Products Derived from Active Pectin in Juzen-taiho-to

| Sample | Treatment | Mitogenic Activity (absorbance at 540 nm) Concentration (μg/ml) | |
		30	100
Control		0.189 ± 0.006	
Polygalacturonic acid from orange		0.200 ± 0.009	0.198 ± 0.001
Active pectin	No treatment	0.397 ± 0.012[b]	0.196 ± 0.001
	Endo-PGase[a]	0.193 ± 0.012	0.196 ± 0.001
"Ramified" region from Active pectin		0.244 ± 0.006[b]	0.268 ± 0.006[b]
Oligogalacturonide fraction from Active pectin		0.204 ± 0.001	0.194 ± 0.004

[a] Endo-α-D-(1$\rightarrow$4)-polygalacturonase digestion sample.
[b] $p < 0.05$.

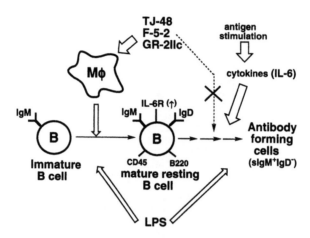

FIGURE 7.5 Mechanism of mitogenic activity of Juzen-taiho-to (TJ-48), active pectin fraction (F-5-2) from TJ-48, and active pectin from Glycyrrhizae Radix.

Comparison of the chemical property between mitogenic polysaccharide from Glycyrrhizae Radix and the most potent mitogenic active polysaccharide from TJ-48 indicated that the component sugars of these polysaccharides were very similar, and the mechanism of mitogenic activity of these polysaccharides (F-5-2 from TJ-48; GR-2IIc from Glycyrrhizae Radix) was also suggested to be the same (Figure 7.5) (Zhao et al., 1991a; Takemoto et al., 1994). All the ramified regions of GR-2IIc acted as active sites for the expression of the activities, the same as the active polysaccharides from TJ-48 (Zhao et al., 1991b). When rhamnogalacturonan cores of the ramified region of GR-2IIa-IIc were destroyed by lithium degradation, the complement-activating and mitogenic activities were reduced significantly; however, the degradation products still showed statistically significant activity (Kiyohara et al., 1996).

The products contained various sizes of neutral oligosaccharide-alditols, which were derived from the side chains of ramified regions of GR-2IIc. The long oligosaccharide-alditol fraction and shorter oligosaccharide-alditol fraction containing di- to tetrasaccharide-alditol had relatively potent complement-activating activity; all the oligosaccharide-alditol fractions expressed weak but significant mitogenic activity. Because standard glucosyl oligosaccharide-alditols did not have any complement-activating and mitogenic activities, the active oligosaccharide-alditols derived from GR-2IIc possessed the essential structure for the expression of the activities. These suggest that immunopharmacological activities of the active pectic polysaccharides in Juzen-taiho-to are due to the side chains in their ramified regions and that different activities depend on the different structures of the side chains (Figure 7.6).

Although the prescriptions consist of some different component crude drugs (see Chapter 5), Ninjin-yoei-to is another Kampo medicine used for the treatment of similar diseases to those for which Juzen-taiho-to is used. In the aspect of immunopharmacological activity on systemic immune system, IL-6 and TNF-α production-enhancing activities of a commercially available dried extract of Ninjin-yoei-to (pharmaceutical preparation, TJ-108 by Tsumura & Co) were analyzed. Ninjin-yoei-to stimulated IL-6 and TNF-α production of spleen cells *in vitro*, and these stimulating activities were expressed by polysaccharide molecules of Ninjin-yoei-to.

When the activities of the polysaccharide fraction from Ninjin-yoei-to were compared with those from Juzen-taiho-to, the fraction from Ninjin-yoei-to stimulated IL-6 and TNF-α productions of spleen cells, whereas the fraction from Juzen-taiho-to enhanced only IL-6 production (Figure 7.7). The active polysaccharides were purified from the polysaccharide fraction of Ninjin-yoei-to for both activities, and two pectic polysaccharides were shown to be active ingredients

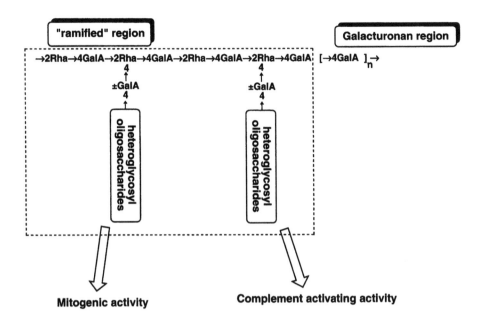

FIGURE 7.6 Structural requirements for complement activating and mitogenic activities in pectic polysaccharides from Glycyrrhizae Radix.

in Ninjin-yoei-to. Pectic polysaccharides are known to be general primary metabolites in plant cell walls, and all decoctions and commercially available dried extracts of Kampo medicines commonly contain several kinds of pectic polysaccharides that have different structures as their ingredients. From the results of Ninjin-yoei-to, it is expected that pectic polysaccharides in Kampo medicines play important roles as prescription-specific active ingredients for expression of clinical effects through the systemic immune system of the respective Kampo medicines.

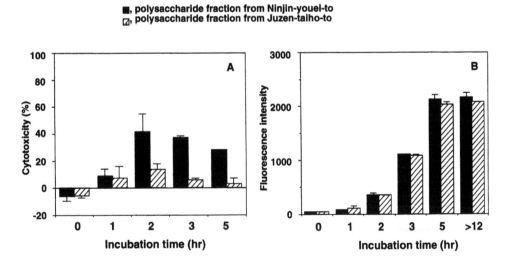

FIGURE 7.7 Effect of polysaccharide fractions from Juzen-taiho-to or Ninjin-youei-to *in vitro* TNF-α (A) and IL-6 productions (B).

7.3 ACTIVE INGREDIENTS RELATED TO THE EFFECT OF JUZEN-TAIHO-TO ON THE MUCOSAL IMMUNE SYSTEM

In Chapter 4, Juzen-taiho-to was shown to stimulate immune function of lymphocytes in Peyer's patches, the interepitherial region, and mesenteric lymph nodes of the intestinal immune system. The intestinal immune system-modulating ingredients in Juzen-taiho-to were clarified by monitoring the stimulatory activity of the ingredients on the production of growth factors for bone marrow cells from Peyer's patch cells (details of the assay procedure are in Chapter 4).

Among the subfractions prepared from Juzen-taiho-to (TJ-48), as described previously, polysaccharide fraction (F-5) and water-soluble and dialyzable fraction containing ingredients of intermediate size (F-3) enhanced proliferation of bone marrow cells mediated by Peyer's patch cells *in vitro* (activity to modulate the intestinal immune system) (Hong et al., 1998; Kiyohara et al., 2000). Oral administration to mice of F-3 and F-5 also showed activity modulating the intestinal immune system, indicating that the ingredients in F-3 and F-5 act as active principles in Juzen-taiho-to for modulation of this system (Kiyohara et al., 2002). The activity of F-5 was significantly decreased by periodate oxidation while the activity of F-3 was reduced by periodate oxidation as well as perchloric acid treatment, which destroys the lignin moiety.

Two active ingredients were isolated from F-3; these consisted of about 20 to 70% carbohydrate, mainly comprising arabinose, galactose, glucose, and galacturonic acid. The active ingredients also gave vanilin, syringaldehyde, and *p*-hydroxybenzaldehyde as oxidation products by alkaline nitrobenzene oxidation, indicating that these also contained the soft glass-type lignin moiety (Table 7.4). By

TABLE 7.4
Chemical Properties of Intestinal Immune System Modulating Lignin–Carbohydrate Complexes from Juzen-taiho-to

	Lignin–Carbohydrate Complex	
	A	B
Molecular weight	12,000	3000
Neutral sugar (%)	20.2	26.0
Uronic acid (%)	19.7	49.3
Bradford method-positive material (%)	54.9	51.3
Protein[a] (%)	13.9	n.d.[b]
Lignin (%)	11.0	19.0
Vanillin	+	+
Syringaldehyde	+	+
p-Hydroxybenzaldehyde	+	+
Component sugar (mol.%)		
Arabinose	16.5	17.9
Rhamnose	7.0	4.6
Fucose	1.8	1.9
Xylose	4.2	4.4
Glucuronic acid	3.4	5.7
Galacturonic acid	11.5	23.1
Mannose	5.8	4.2
Galactose	15.6	19.6
Glucose	32.4	15.4
4-Methylglucuronic acid	1.8	3.3

[a] Elemental analysis.
[b] Not determined.

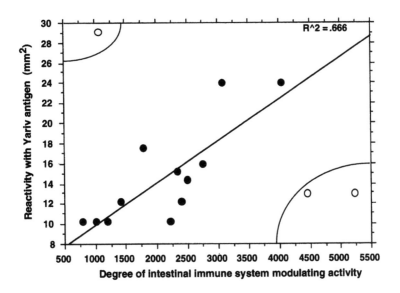

FIGURE 7.8 Relationship between reactivity with Yariv-antigen and intestinal immune system modulating activity of polysaccharides purified from Juzen-taiho-to.

hydrophobic interaction chromatography, the carbohydrate moiety and the lignin moiety could not be separated from each other. Therefore, it was suggested that the lignin and carbohydrate moieties were covalently linked to form a complex in those active ingredients.

Also purified from F-5 were 12 kinds of intestinal immune system-modulating polysaccharides composed mainly of rhamnose, arabinose, galactose, and galacturonic acid; it was suggested that they be grouped into pectic polysaccharides. Beta-glucosyl-Yariv antigen, which is a glucosylated phenylazo derivative, is known to interact with arabino-3, 6-galactan structure and to form a dye complex. Single radial gel diffusion assay using the Yariv antigen indicated that some of the polysaccharides modulating the intestinal immune system contained arabino-3, 6-galactan moiety. Correlation between the content of arabino-3,6-galactan moiety and the degree of activity was good (Figure 7.8). This observation suggests that some of these polysaccharides in Juzen-taiho-to express their activity by the arabino-3, 6-galactan moiety (Kiyohara et al., 2002).

Contributions of component crude drugs of Juzen-taiho-to were compared on the expression of the intestinal immune system modulating activity. Atractylodis Lanceae Rhizoma was suggested to act as the most important component for the activity; among its ingredients, polysaccharide fractions showed the most potent intestinal immune system-modulating activity (Yu et al., 1998). Three active polysaccharides were purified from Atractylodis Lanceae Rhizoma.

One of the active polysaccharides (ALR-5IIa-1-1) was grouped into arabino-3,6-galactan and the other two (ALR-5IIb-2-2 and ALR-5IIc-3-1) were classified to pectic polysaccharide. Because specific glycosidase digestions for arabino-3, 6-galactan decreased ALR-5IIa-1-1 activity in modulating the intestinal immune system, it was proved that β-D-(1→3,6)-galactan moiety of ALR-5IIa-1-1 contributed to the expression of the activity. Analysis of the enzymic digestion products from ALR-5IIa-1-1 indicated that ALR-5IIa-1-1 had the novel structural feature as arabino-3, 6-galactan type polysaccharides. It consisted of β-D-(1→3)-linked galacto-oligosaccharides linked to each other through unknown glycosyl residues to form the backbone; various sizes of partially arabinosylated galactosyl side chains were attached to position 6 of β-D-(1→3)-galactosyl residues in the backbone (Yu et al., 2001a; Taguchi et al., 2004).

Comparison of the activity of the products obtained by the sequential enzymic digestions suggests that the galactosyl side chains located in the outer portion of the ALR-5IIa-1-1 molecule

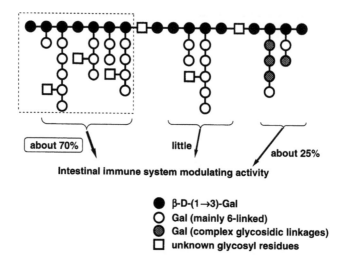

FIGURE 7.9 Active carbohydrate site for expression of intestinal immune system modulating activity in arabinogalactan (ALR-5lla-1-1) purified from Atractylodis Lanceae Rhizoma.

are responsible for activity modulating the intestinal immune system (Figure 7.9). The other active polysaccharide (ALR-5IIc-3-1) from Atractylodis Lanceae Rhizoma comprised galacturonan region (polymerized α-(1→4)-linked galacturonic acid), ramified region (rhamnogalacturonan core possessing side chains rich in neutral sugars), and another uncommon region that resembles rhamnogalacturonan II (RG-II) (Albersheim et al. 1994). This polysaccharide consisted of 2-methylxylose, 2-methylfucose, 3-C-carboxy-5-deoxy-L-xylose (aceric acid); 3-deoxy-D-*manno*-2-octulosonic acid (Kdo); and 3-deoxy-D-*lyxo*-2-heptulosaric acid (Dha) as the characteristic component sugars (Yu et al., 2001b).

Comparison of the activity of these regions indicated that this uncommon region acted as an active site for expression of activity of ALR-5IIc-3-1. Structural analysis indicated that the third active polysaccharide (ALR-5IIb-2-2) also had similar chemical properties to those of the uncommon region in ALR-5IIc-3-1 and RG-II, and the molecular weight was very small (about 3000.) (Yu et al., 2001b). Some of the intestinal immune system-modulating polysaccharides isolated from Juzen-taiho-to comprised similar characteristic component sugars, such as 2-methylfucose, 2-methylxylose, and apiose, suggesting that the uncommon structural region in the active polysaccharides also contributes to the expression of the intestinal immune system-modulating activity of Juzen-taiho-to.

7.4   ACTIVE INGREDIENTS RELATED TO THE EFFECT OF JUZEN-TAIHO-TO ON THE HEMATOPOIETIC SYSTEM

Juzen-taiho-to has been used for treatment of anemia and applied to protect against adverse effects of anticancer drugs and radiation therapy. It was found that Juzen-taiho-to (used pharmaceutical preparation TJ-48 by Tsumura & Co.) enhanced peripheral blood counts in cancer patients who were administered phase-specific drugs and/or received radiation therapy (details described in Chapter 4) (Nabeya and Ri, 1983). Juzen-taiho-to has also been found to enhance day-14 spleen colony forming unit (CFU-S) counts to stimulate proliferation and self-renewal of bone marrow stem cells (Ohnishi et al., 1990).

It has been suggested that not only low molecular weight ingredients but also high molecular weight ingredients stimulate proliferation of bone marrow stem cells in *in vitro* hematopoietic sytem

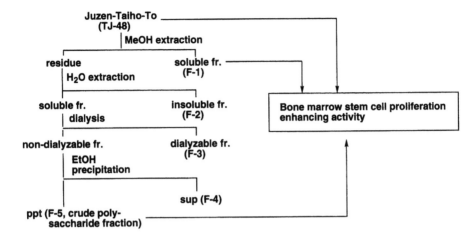

FIGURE 7.10 Bone marrow stem cell proliferation enhancing activity of Juzen-taiho-to and its subfractions.

cell (HSC) assays (Sakurai et al., 1996a) (Figure 7.10); some C18-fatty acids such as oleic and linoleic acids have been found to be low molecular weight active ingredients in Juzen-taiho-to (Hisha et al., 1997; details described in Chapter 4). When fatty acid composition of Juzen-taiho-to was compared with other Kampo medicines, the same active fatty acids were detected even in other Kampo prescriptions not found to accelerate recovery from hemopoietic injury, but in different ratios. Although not all Kampo medicines tested showed the stimulatory activity, their fatty acid fractions did. These suggest that hemopoietic stimulation of Juzen-taiho-to may be the result of the combined effect of the active unsaturated fatty acids and other hydrophilic ingredients (Sakurai et al., 1996a).

The polysaccharide fraction of Juzen-taiho-to (F-5) also enhanced proliferation of bone marrow stem cells *in vitro* (Figure 7.10). F-5 was shown to contain pectin, pectic arabinogalactan, and pectic heteroglycan as pectic polysaccharides from the study for active ingredients against the systemic immune system; the active pectin-containing fraction had the most potent hemopoietic stimulative activity. Two kinds of the active polysaccharides were purified from the pectin-containing fraction, and both were classified as pectin type polysaccharide. These results suggest that the combination of the active unsaturated fatty acids and pectic polysaccharides contributes to the expression of the hemopoietic effect of Juzen-taiho-to.

7.5 ACTIVE INGREDIENTS RELATED TO THE PROTECTIVE EFFECT OF JUZEN-TAIHO-TO ON ADVERSE REACTIONS TO ANTICANCER DRUGS

Because Juzen-taiho-to has been used for treatment of anemia or fatigue, it was tried on cancer patients who were administered phase-specific anticancer drugs and/or received radiation therapy; it protected hemopoietic failure of the patients (Nabeya and Ri, 1983). *cis*-Diaminedichloroplatinum (*cis*-DDP) is one of the most effective cancer chemotherapeutic agents. However the clinical use of *cis*-DDP is limited by severe adverse effects such as nephrotoxicity. Commercially available dried extract of Juzen-taiho-to (TJ-48) was found to prevent the increase of blood urea nitrogen (BUN) and serum creatinin level, the decrease in white blood cells and platelet counts in mice administered *cis*-DDP by combined pre- and postadministrations of Juzen-taiho-to (Sugiyama et al., 1993, 1995). Juzen-taiho-to did not change the antitumor effect of *cis*-DDP.

When these protective effects were compared for fractions prepared from Juzen-taiho-to, methanol-soluble fraction (F-1), polysaccharide fraction (F-5), and methanol-insoluble and water-soluble fraction decreased the BUN level in mice treated with *cis*-DDP (Kiyohara et al., 1995b). The protective effects of hot water extracts of component herbs of Juzen-taiho-to were tested on renal toxicity of *cis*-DDP and that from Angelicae Radix (dried roots of *Angelica sinensis* Diels) showed the most potent protective activity (Sugiyama et al., 1994). The active ingredient, sodium malate, was obtained (details was described in Chapter 4). In Japan, the dried roots of *Angelica acutiloba* Kitagawa (Japanese name: Yamato-toki) are usually used for the component crude drug as Angelicae Radix rather than the root of *Angelica sinensis*. However it is not known whether *Angelica acutiloba* is responsible for the protective activity nor if it contains sodium malate.

When mice were treated with *cis*-DDP intraperitoneally and also orally administered fractions from Juzen-taiho-to, the survival rate of mice administered polysaccharide fraction of Juzen-taiho-to or methanol-insoluble fraction from Juzen-taiho-to increased, compared with that of mice that were administered water only (Kiyohara et al., 1995b). However, methanol-soluble fraction of Juzen-taiho-to did not change the mortality of *cis*-DDP-treated mice. Among the polysaccharides in F-5, the pectin-containing fraction significantly reduced BUN level in *cis*-DDP-treated mice on oral administration similar to F-5, whereas pectic arabinogalactan-containing fraction did not affect the renal toxicity of *cis*-DDP.

Although it is not known whether sodium malate is contained in the roots of *Angelica acutiloba*, as it is in those of *Angelica sinensis*, one possibility is that the protective effect of Juzen-taiho-to on renal toxicity against *cis*-DDP may be expressed by the combination effects of sodium malate and pectin-type polysaccharides in the prescription. It has also been indicated that polysaccharides mainly contribute to the expression of the protective effect of Juzen-taiho-to against the lethal activity of *cis*-DDP.

7.6 INTERACTION OF ACTIVE INGREDIENTS IN JUZEN-TAIHO-TO

7.6.1 Combination Effect of Tetramethylpyrazine and Ferulic Acid on Spontaneous Movement of the Uterus

Tetramethylpyrazine (TMP) is one of the alkaloids contained in Cnidii Rhizoma (dried rhizomes of *Cnidium officinale* Makino). Ferulic acid (FA) is a phenolic compound contained in Angelica Radix (dried roots of *Angelica sinensis* Diels) and in Cnidii Rhizoma. These crude drugs are used as ingredients in Juzen-taiho-to. TMP and FA showed an inhibitory effect on uterine movement when they were given perorally and intravenously to rats (Ozaki and Ma, 1990). The combination of both compounds synergistically inhibited uterine contraction at doses individually insufficient to inhibit, suggesting that the combination of TMP and FA potentiates their inhibiting activities for uterine contraction. Therefore, these active substances may express the combination effect in the decoction of Juzen-taiho-to.

7.6.2 Effect of Other Component Herbs on Extraction Rate of Glycyrrhizin from Glycyrrhizae Radix

Because Kampo medicines generally comprise more than two kinds of component crude drugs in each prescription, it is expected that the content of active ingredients in decoctions is affected by the coexistence of other components. Glycyrrhizin is one of the important ingredients in Glycyrrhizae Radix that is contained in many Kampo medicines. Tomimori and Yoshimoto have studied the influence of coexistence of another crude drug on the extraction rate of glycyrrhizin during decoction (1980). Among the component herbs of Juzen-taiho-to, the coexistence of Poria, Cnidii Rhizoma, Rehmanniae Radix, Angelicae Radix, and Astragali Radix with Glycyrrhizae Radix did not influence the content of glycyrrhizin in the extract (Table 7.5). The glycyrrhizin content in the decoction was slightly decreased by the coexistence of Ginseng Radix, and Atractilodis Rhizoma

TABLE 7.5
Glycyrrhizin Content in Decoction of Glycyrrhizae Radix Mixed with Other Component Herbs of Juzen-taiho-to

Addition		Glycyrrhizin in Decoction Content (%)	Relative Content[a]
Glycyrrhizae Radix	Alone	6.2	100
	+ Poria	7.0	113
	+ Cnidii Rhizoma	6.0	97
	+ Rehmanniae Radix	6.0	97
	+ Angelicae Radix	5.7	92
	+ Astragali Radix	5.5	89
	+ Ginseng Radix	5.0	81
	+ Atractylodis Rhizoma[b]	5.0	81
	+ Atractylodis Lanceae Rhizoma[b]	4.8	77
	+ Paeoniae Radix	3.8	61
	+ Cinnamomi Cortex	3.7	60

[a] Content of glycyrrhizin in decoction prepared from only Glycyrrhizae Radix was expressed as 100.
[b] Atractylodis Lanceae Rhizoma is often used as a component herb for Juzen-Taiho-To instead of Atractylodis Rhizoma.

or Atractylodis Lanceae Rhizoma, whereas the coexistence of Paeoniae Radix and Cinnamomi Cortex significantly decreased the content of glycyrrhizin (about 60%).

These observations suggest that the contents of some other low molecular weight ingredients as well as that of glycyrrhizin are affected by the coexistence of the components of Juzen-taiho-to during decoction. These phenomena are thought to have occurred because of simple solubility factors and/or the formation of water-insoluble complexes among the different ingredients, and also reabsorption of the extracted ingredients to organic materials during decoction, etc.

7.6.3 COMBINATION EFFECTS ON EXPRESSIONS OF COMPLEMENT-ACTIVATING ACTIVITY AND MITOGENIC ACTIVITY

Juzen-taiho-to consists of ten kinds of crude drugs. In order to determine the contribution of each component, the complement-activating activity and mitogenic activity were compared with the hot water extract of each single component and the hot water extract from the corresponding incomplete Juzen-taiho-to.

The hot water extracts of Astragali Radix, Glycyrrhizae Radix, and Atractylodis Lanceae Rhizoma showed potent mitogenic activity. Those from Ginseng Radix expressed more moderate activity and those from Angelicae Radix, Rehmanniae Radix, and Paeoniae Radix showed weak or no activity. Those from Cinnamomi Cortex and Cnidii Rhizoma reduced the proliferation of lymphocytes compared with the control (Yamada et al., 1992).

Decoctions of Juzen-taiho-to with the omission of a single component at a time were prepared, and the mitogenic activity was measured. The decoction in which Astragali Radix, Atractylodis Lanceae Rhizoma, Glycyrrhizae Radix, Cinnamomi Cortex, or Ginseng Radix was omitted from Juzen-taiho-to resulted in decreased mitogenic activity compared with Juzen-taiho-to (Table 7.6). The omission of Atractylodis Lanceae Rhizoma, Glycyrrhizae Radix, Ginseng Radix, or Cinnamomi Cortex from Juzen-taiho-to produced the most decreased activity. When Paeoniae Radix was omitted from Juzen-taiho-to, the mitogenic activity was better than Juzen-taiho-to.

TABLE 7.6
Summary of Individual and Combination Effects of Herbal Components in Juzen-taiho-to

Class	Activity of Component Herb A	Activity of Concoction, Omitted Herb A from Juzen-taiho-to	Complement Activating Activity	Mitogenic Activity
I	Potent	↓	Glycyrrhizae Radix Ginseng Radix	Astragali Radix Glycyrrhizae Radix Atractylodis Lanceae Rhizoma
II	Weak or none	→	Angelicae Radix	Angelicae Radix Rehmanniae Radix Poria
III	Potent	↑ or →	Atractylodis Lanceae Rhizoma Cnidii Rhizoma Cinnamomi Cortex	
IV	Moderate, weak or none	↓	Astragali Radix Rehmanniae Radix Paeoniae Radix	Ginseng Radix
V	None	↑		Paeoniae Radix
VI	Suppressive	↓ or →		Cinnamomi Cortex Cnidii Rhizoma

Hot water extract of Glycyrrhizae Radix, Ginseng Radix, Atractylodis Lanceae Rhizoma, Cnidii Rhizoma, and Cinnammomi Cortex showed potent complement-activating activity; those from Astragali Radix, Rehmanniae Radix, Poria, and Angelicae Radix expressed weak activity and Paeoniae Radix expressed no activity. When Glycyrrhizae Radix or Astragali Radix was omitted from Juzen-taiho-to, the resulting decoctions had a very weak complement-activating activity; however, when Atractylodis Lanceae Rhizoma or Angelicae Radix was omitted, the resulting decoctions did not lose activity. The removal of Cnidii Rhizoma resulted in a higher activity than Juzen-taiho-to, whereas the removal of Ginseng Radix, Poria, Paeoniae Radix, or Rehmanniae Radix also gave a lower activity than Juzen-taiho-to. Although Cinnamomi Cortex on its own showed the highest complement-activating activity, the decoction that omitted Cinnamomi Cortex had a higher activity than Juzen-taiho-to.

These results suggest that complement-activating and mitogenic activities of Juzen-taiho-to are caused by the combination effects of its several crude drugs. Component crude drugs of Classes IV, V, and VI in Table 7.6 are especially thought to contribute to the combination effect. In Class IV, the extract shows potent activity on its own. When that crude drug was omitted from Juzen-taiho-to, the overall activity increased or did not change. In Class V, the extract shows moderate, weak, or no activity. If this crude drug is omitted from Juzen-taiho-to, the resulting activity decreases significantly. In Class VI, although the extract shows suppressive activity on its own, when that crude drug was omitted from Juzen-taiho-to, the overall activity decreased or did not change.

The crude drugs in Class IV cause interaction with other components that enables the given decoction to reveal a higher activity, while the crude drug in Class V causes an increase of something by its absence because of the suppressive substance in that drug. The crude drug in Class VI causes an increase of something by its presence. Not only the complement-activating activity but also mitogenic activity of Juzen-taiho-to is shown to be expressed by pectic polysaccharides (Yamada et al., 1990; Kiyohara et al., 1991), and the activities of some component herbs are also thought to be expressed by polysaccharide molecules (Yamada et al., 1992). These observations suggest that interaction between polysaccharide molecules or between polysaccharide and low molecular weight ingredients contributes to the activity in the decoction of Juzen-taiho-to.

7.6.4 COMBINATION EFFECT ON EXPRESSION OF ACTIVITY MODULATING THE INTESTINAL IMMUNE SYSTEM

Juzen-taiho-to was shown to modulate the intestinal immune system and to activate T-cells in Peyer's patch cells, resulting in potentiation of growth factor production for proliferation of bone marrow cells (Hong et al., 1998). Pectic polysaccharides and lignin–carbohydrate complexes were clarified as active ingredients of Juzen-taiho-to for intestinal immune system-modulating activity through Peyer's patch cells (Kiyohara et al., 2000, 2002).

In order to clarify the active ingredients activity in component herbs of Juzen-taiho-to for modulation of the intestinal immune system, the activities of component herbs were compared. The active lignin–carbohydrate complexes were fractionated into water-soluble and dialyzable fractions (F-3) of Juzen-taiho-to (Figure 7.2). F-3 fractions from components were tested for intestinal immune system-modulating activity. However, all F-3 fractions prepared from the respective ten crude drugs of Juzen-taiho-to did not show the activity, although the F-3, which was prepared from the decoction of Juzen-taiho-to, expressed the potent activity (Kiyohara et al., 2004).

It is generally thought that Juzen-taiho-to comprises two kinds of formulas Shimotsu-to [Si-Wu-Tang in Chinese], composed of Angelicae Radix, Cnidii Rhizoma, Paeoniae Radix, and Rehmanniae Radix, and Shikunshi-to [Si-Jun-Zi-Tang in Chinese], composed of Atractylodis Lanceae Rhizoma, Ginseng Radix, Poria, and Glycyrrhizae Radix) in addition to Astragali Radix and Cinnamomi Cortex. However, F-3 fractions prepared from decoctions of Shimotsu-to or Shikunshi-to did not show the activity. When F-3 was prepared from a concoction that decocted from Shimotsu-to plus Astaragali Radix or Cinnamomi Cortex, no activity was observed. However, F-3 prepared from a concoction decocted from Shikunshi-to plus Astaragali Radix showed significant activity (Figure 7.11).

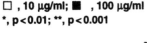

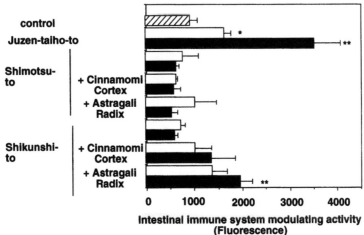

FIGURE 7.11 Intestinal immune system modulating activity of methanol-insoluble and water-soluble fraction (F-3) prepared from Juzen-taiho-to and its basic concoctions (Shimotsu-to and Shikunshi-to).

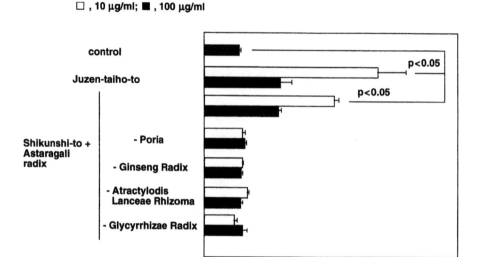

FIGURE 7.12 Changes of intestinal immune system modulating activity of methanol-insoluble and water-soluble fraction (F-3) prepared from Shikunshi-to mixed with Astragali Radix by omission of the component herbs.

Further addition of Cinnammomi Cortex into Shikunshi-to plus Astaragali Radix increased the activity of the resulting F-3 from the formula compared with F-3, which was prepared from concoction of Shikunshi-to plus Astaragali Radix. When each single crude drug was omitted from the formula of Shikunshi-to plus Astaragali Radix, the resulting F-3 showed no activity (Figure 7.12). When F-3 separately prepared from each crude drug of Juzen-taiho-to was mixed and the activity of the mixture tested, no activity was observed.

These observations suggest that F-3 modulating activity through Peyer's patch cells is caused by the combination effect of the "restricted ingredients" from the six crude drugs (Angelicae Radix, Atractylodis Lanceae Rhizoma, Ginseng Radix, Poria, Astragali Radix, and Cinnamomi Cortex) during the decoction (Figure 7.13). Because the mixture of F-3 fractions from these six essential

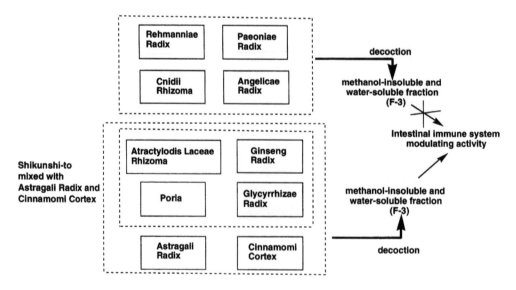

FIGURE 7.13 Basic concoctions of Juzen-taiho-to and their contribution for expression of intestinal immune system modulating activity.

crude drugs did not express the activity, something may have occurred during the extraction procedure of the five crude drugs, such as changes of extraction rate of active ingredients or chemical reaction of ingredients to produce the active ingredients.

7.7 CONCLUSION

In this chapter some of the complex interactions occurring between the constituents of the ten crude drugs during the preparation of the decoction of Juzen-taiho-to have been explored in relation to the body's subsequent complex metabolism of the many compounds.

Activity-guided clarification of the active ingredients in the decoction of the pharmaceutical dried extract preparation of Juzen-taiho-to (TJ-48) leads to the conclusion that low molecular weight ingredients (C_{18}-unsaturated fatty acid such as oleic acid and linoleic acid, and sodium malate) and high molecular weight ingredients (pectic polysaccharides, lignin–carbohydrate complexes) are some of the important active ingredients for pharmacological activity of Juzen-taiho-to for the systemic immune system, intestinal immune system, and hemopoietic system. From these active ingredients, it is thought that parts of the pharmacological activity of Juzen-taiho-to, such as immunostimulatory activity, hemopoietic-stimulating activity, and prevention of adverse effects of anticancer drugs can be explained. Although Juzen-taiho-to is also applied for treatment of autoimmune diseases and prevention of rebound reaction of steroid for chronic atopic skin disease, the active ingredients for these activities are not known; clarification of the active ingredients awaits further study.

The high molecular weight ingredients, pectic polysaccharides, and lignin–carbohydrate complexes, were shown to be largely responsible for the expression of some of the pharmacological activities of Juzen-taiho-to; however, their mechanisms for *in vivo* activities by oral administration are mostly unknown. It has been suggested that pharmacologically active pectin is absorbed partly from the digestive system by oral administration (Sakurai et al., 1996b). It is assumed that the active pectic polysaccharides and lignin–carbohydrate complexes may express their *in vivo* actions by their direct absorption through the digestive system or by stimulation of the intestinal immune system.

Pharmacological evaluation of the combination effects of the crude drugs or ingredients of the crude drug suggests that several complex combination effects occur during the extraction process and before the extract is taken into the body. Interactions are expected to occur not only among low molecular weight ingredients but also between low molecular weight and high molecular weight ingredients, and among the high molecular weight ingredients. This information, along with that in the other chapters, attempts to give explanations for the wide range of beneficial clinical activity available from Juzen-taiho-to.

REFERENCES

Abdel–Hafez, A.A., Meselhy, M.R., Nakamura, N., Hattori, M., Watanabe, H., Murakami, Y., El-Gendy, M.A., Mahfouz, N.M. and Mohamed, T.A. (1998) Effects of paeoniflorin derivatives on scopolamine-induced amnesia using a passive avoidance task in mice; Structure–activity relationship. *Biol. Pharm. Bull.*, 21, 1174–1179.

Aida, K., Tawata, M., Shindo, H., Onaya, T., Sasaki, H., Yamaguchi,T., Chin, M. and Mitsuhashi, H. (1990) Isoliquiritigenin : a new aldose reductase inhibitor from Glycyrrhizae Radix. *Planta Med.*, 56, 254–258.

Albershein, P., An, J., Freshour, G., Fuller, M.S., Guillen, R., Ham, K.-S., Hahn, M.G., Huang, J., O'Neill, M., Whitcombe, A., Williams, M.V., York, W.S. and Darvill, A. (1994) Structure and function studies of plant cell wall polysaccharides. *Biochem. Soc. Trans.*, 22, 374–378.

Baba, M. and Shigeta, S. (1987) Antiviral activity of glycyrrhizin against varicella-zoster virus *in vitro. Antiviral Res.*, 7, 99–107.

Badam, L. (1997) *In vitro* antiviral activity of indigenous glycyrrhizin, licorice and glycyrrhizic acid (Sigma) on Japanese encephalitis virus. *J. Commun. Dis.,* 29, 91–99.

Chen, Y.F., Tsai, H.Y. and Wu, T.S. (1995) Anti-inflammatory and analgesic activities from roots of *Angelica pubescens. Planta Med.,* 61, 2–8

Claeson, P., Panthong, A., Tuchinda, P., Reutrakul, V., Kanjanapothi, D. and Taylor, W.C. (1993) Three non-phenolic diarylheptanoids with anti-inflammatory activity from *Curcuma xanthorrhiza. Planta Med.,* 59, 451–454.

Crance, J.M., Gratier, D., Guimet, J. and Jouan, A. (1997) Inhibition of sandfly fever Sicilian virus (Phlebo-virus) replication *in vitro* by antivial compounds. *Res. Virol.,* 148, 353–365.

Francischetti, I.M., Monteiro, R.Q. and Guimaraes, J.A. (1997) Identification of glycyrrhizin as a thrombin inhibitor. *Biochem. Biophys, Res. Commun.,* 235, 259–263.

Furuya, T., Yamagata, S., Shimoyama, Y., Fujihata, M., Morishima, N. and Ohtsuki, K. (1997) Biochemical characterization of glycyrrhizin as an effecttive inhibitor for hyaluroindases from bovine testis. *Chem. Pharm. Bull.,* 20, 973–977.

Haraguchi, H., Ishikawa, H., Mizutani, K., Tamura, Y. and Kinoshita, T. (1998) Antioxidative and superoxide scavenging activities of retrochalcones in *Glycyrrhiza inflata. Bioorg. Med. Chem.,* 6, 339–347.

Hase, K., Li, J., Basnet, P., Xiong, Q., Takamura, S., Namba, T. and Kadota, S. (1997) Hepatoprotective principles of *Swertia japonica* Makino on D-galactosamine/lipopolysaccharide-induced liver injury in mice. *Chem. Pharm. Bull.,* 45, 1823–1827.

Hisha, H., Yamada, H., Sakurai, M.H., Kiyohara, H.,Li, Y., Yu, C.Z., Takemoto, N., Kawamura, H., Yamaura, K., Shinohara, S., Komatsu, Y., Aburada, M. and Ikehara, S. (1997) Isolation and identification of hematopoietic stem cell-stimulating substances from Kampo (Japanese herbal) medicine, Juzen-taiho-to. *Blood,* 90, 1022–1030.

Hong, T., Matsumoto, T., Kiyohara, H. and Yamada, H. (1998) Enhanced production of hematopoietic growth factors through T-cell activation in Peyer's patches by oral administration of Kampo (Japanese herbal) medicines, Juzen-taiho-to. *Phytomedicine,* 5, 353–360.

Hosoya, E. Scientific reevaluation of Kampo prescriptions using modern technology. In *Recent Advances in The Pharmacology of Kampo Medicines,* Eds. Hosoya, E. and Yamamura, Y., Excerpta Medica, Tokyo, pp. 17–37 (1988).

Hsu, F.L., Lai, C.W. and Cheng, J.T. (1997) Antihyperglycemic effects of paeoniflorin and 8-debenzoylpae-oniflorin, glucosides from the roots of *Paeonia lactiflora. Planta Med.,* 63, 323–325.

Imai, T., Sakai, M., Ohtake, H., Azuma, H. and Otagiri, M. (1991) *In vitro* and *in vivo* evaluation of the enhancing activity of glycyrrhizin on the intestinal absorption of drugs. *Pharm. Res.,* 16, 80–86.

Ito, M., Nakashima, H., Baba, M., Pauwels, R., De Clercq, E., Shigeta, S. and Yamamoto, N. (1987) Inhibitory effect of glycyrrhizin in the *in vitro* infectivity and cytopathic activity of the human immunodeficiency virus [HIV (HTLV-III/LAV9)]. *Antiviral. Res.,* 7, 127–137.

Itoh, K., Hara, T., Shiraishi, T., Taniguchi, K., Morimoto, K. and Onishi, T. (1989) Effects of glycyrrhizin and glycyrrhetinic acid on N+-K+-ATPase of renal basolateral membranes *in vitro. Biochem. Int.,* 18, 81–89.

Juniewicz, P.E., Pallante Morell, S., Mosen, A. and Ewing, L.L. (1988) Identification of phytoestrogens in the urine of male dogs. *J. Steroid Biochem.,* 31, 987–994.

Kageyama, Y., Suzuki, H. and Saruta, T. (1992) Glycyrrhizin induces mineral corticoid activity through alterations in cortisol metabolism in the human kidney. *J. Endocrinol.,* 135, 147–152.

Kase, Y., Saitoh, K., Makino, B., Hashimoto, K., Ishige, A. and Komatsu, Y. (1999) Relationship between the antidiarrhoeal effects of Hange-shashin-to and its active components. *Phytother. Res.,* 13, 468–473.

Kawamori, T., Tanaka, T., Hara, A., Yamahara, J. and Mori, H. (1995) Modifying effects of naturally occurring products on the development of colonic aberrant crypt foci induced by azoxymethane in F344 rats. *Can. Res.,* 15, 1277–1282.

Keung, W.M. (1993) Biochemical studies of a new class of alcohol dehydrogenase inhibitors from *Radix puerariae, Alcohol Clin. Exp. Res.,* 17, 1254–1260.

Kiyohara, H., Takemoto, N., Komatsu, Y., Kawamura, H., Hosoya, E. and Yamada, H. (1991) Characterization of mitogenic pectic polysaccharides from Kampo (Japanese herbal) medicine "Junzen-taiho-To." *Planta Med.,* 57, 254–259.

Kiyohara, H., Yamada, H., Takemoto, N., Kawamura, H., Komatsu, Y. and Oyama, T. (1993) Characterization of *in vitro* IL-2 production enhancing and anticomplementary pectic polysaccharides from Kampo (Japanese herbal) medicine "Juzen-taiho-to." *Phytother. Res.,* 7, 367–375.

Kiyohara, H., Matsumoto, T., Takemoto, N., Kawamura, H., Komatsu, Y. and Yamada, H. (1995a) Effect of oral administration of a pectic polysaccharide fraction from a Kampo (Japanese herbal) medicine "Juzen-taiho-to" on antibody response of mice. *Planta Med.,* 61, 429–434.

Kiyohara, H., Matsumoto, T., Komatsu, Y. and Yamada, H. (1995b) Protective effect of oral administration of a pectic polysaccharide fraction from a Kampo (Japanese herbal) medicine "Juzen-Taiho-To" on adverse effects of *cis*-diaminedichloroplatinum. *Planta Med.,* 61, 531–534.

Kiyohara, H., Takemoto, N., Zhao, J.F., Kawamura, H. and Yamada, H. (1996) Pectic polysaccharide from roots of *Glycyrrhiza uralensis*: possible contribution of neutral oligosaccharide in the galacturonase-resistant region to anti-complementary and mitogenic activities. *Planta Med.,* 62, 14–19.

Kiyohara, H., Matsumoto, T. and Yamada, H. (2000) Lignin–carbohydrate complexes: intestinal immune system modulating ingredients in Kampo (Japanese herbal) medicine, Juzen-taiho-to. *Planta Med.,* 66, 20–24.

Kiyohara, H., Matsumoto, T. and Yamada, H. (2002) Intestinal immune system modulating polysaccharides in a Japanese herbal (Kampo) medicine, Juzen-taiho-to. *Phytomedicine,* 9, 614–624.

Kiyohara, H., Matsumoto, T., and Yamda, H. (2004) Combination effects of herbs in a multi-herbal formula: expression of Juzen-taiho-to's immunomodulatory activity on the intestinal immure system. *Evidence-Based Complementary Alternative Med.* (eCAM), 1, 83–91.

Kobayashi, M., Schmitt, D.A., Utsunomiya, T., Pollard, R.B. and Suzuki, F. (1993a) Inhibition of burn-associated suppressor cell generation by glycyrrhizin through the induction of contrasuppressor T-cells. *Immunol. Cell. Biol.,* 71, 181–189.

Kobayashi, S., Mimura, Y., Naitoh, T., Kimura, T. and Kimura, M. (1993b) Chemical structure–activity of Cnidium rhizome-derived phthalides for the competence inhibition of proliferation in primary cultures of mouse aorta smooth muscle cells. *Jap. J. Pharmacol.,* 63, 353–359.

Kobayashi, S., Miyamoto, T., Kimura, I. and Kimura, M. (1995) Inhibitory effect of isoliquiritin, a compound in licorice root, on angiogenesis *in vivo* and tube formation *in vitro*. *Biol. Pharm. Bull.,* 18, 1382–1386.

Koh, W.S., Yoon, S.Y., Kwon, B.M., Jeong, T.C., Nam, K.S. and Han, M.Y. (1998) Cinnamaldehyde inhibits lymphocyte proliferation and modulates T-cell differentiation. *Int. J. Immunopharmacol.,* 20, 643–660.

Kusano, A., Nikaido, T., Kuge, T., Ohmoto, T., Delle Monache, G., Botta, B., Botta, M. and Saito, T. (1991) Inhibition of adenosine 3′5′-cyclic monophosphate phosphodieterase by flavonoids from licorice roots and 4-arylcoumarins. *Chem. Pharm. Bull.,* 39, 930–933.

Kwon, B.M., Lee, S.H., Choi, S.U., Park, S.H., Lee, C.O., Cho, Y.K., Sung, N.D. and Bok, S.H. (1998) Synthesis and *in vitro* cytotoxicity of cinnamaldehyde to human solid tumor cells. *Arch. Pharm. Res.,* 21, 147–152.

Matsui, M., Sato, Y., Bando, H., Murayama, M., Osawa, T., Miura, T. and Oshima, Y. (1998) Studies on constituents of "Kako-bushi-matsu." *Nat. Med.,* 52, 232–235.

Matsumoto, K., Kohno, S., Ojima, K., Tezuka, Y., Kadota, S. and Watanabe, H. (1998) Effects of methylenechloride-soluble fraction of Japanese Angelica root extract, ligustilide and butylidenephthalide, on pentobarbital sleep in group-housed and socially isolated mice. *Life Sci.,* 62, 2073–2082.

Mitscher, L. A., Park, Y.H., Clark, D. and Beal, J. (1980) Anti-microbial agents from higher plants. Antimicrobial isoflavonoid and related substances from *Glycyrrhiza glabra* L. var. typica. *J. Nat. Products,* 43, 259–269.

Nabeya, K. and Ri, S. (1983) Effect of oriental medical herbs on the restoration of the human body before and after operation. *Proc. Symp. WAKAN-YAKU.* 16, 201.

Nagai, T., Egashira, T., Yamanaka, Y. and Kohno, M. (1991) The protective effect of glycyrrhizin against injury of the liver caused by ischemia-reperfusion. *Arch. Environ. Contam. Toxicol.,* 20, 432–436.

Ohnishi, Y., Yasumizu, R., Fan, H., Liu, J., Takao-Lin, F., Komatsu, Y., Hosoya, E., Good, R.A. and Ikehara, S. (1990) Effect of Juzen-taiho-to (TJ-48), a traditional Oriental medicine, on hematopoietic recovery from radiation injury in mice. *Exp. Hematol.,* 18, 18.

Ohtsuki, K., Oh-Ishi, M. and Nagata, N. (1992) The stimulatory and inhibitory effects of glycyrrhizin and a glycyrrhetinic acid derivative on phosphorylation of lipocortin I by A-kinase *in vitro*. *Biochem. Int.,* 28, 1045–1053.

Ohta, T., Watanabe, K., Moriya, M., Shirasu, Y. and Kada, T. (1983) Analysis of the antimutagenic effect of cinnamaldehyde on chemically induced mutagenesis in *Escherichia coli*. *Mol. Gen. Genet.,* 192, 309–315.

Ohuchi, K., Kamada, Y., Levine, L. and Tsurufuji, S. (1981) Glycyrrhizin inhibits prostaglandin E2 production by activated peritoneal macrophages from rats. *Prostaglandins Med.,* 7, 457–463.

Ozaki, Y. and Ma, J.P. (1990) Inhibitory effects of tetramethylpyrazine and ferulic acid on spontaneous movement of rat uterus *in situ. Chem. Pharm. Bull.*, 38, 1620–1623.

Ruiz–Larrea, M.B., Mohan, A.R., Paganga, G., Miller, N.J., Bolwell, G.P., and Rice–Evans, C.A. (1997) Antioxidant activity of phytoestrogenic isoflavones. *Free Radical Res.*, 26, 63–70.

Sakurai, M., Kiyohara, H., Yamada, H., Hisha, H., Li, Y., Takemoto, N., Kawamura, H., Yamamura, K., Shinohara, S., Komatsu, Y., Aburada M., and Ikehara, S. (1996a) Hematopoietic stem cell-stimulating ingredients in Kampo (Japanese herbal) medicine, "Juzen-Taiho-To." In *Bone Marrow Transplantation—Basic and Clinical Studies*, Ikehara, S., Takaku, F. and Good, R.A., Eds., 64–67, Springer–Verlag, Tokyo.

Sakurai, M., Matsumoto, T., Kiyohara, H. and Yamada, H. (1996b) Detection and tissue distribution of anti-ulcer pectic polysaccharides from *Bupleurum falcatum* by polyclonal antibody. *Planta Med.*, 62, 341–346.

Sharma, R.D. (1979) Isoflavones and hypercholesterolemia in rats. *Lipids*, 14, 535–539

Shiki, Y., Sasaki, N., Shirai, K., Saito, Y., Yoshida, S. (1986) Effect of glycyrrhizin on stability of lysosomes in the rat arterial wall. *Am. J. Chin. Med.*, 14, 131–137.

Sugaya, A., Suzuki, T., Sugaya, E., Yuyama, N., Yasuda, K., and Tsuda, T. (1991) Inhibitory effect of peonyl root extract on pentylanetetrazol-induced EEG power spetrum changes and extracellular calcium concentration changes in rat cerebral cortex. *J. Ethnopharmacol.*, 33, 159–167.

Sugiyama, K., Yokota, M., Ueda, H., and Ichio, Y. (1993) Protective effects of Kampo medicines against cis-diaminedichloroplatinum (II)-induced nephrotoxicity and bone marrow toxicity in mice. *J. Med. Pharm. Soc. WAKAN-YAKU*, 10, 76–85.

Sugiyama, K., Ueda, H., Suhara, Y., Kajima, Y., Ichio, Y., and Yokota, M. (1994) Protective effect of sodium L-malate, an active constituent isolated from Angelicae Radix, on *cis*-diaminedichloroplatinum (II)-induced toxic side effect. *Chem. Pharm. Bull.*, 42, 2565–2568.

Sugiyama, K., Ueda, H., Ichio, Y., and Yokota, M. (1995) Improvement of cisplatin toxicity and lethality by Juzen-taiho-to in mice. *Biol. Pharm. Bull.*, 18, 53–58.

Suzuki, F., Schmitt, D.A., Utsunomiya, T., and Pollard, R.B. (1992) Stimulation of host resistance against tumors by glycyrrhizin, an active component of licorice roots. *In Vivo*, 6, 589–596.

Taguchi, I., Kiyohara, H., Matsumoto, T., Yamada, H. (2004) Structure of oligosaccharide side chains of an intestinal immune system modulating arabinogalactan isolated from *atractylodes lancea* DC. *Carbohydr. Res.*, 339, 763–770.

Takemoto, N., Kiyohara, H., Maruyama, H., Komatsu, Y., Yamada, H., and Kawamura, H. (1994) A novel type of B-cell mitogen isolated from Juzen-taiho-to (TJ-48), a Japanese traditional medicine. *Int. J. Immunopharmacol.*, 16, 919–929.

Takeuchi, T., Nishii, O., Okamura T., and Yaginuma, T. (1991) Effect of paeoniflorin, glycyrrhizin and glycyrrhetic acid on ovarian androgen production. *Am. J. Chin. Med.*, 19, 73–78.

Tamaya, T., Sato, S. and Okada, H. (1986) Inhibition by plant herb extracts of steroid bindings in uterus, liver and serum of the rabbit. *Acta Obset. Gynecol. Scand.*, 65, 839–842.

Tanaka, M., Mano, N., Akazai, E., Narui, Y., Kato, F., and Koyama, Y. (1987) Inhibition of mutagenicity by glycyrrhiza extract and glycyrrhizin. *J. Pharmacobiodyn.*, 10, 685–688.

Tawata, M., Aida, K., Noguchi, T., Ozaki, Y., Kume, S., Sasaki, H., Chin, M., and Onaya, T. (1992) Anti-platelet action of isoliquiritigenin, an aldose reductase inhibitor in licorice. *Eur. J. Pharmacol.*, 212, 87–92.

Tomimori, T. and Yoshimoto, M. (1980) Quantitative variation of glycyrrhizin in the decoction of Glycyrrhizae Radix mixed with other crude drugs. *Shoyakugaku-Zasshi*, 34, 138–144 (in Japanese).

Ueda, H., Sugiyama, K., Kajima, Y., and Yokota, M. (1997) Role of sodium malate in the inhibitory effect of Juzen-taiho-to against cisplatin-induced toxic side effect. *J. Trad. Med.*, 14, 199–203

Wegener, J.W. and Nawrath, H. (1997a) Cardiac effects of isoliquiritigenin. *Eur. J. Pharmacol.*, 326, 37–44.

Wegener, J.W. and Nawrath, H. (1997b) Differential effects of isoliquritigenin and YC-1 in rat aortic smooth muscle. *Eur. J. Pharmacol.*, 323, 89–91.

Yamada, H., Kiyohara, H., Cyong, J.C., Takemoto, N., Komatsu, Y., Kawamura, H., Aburada, M., and Hosoya, E. (1990) Fractionation and characterization of mitogenic and anti-complementary active fractions from Kampo (Japanese herbal) medicine "Juzen-taiho-to." *Planta Med.*, 56, 386–391.

Yamada, H., Kiyohara, H., Takemoto, N., Zhao, J.F., Kawamura, H., Komatsu, Y., Cyong, J.C., Aburada, M., and Hosoya, E. (1992) Mitogenic and complement activating activities of the herbal components of Juzen-taiho-to. *Planta Med.*, 58, 166–170.

Yamamoto, S., Aizu, E., Jiang, H., Nakadate, T., Kiyoto, I., Wang, J.-C., and Kato, R. (1991) The potent anti-tumor-promoting agent isoliquiritigenin. *Carcinogenesis,* 12, 317–323.

Yamamura, Y., Kotake, H., Tanaka, N., Aikawa, T., Sawada, Y., and Iga, T. (1997) The pharmacokinetics of glycyrrhizin and its restorative effect on hepatic function in patients with chronic hepatitis and in chronically carbontetrachloride-intoxicated rats. *Biopharm. Drug Dispos.,* 18, 717–725.

Yoshikawa, M., Matsui, Y., Kawamoto, H., Umemoto, N., Oku, K., Koizumi, M., Yamao, J., Kuriyama, S., Nakano, H., Hozumi, N., Ishizaka, S., and Fukui, H. (1997) Effects of glycyrrhizin on immune-mediated cytotoxicity. *J. Gastroenterol. Hepatol.,* 12, 243–248.

Yu, K.W., Kiyohara, H., Matsumoto, T., Yang, H.C., and Yamada, H. (1998) Intestinal immune system modulating polysaccharides from rhizomes of *Atractylodes lancea. Planta Med.,* 64, 714–719.

Yu, K.W., Kiyohara, H., Matsumoto, T., Yang, H.C., and Yamada, H. (2001a) Structural characterization of intestinal immune system modulating new arabino-3,6-galactan from rhizomes of *Atractylodes lancea* DC. *Carbohydr. Polyms.,* 46, 147–156.

Yu, K.W., Kiyohara, H., Matsumoto, T., Yang, H.C., and Yamada, H. (2001b) Characterization of pectic polysaccharides having intestinal immune system modulating activity from rhizomes of *Atractylodes lancea* DC. *Carbohydr. Polyms.,* 46, 125–134.

Zhao, J.F., Kiyohara, H., Sun, X.B., Matsumoto, T., Cyong, J.C., Yamada, H., Yakemoto, N., and Kawamura, H. (1991a) *In vitro* immunostimulating polysaccharide fractions from roots of *Glycyrrhiza uralensis* Fisch. et DC. *Phytother. Res.,* 5, 206–210.

Zhao, J.F., Kiyohara, H., Yamada, H., Takemoto, N., and Kawamura, H. (1991b) Heterogeneity and characterization of mitogenic and anti-complementary pectic polysaccharides from the roots of *Glycyrrhiza uralensis* Fisch. et DC. *Carbohydr. Res.,* 219, 149–172.

8 Effects of Juzen-taiho-to on Carcinogenesis, Tumor Progression, and Metastasis *In Vivo*

Ikuo Saiki

CONTENTS

8.1 INTRODUCTION

The development of cancer (*carcinogenesis*) generally involves many steps, each governed by multiple factors; some depend on the genetic constitution of the individual and others on the environment and way of life. Most cancers are probably initiated by a change in the cell's DNA sequences, but a single mutation is not enough to cause cancer. *Tumor progression* is the process by which an initial population of slightly abnormal cells evolves from bad to worse through successive cycles of mutation and natural selection. *Metastasis*, one of the major causes of mortality

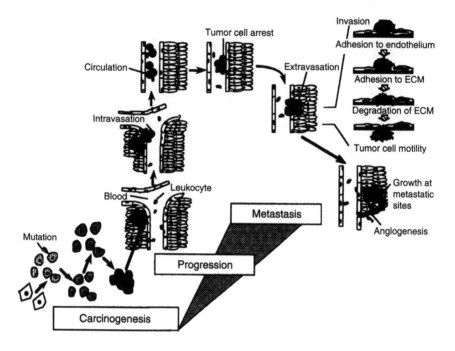

FIGURE 8.1 A schematic representation of various steps of tumor development (carcinogenesis), malignant progression, and metastasis.

in cancer, is a complex cascade of events involving tumor dissemination from the primary site of growth to distant organs. Following tumor development and progression, the pathogenesis of metastases can be subdivided into a variety of sequential steps (Figure 8.1) (Fidler, 1984; Nicolson, 1987; Hart, 1982; Liotta et al., 1983):

1. Release from the primary tumor and invasion of the surrounding tissues
2. Entry into the vascular or lymphatic circulation
3. Transit in the circulation
4. Arrest in the capillary bed of a distant organ
5. Extravasation from the circulation
6. Growth at apparently selected sites distant from the original tumor site

Few cells in a primary tumor can complete all the steps necessary to achieve metastasis. Specific tumor interactions with host cells or components are therefore fundamental events in preferential organ colonization whereby metastases occur in specific organs and not randomly.

Despite advances in diagnostic techniques for the early detection of various cancers and significant improvement in surgical procedures, the mortality rate of cancer has been increasing steadily (Eisenberg et al., 1982; Galandivk et al., 1992; Gastrointestinal Tumor Study Group, 1985), and metastasis is a frequent cause of death by cancer. For instance, the liver is the most common target of the hematogenous metastasis in gastrointestinal tract cancer, especially colon cancer; the prognosis for cases with liver metastasis is extremely poor (Bengmark and Hafstrom, 1969; Gavowski et al., 1994). If occult micrometastases that had been established at the time of surgery could be inhibited, the prognosis of patients with colon carcinoma would improve.

Juzen-taiho-to was introduced from China to Japan in the Kamakura dynasty (AD 1192–1333) and since then has been used as a cure for consumption, general debility, deficiency, and impairments of "In" (Yin in Chinese), "Yo" (Yang in Chinese), "Ketsu" (Xue in Chinese; a concept referring to blood, hormones, the autonomic nervous system, and other regulatory functions of the body's internal environment) or vital energy, "Ki" (Qi in Chinese) in the viscera or bowels, and lack of appetite. It is one of the nourishing agents, so-called "Hozai" (Buji in Chinese), for improving deficiency syndrome, "Kyo-sho" (Xu-zheng in Chinese), and disturbances and imbalances in the homeostatic condition of the body diagnosed by Kampo medicine, and is currently administered to patients weakened by long illness, fatigue, loss of appetite, night sweats, circulatory problems, and anemia. It is also used for cancer patients.

Clinically, Juzen-taiho-to is known to improve the general systemic condition of cancer patients and to reduce the adverse effects of chemotherapy, radiation therapy, and surgical treatment, which may lead to the induction of "Kyo-sho"(deficiency of Ki and Ketsu). It has also been used for the treatment of rheumatoid arthritis, atopic dermatitis, chronic fatigue syndrome, ulcerous colitis, and so on. Pharmacological studies of Juzen-taiho-to have been carried out to determine the clinical efficacy of Juzen-taiho-to by testing effects such as

- Enhancement of phagocytosis (Maruyama et al., 1988)
- Cytokine induction (Haranaka et al., 1985; Kubota et al., 1992)
- Antibody production (Hamada et al., 1988)
- Mitogenic activity of spleen cells (Kiyohara et al., 1991)
- An antitumor effect when combined with surgical excision (Maruyama et al., 1993)
- Antitumor activity with or without other drugs (Haranaka et al., 1988; Sugiyama et al., 1995a)
- Protection from the deleterious effects of anti-cancer drugs (Sugiyama et al., 1995b)
- Radiation-induced immuno-suppression and bone marrow toxicity (Kawamura et al., 1989; Ohnishi et al., 1990)

This chapter describes the effect of Juzen-taiho-to on carcinogenesis, tumor progression and metastasis *in vivo*, and the inhibitory mechanism of action.

8.1.1 JUZEN-TAIHO-TO AND ITS CONSTITUENTS

Juzen-taiho-to is composed of ten herbs/crude drugs (shown in the previous chapters), of which the quality is controlled by *Japanese Pharmacopoeia XIII*. Juzen-taiho-to is prepared as follows: a mixture of Astragli Radix (3.0 g), Cinnamomi Cortex (3.0 g), Rehmanniae Radix (3.0 g), Paeoniae Radix (3.0 g), Cnidii Rhizoma (3.0 g), Atractylodis Lanceae Rhizoma (3.0 g), Angelicae Radix (3.0 g), Ginseng Radix (3.0 g), Poria (3.0 g), and Glycyrrhizae Radix (1.5 g) is added to 285 ml of water and extracted at 100°C for 1 h. The extracted solution is filtrated and spray-dried to obtain the dry extract powder (2.3 g). In Japan, Juzen-taiho-to is now available as a decoction or extract preparation, e.g., TJ-48.

8.2 EFFECT ON PROCESSES OF CARCINOGENESIS, MALIGNANT PROGRESSION, AND METASTASIS

8.2.1 INHIBITION OF CARCINOGENESIS

Tatsuta et al. (1994) investigated the effect of Juzen-taiho-to on hepatocarcinogenesis induced by N-nitrosomorpholine (NNM) in Sprague–Dawley rats. Administered in the diet for 15 weeks, 2.0 or 4.0% Juzen-taiho-to significantly reduced the size, volume, and/or number of glutathione-S-transferase (GST-P)-positive and γ-glutamyl transpeptidase (GGT)-positive hepatic lesions. This treatment also caused a significant increase in the proportion of IL-2 receptor-positive lymphocytes among the lymphocytes infiltrating the tumors, as well as a significant decrease in the labeling index of preneo-plastic lesions. This suggests that the inhibition of hepatocarcinogenesis by Juzen-taiho-to is in part due to activation of the immune system. Similarly, Juzen-taiho-to significantly inhibited the formation of GAT-P positive foci treated with diethylnitrosamine and two thirds partial hepatectomy.

The inhibitory effects of Juzen-taiho-to on the development of bladder cancer induced by *N*-butyl-*N*-(4-hydroxybutyl)nitrosamine (BBN) were also evaluated in C57BL/6 strain mice (Watanabe et al., 1997). The group administered Juzen-taiho-to orally showed a lower rate of bladder cancer incidents, compared to a control group. Furthermore, neither normal epithelia nor simple hyperplasia was observed in the control group at 12 weeks after BBN administration, whereas 70% of the treated group retained normal epithelia or had simple hyperplasia of the bladder tissue. This suggests that Juzen-taiho-to is likely to be effective in preventing the early neoplastic changes that occur in the mouse urinary bladder cancer induced by BBN.

Sakamoto et al. (1994) have reported that Juzen-taiho-to and Sho-saiko-to (Xiao-Chai-Hu-Tang in Chinese) suppressed the activities of thymidylate synthetase and thymidine kinase involved in *de novo* and salvage pathways for pyrimidine nucleotide synthesis, respectively, in spontaneous mammary tumors of SHN mice with a reduced level of serum prolactin. This indicates that these Kampo medicines may have antitumor effects on mammary tumors.

8.2.2 PREVENTION OF MALIGNANT TUMOR PROGRESSION

Malignant tumor progression is the process by which tumor cells acquire a more malignant phenotype, such as enhancement of the ability to proliferate, to invade, or to metastasize, and is affected by various factors. However, few studies have been conducted on the mechanisms, facilitating factors, and inhibitors of progression because of the lack of a suitable experimental model. In one tumor progression model (Figure 8.2), spontaneously regressive QR-32 tumor cells, when coimplanted s.c. with a foreign body (a gelatin sponge), irreversibly acquired the ability to grow progressively at the inoculated sites—even without a gelatin sponge (Okada et al., 1992). In contrast, QR-32 cells alone gradually grew over 15 days after inoculation and thereafter regressed for up to 25 days.

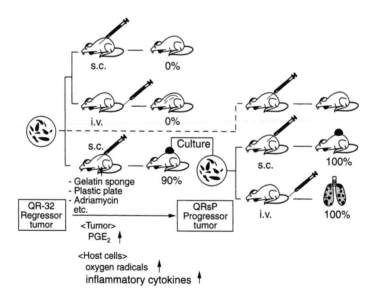

FIGURE 8.2 Malignant progression model by regressor tumor. A weakly malignant QR-32 fibrosarcoma undergoes spontaneous regression in normal C57BL/6 mice unless extremely large numbers of the cells were inoculated i.v. or s.c. Coimplantation of QR-32 cells with a foreign body, gelatin sponge (in some cases, small fragments of plastic plate or a low dose of adriamycin), permits the progressive growth of the tumors that develop *in vivo*. Inoculation of the resultant progressor tumor (QRsP) without gelatin sponge into fresh mice resulted in lethal growth.

It has also been shown that such progressive growth of tumors was provoked through the enhancement of PGE_2 production in the tumors or by oxygen radicals and inflammatory cytokines produced by host cells reactive with the gelatin sponge (Okada et al., 1992, 1993, 1994). This seems analogous to the fact that malignant progression of tumors followed by metastasis is clinically observed to be elicited by various factors and circumstances, including stresses, anticancer drugs, inflammation, etc. The progressive growth after s.c. coimplantation of QR-32 cells with a gelatin sponge was prevented by endogenous induction of antioxidative enzymes or scavengers such as manganese superoxide-dismutase (Mn-SOD) or metallothionein at the tumor sites by orally administered bismuth subnitrate and lipophilic vitamin C (Okada et al., 1995).

The inhibitory effect of oral administration of Juzen-taiho-to (pharmaceutical preparation, TJ48, by Tsumura & Co) on progressive growth of this mouse fibrosarcoma was investigated in this tumor progression model. Oral administration of Juzen-taiho-to caused significant inhibition of the progressive growth of QR-32 regressor tumors after coimplantation with gelatin sponge (Figure 8.3) and prolonged the survival of tumor-bearing mice. These results indicate that Juzen-taiho-to may be effective for preventing weakly malignant tumors from growing progressively upon coimplantation with a gelatin sponge (Ohnishi et al., 1996).

Such progressive growth, however, may not necessarily be equivalent to malignant progression because even cells that do not acquire a more malignant phenotype can show transient proliferation depending on the host circumstances and the implantation conditions. Resultant progressive tumor (QRsP) showed gelatin sponge-independent growth after reinoculation (six out of six mice, Figure 8.3), as also demonstrated previously (Okada et al., 1992). This phenomenon could be regarded

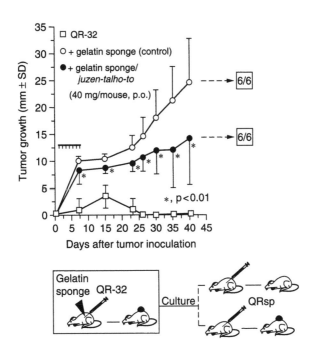

FIGURE 8.3 Effect of Juzen-taiho-to on the growth of QR-32 fibrosarcoma after coimplantation with gelatin sponge. Six C57BL/6 mice per group were inoculated s.c. with QR-32 cells (10^5) with or without gelatin sponge (10- × 5- × 3-mm size). Juzen-taiho-to (40 mg/day/mouse) was administered orally for 7 days after tumor inoculation. Tumor growth was measured as a time of function after the implantation and calculated as average diameter (mm) of long and short axes. (□), QR-32 alone; (○), QR-32+gelatin sponge; (●), QR-32+gelatin sponge + Juzen-taiho-to; *, $p < 0.01$ compared with control (QR-32 + gelatin sponge) by Student's two-tailed *t*-test.

as progression. Actually, tumors obtained from the group treated with Juzen-taiho-to did not grow progressively after reinoculation into syngeneic mice without a gelatin sponge (zero out of six mice, Figure 8.3).

On the other hand, oral administration of Juzen-taiho-to for 7 days after inoculation of QRsP progressive cells resulted in a significant reduction of the tumor growth and enhancement of the survival rate in tumor-bearing mice compared with the control (data not shown, Ohnishi et al., 1996). Because Juzen-taiho-to has been reported to possess antitumor effects based on the activation of macrophages (Maruyama et al., 1988), cytokine induction (Haranaka et al., 1985; Kubota et al., 1992), augmentation of NK cell activity (Takahashi and Nakazawa, 1995), etc., the inhibitory mechanisms of the progressive growth of QR-32 regressor cells and the growth of the resultant QRsP progressor cells may be in part associated with induction of host-mediated immune surveillance by Juzen-taiho-to.

However, oral administration of bismuth subnitrate, which induces metallothionein as a scavenger of oxygen radicals in the tumor tissue (Satoh et al., 1989), resulted in a significant inhibition of gelatin sponge-elicited progressive growth (data not shown, Ohnishi et al., 1996). These results suggest that Juzen-taiho-to may act to induce antioxidants and to reduce PGE_2 production (Okada et al., 1990) during tumor progression, in addition to augmenting the host-mediated immune responses.

8.2.3 Inhibition of Tumor Growth and Metastasis and the Inhibitory Mechanism

The antitumor activity of Kampo medicines including Juzen-taiho-to has been investigated using various tumor models (Table 8.1). When Juzen-taiho-to (150 or 300 mg/kg) was administered orally twice a day for 10 days after i.p. inoculation of Ehrlich ascites tumor in ICR mice, it significantly suppressed the tumor growth and also prolonged the survival time of tumor-bearing mice (Itoh and Shimura, 1985a). Sho-saiko-to, Chorei-to (Zhu-Ling-Tang in Chinese), and Juzen-taiho-to showed antitumor activities against Ehrlich ascites tumor. However, Juzen-taiho-to was not effective or was only slightly effective in inhibiting the growth of P388 leukemia, Lewis lung carcinoma (LLC), and sarcoma-180 ascites tumor (Aburada et al., 1983; Itoh, 1989; Komiyama et al., 1988).

Oral administration of Juzen-taiho-to or Sho-saiko-to resulted in a significant increase in relative organ weights of the spleen, thymus, and liver and also enhancement of reticuloendothelial cell function such as the phagocytic index, compared with the control (Itoh and Shimura, 1985b). The lung metastasis of LLC was inhibited by i.v. injection of peritoneal macrophages activated with Juzen-taiho-to (Itoh, 1989). Thus, the mechanism of antitumor activity of Juzen-taiho-to may be partly due to the stimulation of the reticuloendothelial system, C3 activation, and depression of the liver microsomal drug-metabolizing enzymatic system.

Oral administration of Juzen-taiho-to significantly suppressed the growth of human U-87MG glioma cells transplanted s.c. into BALB/c nude mice and prolonged the survival time of the mice (Takahashi and Nakazawa, 1995). Endogenous TNF production was augmented by administration of Juzen-taiho-to without a secondary stimulus. Similar results were also obtained by using murine 203-glioma cells in syngeneic mice. The antitumor mechanism of Juzen-taiho-to appears in part to involve the ability to induce endogenous TNF production.

We also examined the effect of oral administration of Juzen-taiho-to on liver metastasis resulting from intraportal vein injection of colon 26-L5 carcinoma cells *in vivo* (Ohnishi et al., 1997) and the role of the immune system after the administration. Oral administration of Juzen-taiho-to before tumor inoculation resulted in the dose-dependent inhibition of liver metastasis of colon 26-L5 carcinoma cells and a significant enhancement of survival rate compared to the untreated control (Ohnishi et al., 1998a) (Figure 8.4 through Figure 8.6). *cis*-Diamminedichloroplatinum II (CDDP) significantly inhibited liver metastasis at 80 µg/mouse (Sugiyama et al., 1995a, b), but it produced severe adverse effects

TABLE 8.1
Antitumor and Antimetastatic Effects of Juzen-taiho-to in Combination with or without Other Treatment Modalities

Tumor	Combined with	Efficacy	Ref.
Ehrlich ascites tumor	—	Growth ↓/survival↑	Ito and Shimura, 1985a, b; Sakagami et al., 1988
	+ OK432	Effective	Haranaka et al., 1985, 1988, 1995
	+ x-Irradiation (20 Gy)	Survival↑	Hosokawa, 1993
B-1F Leydig cell tumor	—	Growth↓/survival↑	Nishizawa et al., 1995
203 or U-87MG glioma	—	Growth↓/survival↑	Takahashi and Nakazawa, 1995
203 Gl glioma	+ ACNU	Growth↓	Nakamura et al., 1994
P388 leukemia	—	—	Aburada et al., 1983
	+ Mitomycin C	Survival↑	Sugiyama, 1996
	+ CDDP	Survival↑	Sugiyama, 1996
LLC carcinoma	—	—	Itoh, 1989
	+ Cyclophosphamide	Metastasis↓	Itoh, 1989
EL4 lymphoma	+ Cyclophosphamide	Survival↑	Yamaguchi et al., 1991
		Liver metastasis↓	Yamaguchi et al., 1991
Sarcoma-180	+ Mitomycin C	Growth↓	Komiyama et al., 1988
Meth A fibrosarcoma	+ Cytoxan	Growth↓	Komiyama et al., 1988
B16 melanoma	+ Adriamycin	Growth↓	Komiyama et al., 1988
	+ 5-FU	Growth↓	Komiyama et al., 1988
Sarcoma-180 or B16 melanoma	+ Mitomycin C/hyperthermia	Growth↓	Komiyama et al., 1989
Sarcoma-180	+ CDDP	Growth↓	Sugiyama, 1996
BMT-2 bladder tumor	+ CDDP	Growth↓/survival↑	Ebisuno et al., 1989
Meth A fibrosarcoma	+ CDDP	Growth↓	Ikekawa et al., 1991
	+ Surgery	Growth↓	Maruyama et al., 1993
QR-32 fibrosarcoma (Regressor tumor)	—	Malignant progression↓	Ohnishi et al., 1996
		Survival↑	Ohnishi et al., 1996
Colon 26-L5 carcinoma	—	Liver metastasis↓	Ohnishi et al., 1998a
		Survival↑	Saiki et al., 1999
		Cachexia↓	Sairenji et al., 1992
B16-BL6 melanoma	—	Lung metastasis↓	Ohnishi et al., 1998b

Notes: ↓: inhibition of tumor growth or metastasis; ↑: enhancement of tumor growth, metastasis, and survival.

such as decrease of body weight and a 50% death rate of the mice. Juzen-taiho-to did not produce any side effects, nor did it directly affect the tumor cells *in vitro*. Thus, Juzen-taiho-to may be a biological response modifier that inhibits micrometastasis and differs from chemotherapeutic agents. Similarly, oral administration of Juzen-taiho-to before i.v. inoculation of B16-BL6 melanoma cells resulted in marked inhibition of lung metastasis without causing any loss of body weight (data not shown, Ohnishi et al., 1998b).

Because metastasizing tumor cells interact with host cells such as lymphocytes, NK cells, and monocytes, which are important in the destruction of tumor cells (Fidler, 1985; Hanna, 1982), we investigated whether Juzen-taiho-to could stimulate immune cells to inhibit tumor metastasis. Antiasialo GM_1 serum can selectively eliminate NK cells (Habu et al., 1981; Saiki et al., 1989) and 2-chloroadenosine can eliminate macrophages (Saiki et al., 1989; Saito and Yamaguchi, 1981). Figure 8.7 shows that liver metastasis was enhanced in mice pretreated with antiasialo GM1 serum or 2-chloroadenosine, and in T-cell-deficient nude mice, compared to untreated

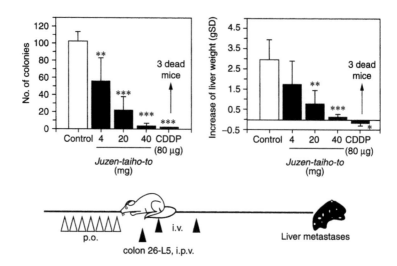

FIGURE 8.4 Effect of oral administration of Juzen-taiho-to on experimental liver metastasis produced by the intraportal vein injection of colon 26-L5 carcinoma cells. Five BALB/c mice per group were inoculated intraportally with colon 26-L5 cells (2×10^4). Juzen-taiho-to at the indicated doses was orally administered for 7 days before tumor inoculation. CDDP was injected intravenously on days 1 and 8 after tumor inoculation. Mice were sacrificed 19 days after tumor inoculation and the number of liver colonies and liver weight were manually counted. *, $p < 0.05$; **, $p < 0.01$; ***, $p < 0.001$ compared to untreated controls by Student's two-tailed t-test. △△△△△△△: Juzen-taiho-to (p.o.) ▲▲: CODP (iv)

normal mice. This indicates that NK cells, macrophages, and T-cells play important roles in the prevention of the metastatic spread of tumor cells. Juzen-taiho-to significantly inhibited the experimental liver metastasis of colon 26-L5 cells in mice pretreated with antiasialo GM1 serum as well as untreated normal mice, buts it did not inhibit the metastasis in 2-chloroadenosine-pretreated mice (Figure 8.7) and T-cell-deficient nude mice.

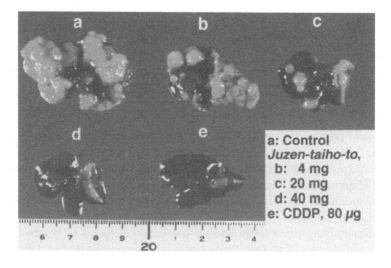

FIGURE 8.5 Macroscopic observation of liver metastasis by colon 26-L5 cells 19 days after tumor inoculation. Five BALB/c mice per group were injected intraportally with colon 26-L5 cells (2×10^4) with or without Juzen-taiho-to or CDDP. Mice were sacrificed 19 days after tumor inoculation. a, Control; b, c, or d, 4, 20, or 40 mg Juzen-taiho-to, respectively; e, 80 μg CDDP.

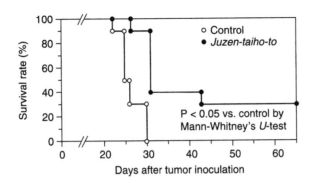

FIGURE 8.6 Effect of Juzen-taiho-to on the survival of mice after the inoculation of colon 26-L5 cells into the portal vein. Ten BALB/c mice per group were orally administered with (●) or without (○) Juzen-taiho-to (40 mg/day) for 7 days before the intraportal inoculation of colon 26-L5 cells (10^4). Survival was monitored as a function of time. $p < 0.05$, Juzen-taiho-to vs. untreated control by the Mann–Whitney U-test.

Juzen-taiho-to was inactive when the contributions of macrophages and T-cells were eliminated from our system, so its inhibitory mechanism is likely to be related to the activation of these cells. We also found that the oral administration of Juzen-taiho-to caused peritoneal exudate macrophages (PEMs) to become cytostatic against tumor cells *in vitro*. Although the exact mechanism responsible for the inhibition of liver metastasis by Juzen-taiho-to is not fully understood, this inhibitory effect is partly associated with the activation of macrophages (Ohnishi et al., 1998a). Further investigation will be needed to determine the detailed mechanisms responsible for the inhibition of tumor metastasis by Juzen-taiho-to. However, it may be difficult to evaluate the results of the *in vitro* study using this formulation, the so-called "Furikake" (in Japanese) study, because the expression of *in vivo* efficacy by the formulation is sometimes mediated by the active (intestinal bacterial) metabolites after the oral administration.

In conclusion, oral administration of Juzen-taiho-to inhibited liver metastasis of colon 26-L5 carcinoma cells and enhanced the survival rate, possibly through the activation of macrophages and T-cells (Ohnishi et al., 1998a). Thus, Juzen-taiho-to may be therapeutically effective for the prevention of cancer metastasis.

8.2.4 COMBINATION WITH OTHER TREATMENT MODALITIES (CHEMOTHERAPY, HYPERTHERMIA, RADIATION, ETC.)

In addition to identifying new types of anticancer agents, it is also important to develop methods or agents that can augment the therapeutic efficacy and/or reduce the side effects of other anticancer agents. Combination therapy with Juzen-taiho-to and other treatment modalities including anti-cancer drugs has also been investigated using various tumor models to examine whether Juzen-taiho-to could potentiate the efficacy. The results are summarized in Table 8.1.

Daily administration of Juzen-taiho-to alone after subcutaneous inoculation of Meth A fibro-sarcoma, sarcoma-180, or B16 melanoma cells resulted in almost no inhibition of tumor growth (Komiyama et al., 1988). In contrast, combination treatment with Juzen-taiho-to (p.o.) and mito-mycin C (i.p.) resulted in more effective inhibition of the growth of Meth A fibrosarcoma or B16 melanoma cells than treatment with mitomycin C or Juzen-taiho-to alone (Komiyama et al., 1988). Treatment with the combination of Juzen-taiho-to and mitomycin C also resulted in a significant enhancement of the survival rate of mice inoculated with P388 leukemia cells and markedly reduced the side effects caused by mitomycin C (Aburada, et al., 1983).

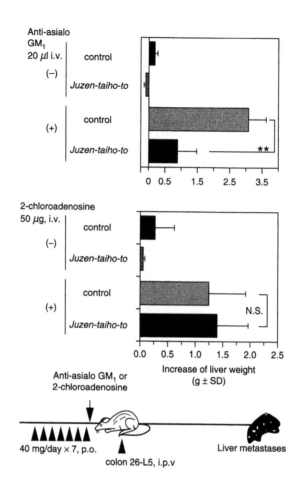

FIGURE 8.7 Effect of antiasialo GM1 serum or 2-chloroadenosine on Juzen-taiho-to-mediated inhibition of experimental liver metastasis produced by the intraportal vein injection of colon 26-L5 cells. Five BALB/c mice per group were orally administered with or without Juzen-taiho-to (40 mg/day) for 7 days before tumor inoculation. Colon 26-L5 cells (10^4) were intraportally injected into groups of control mice or mice pretreated 24 h earlier with antiasialo GM1 serum (20 μl/mouse) or 2-chloroadenosine (50 μg/mouse). Mice were sacrificed 14 days after tumor inoculation and the number of tumor colonies in the liver and liver weight were manually counted. *, $p < 0.05$; **, $P < 0.001$; N.S., not significant compared with an untreated control by Student's two-tailed t-test.

Juzen-taiho-to was given orally (days 1 to 50) to ICR mice inoculated s.c. with sarcoma-180. Mice were administered mitomycin C on various days and treated with hyperthermia (43°C, 30 min). Juzen-taiho-to markedly potentiated the antitumor effect of the combination of mitomycin C and hyperthermia (Komiyama et al., 1989). Similar results using this treatment modality were obtained after the inoculation of B16 melanoma cells into BDF1 mice. In addition to the augmentation of the combined effect of mitomycin C and hyperthermia by Juzen-taiho-to, the immunotoxicity induced by mitomycin C and the subsequent marked growth of the tumors were reduced by the administration of Juzen-taiho-to (Komiyama et al., 1989).

In the MBT-2 bladder tumor-C3H/He mouse model, treatment with a combination of Juzen-taiho-to and cis-diamine dichloroplatinum (CDDP) inhibited tumor growth and prolonged the survival rate of mice more effectively than CDDP alone (Ebisuno et al., 1989). In addition, oral administration of Juzen-taiho-to for 2 weeks significantly ameliorated the adverse effects caused by treatment with a high dose of CDDP, including lethal toxicity, renal and hepatic toxicity, and myelosuppression (Ebisuno et al., 1989).

Administration of Juzen-taiho-to alone exerted no inhibitory effect on the growth of intradermally inoculated Meth A tumor cells. However, Juzen-taiho-to displayed a marked antitumor effect when combined with surgical excision (Maruyama et al., 1993). When oral administration of Juzen-taiho-to (0.5 g/kg/day) was begun on postexcision day 1, the growth of a secondary tumor inoculated on postexcision day 7 was greatly inhibited. Because the antitumor immunity of spleen cells from mice treated with Juzen-taiho-to was abolished by treatment of the spleen cells with anti-L3T4 monoclonal antibody + complement but not with anti-Lyt-2 monoclonal antibody + complement, the effect is mediated by L3T4-positive helper T-cells.

Combination therapy with Kampo preparations (Sho-saiko-to, Juzen-taiho-to, or Cinnamomum cortex) and *Streptococcus pyogenes* products (OK432) strongly inhibited the growth of Ehrlich ascites or Meth A tumor cells through the increased production of endogenous tumor necrosis factor (TNF) (Haranaka et al., 1985, 1988). A significant negative correlation was observed between the TNF activity and tumor weight, and a positive correlation was found between the TNF activity and spleen weight in the Ehrlich ascites tumor-bearing ddY mice receiving Kampo preparations or OK432. Marked lymphocytosis, hyperplasia, and hypertrophy of Kupffer cells in the liver were noted in the tumor-bearing ddY mice receiving the Kampo preparations or OK432.

These results suggest that the antitumor activities and capacity to induce TNF production of the preparations are probably due in part to stimulation of the reticuloendothelial system, including macrophages, and induction of host-mediated antitumor substance(s) like TNF as immunopotentiator(s). Also, antitumor activity of Juzen-taiho-to and OK432 was observed in one patient with hepatocellular carcinoma (Haranaka et al., 1995).

8.2.5 PROTECTION AGAINST DELETERIOUS EFFECTS OF ANTICANCER DRUGS AND RADIATION-INDUCED IMMUNOSUPPRESSION

Consumption of a diet containing 1 or 0.5% Juzen-taiho-to for 2 weeks resulted in significant protection against the adverse effects caused by treatment with a high dose of CDDP, including lethal toxicity, renal and hepatic toxicity, and myelosuppression (Ebisuno et al., 1989). Oral administration of Juzen-taiho-to increased the lowered immune response to the normal level in MBT-2 tumor-bearing mice (Ebisuno et al., 1990). Juzen-taiho-to also protected aged mice (13 to 15 months old) from CDDP-induced damage to the immune function and restored the lowered cytotoxic activity in tumor-bearing mice treated with CDDP.

Sugiyama et al. (1995a) showed that the oral administration of Juzen-taiho-to for 12 days prevented increases in blood urea nitrogen, serum creatinine, serum glutamic-oxaloacetic transaminase, serum glutamic-pyruvic transaminases, and relative stomach weight, as well as decreases in white blood cell count, platelet count, relative spleen and thymus weights, food intake, and body weight caused by CDDP to nearly the control levels without reducing the antitumor activity of CDDP against S-180. Also, preventive effects of Juzen-taiho-to were observed against carboplatin-induced myelosuppression and hepatic toxicity (Sugiyama et al., 1995b).

Iijima et al. (1988) reported that treatment with Juzen-taiho-to for 7 days delayed deaths due to lethal doses of MMC or CDDP and markedly improved the survival curves. Also, Juzen-taiho-to reduced the atrophy of the testis, thymus, and spleen caused by MMC. It also had protective effects against leukopenia, anemia, and body weight loss caused by MMC and against the increases of BUN and creatinine caused by CDDP. These results indicate that combination with Juzen-taiho-to may be a new way to minimize the toxicity of MMC or CDDP.

Kawamura et al. (1989) reported that intraperitoneal injection of MMC resulted in a marked reduction of the numbers of colony-forming units in spleen (CFU-S) and granulocyte-macrophage colony-forming cells (CFU-GM). Administration of Juzen-taiho-to before MMC injection did not protect the mice from damage to hematopoietic function caused by MMC, but remarkably accelerated the recovery of CFU-S and CFU-GM. Juzen-taiho-to was effective when its administration was begun after MMC injection. These results suggest that Juzen-taiho-to has the ability to accelerate hematopoietic recovery from bone marrow injury by MMC.

Continuous oral administration of Juzen-taiho-to 2 to 3 weeks before a dose of x-irradiation causing bone marrow death increased the 30-day survival ratios of x-irradiated mice (Hosokawa, 1993). The administration enhanced the recovery of blood cell counts, especially those of thrombocytes as well as blood forming stem cells (CFUs) in bone marrow. Oral administration of Juzen-taiho-to 7 days after x-irradiation (20 Gy) to the i.m. inoculated Ehrlich tumors significantly prolonged the survival rate of x-irradiated mice. Administration of Juzen-taiho-to for 7 days after irradiation showed radioprotective effects by increasing the number and size of day-14 spleen colony-forming units (CFU-S) (Ohnishi et al., 1990).

Sairenji et al. (1992) investigated the effect of Juzen-taiho-to on the survival of cachectic mice inoculated with colon 26 adenocarcinoma cells. Juzen-taiho-to prevented a decrease in body weight of tumor-bearing mice. Consequently, it prolonged the survival rate of mice bearing colon 26, suggesting that Juzen-taiho-to has the ability to ameliorate the cachexia induced by transplantable colon 26 carcinoma.

8.3 ATTEMPTS TO OBTAIN JUZEN-TAIHO-TO PREPARATIONS WITH CONSTANT EFFICACY

Herbal prescriptions including Kampo medicines have been recognized by the scientific medical system and have become increasingly popular. Because Kampo formulations are generally prepared from the combination of many crude drugs, they may have effects that differ from the sum of the effects of the individual constituent crude drugs. They must have an acceptable efficacy and quality when used as therapeutic medicines. Formulations prepared from crude drugs with different qualities would have different biological activities and efficacies.

Therefore, it is necessary to control the quality of the formulations and their component crude drugs and their processing procedures to obtain reproducibility of the formulation and efficacy because their quality varies with the origins of crude drugs and the time and place of harvest, etc. In Japan, the quality of crude drugs is controlled by *Japanese Pharmacopeia XIII*, which regulates the botanical origin, crude drug test of foreign matter, loss by drying, total ash, acid-insoluble ash, extract content, essential oil content, and microscopic examination. However, to our knowledge, these matters have not been studied in detail in the case of Juzen-taiho-to, although a lot of information about Kampo formulations and their constituents is available in Japanese and Chinese archaic writings.

8.3.1 HPLC Profiles of Juzen-taiho-to and Its Constituent Crude Drugs

For the purpose of obtaining proper formulations with constant quality and efficacy, we have performed HPLC pattern analysis of Juzen-taiho-to by using its chemically defined components as standard references (Saiki et al., 1999). Figure 8.8 shows the HPLC profiles of Juzen-taiho-to by single monitor (220 nm) and contour plot (190 to 400 nm) using a photodiode array system as a detector. The contour plot of the UV absorbance intensity of the compounds shows all of the compounds that have detectable UV absorbance in the extracts from the formulation. The origin of each peak of Juzen-taiho-to was identified by comparison with the retention time and UV spectrum of each extract of crude drug or the chemically defined standard compounds (A–L in Figure 8.8). For example, the peaks of paeoniflorin from Paeoniae Radix and glycyrrhizin from Glycyrrhizae Radix were detected at the D and L positions of the contour plot in Figure 8.8.

Thus, in addition to testing whether the standard compounds have the pharmacological activity of inhibiting tumor metastasis or not, this HPLC pattern analysis—the so-called *fingerprint* method—could provide a useful means of identifying the crude drugs and preparing batches of

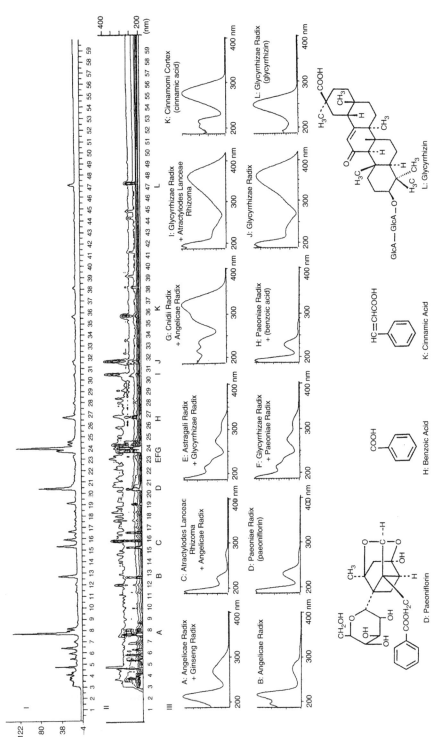

FIGURE 8.8 HPLC profile of Juzen-taiho-to and UV spectra of its constituent crude drugs. I: HPLC pattern analyzed by absorbance at 220 nm; II: contour plot of HPLC pattern by UV absorbance (190 to 400 nm); III: UV spectra of main peaks, origins of peak (A: Angelicae Radix, Ginseng Radix; B: Angelicae Radix; C: Atractylodis Lanceae Rhizoma, Angelicae Radix; D: Paeoniae Radix [paeoniflorin]; E: Astragali Radix, Glycyrrhizae Radix; F: Glycyrrhizae Radix, Paeoniae Radix; G: Cnidii Radix, Angelicae Radix; H: Paeoniae Radix [benzoic acid]; I: Glycyrrhizae Radix, Atractylodes lanceae Rhizoma; J: Glycyrrhizae Radix; K: Cinnamomi Cortex [cinnamic acid]; L: Glycyrrhizae Radix [glycyrrhizin]).

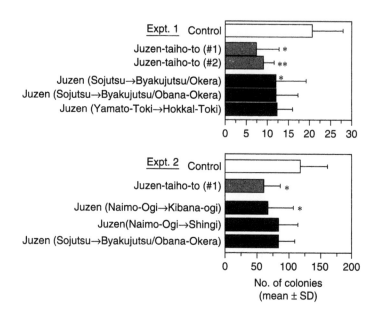

FIGURE 8.9 Effect of oral administration of Juzen-taiho-to and its variant formulations on experimental liver metastasis produced by the intraportal injection of colon 26-L5 carcinoma cells. Five BALB/c mice per group were inoculated intraportally with colon 26-L5 cells (1×10^4 in Experiment 1 or 2×10^4 in Experiment 2). Juzen-taiho-to and its variant formulations (40 mg/mouse) were orally administered for 7 days before tumor inoculation. Mice were sacrificed 19 days after tumor inoculation and the number of liver colonies was manually counted. *, $p < 0.05$; **, $p < 0.01$ compared to untreated controls.

constant formulation. Although the many compounds that have no UV absorbance cannot be detected by this method, a reproducible fingerprint pattern of the formulation may be primarily useful for the assessment of the homogeneity of the formulation, which should lead to constant efficacy.

To evaluate the efficacy of thus prepared Juzen-taiho-to, we examined the antimetastatic effect of oral administration of two Juzen-taiho-to formulations (batches #1 and # 2), which were independently prepared using the same ten crude drugs by the same procedure. The fingerprint analysis of the two batches of Juzen-taiho to showed HPLC profiles similar to that shown in Figure 8.8. Oral administration of the two Juzen-taiho-to preparations (batches #1 and # 2) at the effective dose of 40 mg/day (Ohnishi et al., 1998a, b) significantly reduced the number of colon 26-L5 tumor colonies in the liver. Similar results were observed with both batches of Juzen-taiho-to formulations (Saiki et al., 1999) (Figure 8.9, upper panel).

8.3.2 VARIANT FORMULATIONS AND THEIR ANTIMETASTATIC EFFICACIES

In the usage and preparation of Kampo formulations, some component crude drugs in the formulation are in some cases replaced with related crude drugs. To examine the effect on antimetastatic action when original Juzen-taiho-to constituents were replaced with different crude drugs, we prepared variant formulations of Juzen-taiho-to in which one crude drug was substituted with a related crude drug from different sources or places of origin (Table 8.2). Juzen (Naimo-ogi→ Kibana-ogi) as well as Juzen-taiho-to (#1) significantly inhibited liver metastasis compared with the untreated control (66.7 ± 42.0 and 63.3 ± 28.6 vs. 121.8 ± 55.1 of untreated control, respectively, in lower panel of Figure 8.9). Also, the HPLC pattern of the extract of the root of Astragalus membranaceus (Kibana-ogi) was very similar to that from Astragalus mongholicus (Naimo-ogi) (Figure 8.10).

TABLE 8.2
List of Kampo Formulations and Their Constituent Crude Drugs

Crude Drug (Latin name)	Kampo Medicine (Japanese name)							
	Juzen-taiho-to	Shimotsu-to	Shikunshi-to	Rikkunshi-to	Unsei-in	Ninjin-yoei-to	Toki-shakuyaku-san	Hochu-ekki-to
	+	+	−	−	+	−	−	+
Astragali Radix	△(3)					△(1.5)		△(4)
Cinnamomi Cortex	△(3)					△(2.5)		
Rehmanniae Radix	●(3)	●(3)			●(3)	●(4)		
Paeoniae Radix	●(3)	●(3)			●(3)	●(2)	●(4)	
Cnidii Rhizoma	●(3)	●(3)			●(3)		●(3)	
Angelicae Radix	●(3)	●(3)			●(3)		●(3)	●(3)
Atractylodis Lanceae Rhizoma	■(3)		■(4)	■(4)		■(4)[a]	■(4)	■(4)
Ginseng Radix	■(3)		■(4)	■(4)		■(3)		■(4)
Poria	■(3)		■(4)	■(4)		■(4)	■(4)	
Glycyrrhizae Radix	■(1.5)		■(1)	■(1)		■(1)		■(1.5)
Zingiberis Rhizoma								○(0.5)
Zizyphi Fructus								○(2)
Pinelliae Tuber				○(4)				
Bupleuri Radix								○(2)
Aurantii Nobilis Pericarpium				○(2)		○(2)		○(2)
Polygalae Radix						○(2)		
Schisandrae Fructus						○(1)		
Cimicifugae Rhizoma								○(1)
Scutellariae Radix					○(1.5)			
Phellodendri Cortex					○(1.5)			
Coptidis Rhizoma					○(1.5)			
Gardeniae Fructus					○(1.5)			
Alismatis Rhizoma								

[a] Atractylodis Rhizoma (Byakujutsu) was used in place of Atractylodis Lanceae Rhizoma (Sojutsu). Numbers in parentheses represent the ratios of crude drugs in formulation preparations.

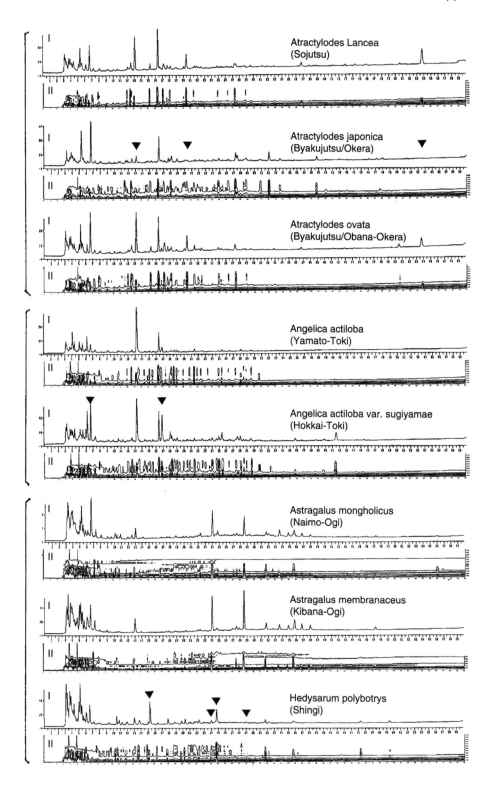

FIGURE 8.10 HPLC profile and UV spectra of substitutable crude drugs of Juzen-taiho-to. I: HPLC pattern analyzed by absorbance at 220 nm; II: contour plot of HPLC pattern by UV absorbance (190 to 400 nm). Some significant peaks were indicated by arrowheads on the chromatograms, compared with original standards.

In contrast, Juzen (Sojutsu→Byakujutsu/okera), Juzen (Sojutsu→Byakujutsu/Obana-okera), Juzen (Yamato-toki→Hokkai-toki), and Juzen (Naimo-ogi→Shingi) had less antimetastatic effect than the original Juzen-taiho-to (#1). Among these four variant formulations with reduced effects, some differences in the HPLC fingerprint patterns were discernible between the original and substituted crude drugs, except in the case of Atractylodis Rhizoma [Byakujutsu/Obana-okera], i.e., the rhizome of *Atractylodes ovata* [Obana-okera]. For example, some additional peaks were seen by arrowheads on the chromatograms compared with those of the original standards (Figure 8.10). This suggests that the reduced effects of the variant formulations may be associated with the marked differences indicated in the fingerprint patterns of the substituted crude drugs.

However, Juzen (Sojutsu→ Byakujutsu/Obana-okera) was less effective in inhibiting the metastasis than the original Juzen-taiho-to (84.1 ± 27.1 and 63.3 ± 28.6, respectively, in lower panel of Figure 8.9), despite the similarity of the HPLC patterns of Sojutsu and Byakujutsu/Obana-okera (Figure 8.10). Therefore, other components in Byakujutsu/Obana-okera that cannot be detected by HPLC analysis, such as polysaccharides or peptides, may be responsible for the reduced efficacy of Juzen (Sojutsu→Byakujutsu/Obana-okera).

In conclusion, we demonstrated differential effects of variant formulations of Juzen-taiho-to on tumor metastasis. The reduced antimetastatic effect of the variant formulations used in this study may be related to the differences in the fingerprint patterns between the original and substituted crude drugs. Thus, HPLC pattern analysis of Kampo medicines may provide a useful method for obtaining their optimal efficacy as well as constant quality of the formulation (Saiki et al., 1999), although such analysis has problems and limitations.

8.4 ANTIMETASTATIC EFFECTS OF FORMULATIONS RELATED TO JUZEN-TAIHO-TO

Juzen-taiho-to is composed of eight crude drugs derived from two prescriptions, Shimotsu-to (Si-Wu-Tang in Chinese) and Shikunshi-to (Si-Jun-Zi-Tang in Chinese), and two other crude drugs (Astragli Radix and Cinnamomi Cortex) as shown in Table 8.2. We therefore examined the effect of Shimotsu-to and Shikunshi-to as well as Juzen-taiho-to on liver metastasis by colon 26-L5 cells in order to clarify which constituents in Juzen-taiho-to are responsible for the antimetastatic effect. Oral administration of Shimotsu-to or Juzen-taiho-to before the tumor inoculation significantly inhibited the number of metastatic colonies and the increase in liver weight compared with the control (Figure 8.11). In contrast, Shikunshi-to did not inhibit liver metastasis effectively. Shimotsu-to is composed of four crude drugs (Rehmanniae Radix, Paconiae Radix, Cnidii Rhizoma, and Angelicae Radix) possessing the ability to improve a deficiency of Ketsu, a concept referring to blood, hormones, the autonomic nervous system, and other regulatory functions of the body's internal environment.

On the other hand, Shikunshi-to is composed of four crude drugs (Atractylodis Lanceae Rhizoma, Ginseng Radix, Poria, and Glycyrrhizae Radix) and used for improving a depression of Ki, a concept that encompasses mental nervous function, especially the appetite for food and the actual process of digesting and absorbing nutrients. These findings suggest that the four constituents in Shimotsu-to (a Ketsu deficiency-improving formulation), rather than the constituents in Shikunshi-to (a Ki depression-improving formulation), are mainly responsible for the antimetastatic effect induced by Juzen-taiho-to (Ohnishi et al., 1998b).

Unsei-in (Wen-Qing-Yin in Chinese), which contains the four Shimotsu-to-derived constituents, was also effective at inhibiting liver metastasis (Table 8.2). Toki-shakuyaku-san (Dang-Gui-Shao-Yao-San in Chinese) and Ninjin-yoei-to (Ren-Shen-Yang-Rong-Tang in Chinese), which include the Shimotsu-to constituents lacking Rehmanniae Radix and Cnidii Rhizoma, respectively, did not show a significant antimetastatic effect. Interestingly, although Ninjin-yoei-to contains nine of the crude drugs (all except Cnidii Rhizoma) in Juzen-taiho-to, its inhibitory effect was much weaker

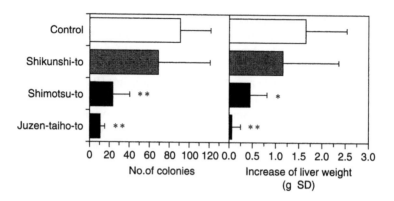

FIGURE 8.11 Effect of oral administration of Juzen-taiho-to, Shimotsu-to, and Shikunshi-to on liver metastasis produced by intraportal injection of colon 26-L5 carcinoma cells. Five BALB/c mice per group were orally administered Juzen-taiho-to, Shimotsu-to, or Shikunshi-to at the dose of 40 mg/day/mouse for 7 days before intraportal vein injection of colon 26-L5 carcinoma cells (2×10^4). Mice were sacrificed 16 days after tumor inoculation and the livers removed. The number of tumor colonies in the livers and liver weights were measured manually. *, $p < 0.01$; **, $p < 0.001$ compared with control.

than that of Juzen-taiho-to. In addition, Rikkunshi-to (Riu-Jun-Zi-Tang in Chinese) and Ninjin-yoei-to, which contain Shikunshi-to constituents, did not affect the inhibition of liver metastasis, as was true of Shikunshi-to. However, because Juzen-taiho-to was more effective at inhibiting tumor metastasis than Shimotsu-to, it would seem likely that some constituents other than those of Shimotsu-to are associated with augmentation of the antimetastatic effect. Thus, Juzen-taiho-to has been shown to be an efficacious formulation with beneficial effects on Ki and Ketsu deficiency.

Hochu-ekki-to (Bu-Zhong-Yi-Qi-Tang in Chinese) as well as Juzen-taiho-to and Ninjin-yoei-to are known to be Hozai (nourishing agents), with an ability to modulate host-mediated immune responses. Hochu-ekki-to also exhibited a significant inhibition of liver metastasis by colon 26-L5 cells, similarly to Juzen-taiho-to (Figure 8.12). The constituents in Hochu-ekki-to are apparently different from those in Juzen-taiho-to (Table 8.2) and some constituents have the ability to improve Ki deficiency. Cho et al. (1991) reported that Hochu-ekki-to was able to stimulate NK cells and thereby to enhance the inhibition of tumor growth. The growth of tumors was clearly inhibited in syngeneic animals given Hochu-ekki-to prophylactically. The splenic NK cells of Hochu-ekki-to-treated WKA rats showed enhanced cytotoxicity against tumors, including NK-sensitive YAC-1 targets. Therefore, the mechanism responsible for the inhibition of liver metastasis by Hochu-ekki-to may be different from that of Juzen-taiho-to and Shimotsu-to.

Our previous studies have shown that Juzen-taiho-to was inactive at inhibiting liver metastasis of colon 26-L5 cells when the contributions of macrophages and T-cells were removed from the model system (Figure 8.7, Ohnishi et al., 1998a). In contrast, Hochu-ekki-to significantly inhibited the liver metastasis of colon 26-L5 cells in mice pretreated with 2-chloroadenosine as well as untreated normal mice, but did not inhibit the metastasis in antiasialo GM1 serum-pretreated mice (Figure 8.12). These results clearly indicate that the antimetastatic mechanism of Hochu-ekki-to is different from that of Juzen-taiho-to and mainly involves the activation of NK cells (Figure 8.13).

In conclusion, the antimetastatic effect induced by Juzen-taiho-to is primarily associated with its Shimotsu-to-derived constituents (Rehmanniae Radix, Paeoniae Radix, Cnidii Rhizoma, and Angelicae Radix). This conclusion is also supported by the evidence that some formulations containing Shimotsu-to constituents are effective inhibitors of tumor metastasis. Because antimetastatic mechanisms by Juzen-taiho-to and Hochu-ekki-to are primarily mediated by the stimulation of macrophages and NK cells, respectively (Figure 8.13), the contributions of these

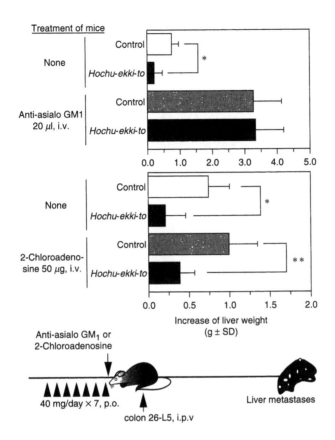

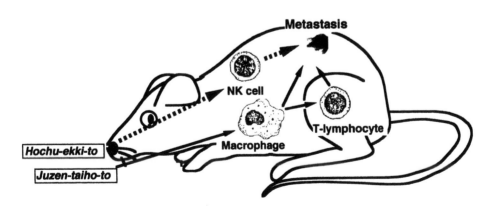

FIGURE 8.12 Effect of antiasialo GM1 serum or 2-chloroadenosine on Hochu-ekki-to-mediated inhibition of experimental liver metastasis produced by the intraportal injection of colon 26-L5 cells. Five BALB/c mice per group were orally administered with or without Hochu-ekki-to (40 mg/day) for 7 days before tumor inoculation. Colon 26-L5 cells (10^4) were intraportally injected into groups of control mice or mice pretreated 24 h earlier with antiasialo GM1 serum (20 µl/mouse) or 2-chloroadenosine (50 µg/mouse). Mice were sacrificed 14 days after tumor inoculation and the number of tumor colonies in the liver and liver weight were manually counted. *, $p < 0.05$; **, $p < 0.001$; N.S., not significant compared with an untreated control by Student's two-tailed t-test.

FIGURE 8.13 Proposed mechanism for the inhibition of tumor metastasis by Juzen-taiho-to and Hochu-ekki-to.

cell populations in antimetastatic actions are partly associated with the improvement of dysfunctions of Ki and Ketsu—dysfunctions representing disturbances in the homeostatic condition of the body.

8.5 ORGAN SELECTIVITY OF JUZEN-TAIHO-TO AND NINJIN-YOEI-TO IN THE EXPRESSION OF ANTIMETASTATIC EFFICACY

The preceding results indicated that Kampo formulations containing four Shimotsu-to constituents as well as Juzen-taiho-to were active in inhibiting liver metastasis of colon 26-L5 carcinoma cells and that Ninjin-yoei-to, which does not include all Shimotsu-to constituents, did not show a significant antimetastatic effect (Ohnishi et al., 1998b). Recently, our preliminary study suggests that mediastinal lymph node metastasis produced by orthotopic implantation of murine Lewis lung carcinoma (Doki et al., 1999) was significantly inhibited by oral administration of Ninjin-yoei-to, but not by Juzen-taiho-to (data not shown). Considering our preliminary study, we investigated whether the difference of antimetastatic effects of Juzen-taiho-to and Ninjin-yoei-to is due to the existence of selective organs and tissues for the expression of the efficacy.

As shown in Figure 8.14, Juzen-taiho-to significantly reduced the number of tumor colonies in the liver ($p < 0.001$) by the intraportal vein (i.p.v.) injection of colon 26-L5 carcinoma cells, but Ninjin-yoei-to had no effect. These results are consistent with our previous report (Ohnishi et al., 1998b). To further investigate the relationship between the efficacy of these formulations and

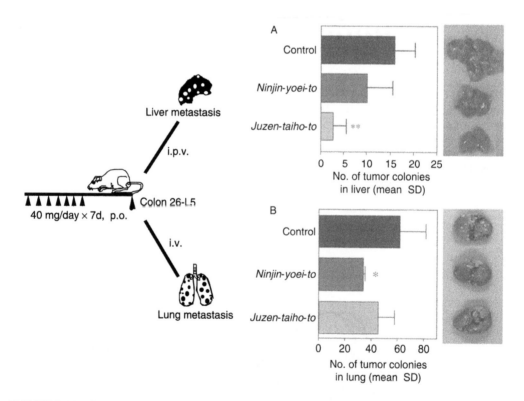

FIGURE 8.14 Effect of oral administration of Juzen-taiho-to and Ninjin-yoei-to on liver or lung metastasis of colon 26-L5 carcinoma cells. Seven BALB/c mice per group were orally administered the indicated Kampo medicines at the dose of 40 mg/day/mouse for 7 days before intraportal vein (A) or intravenous (B) injection of colon 26-L5 cells. The mice were sacrificed 14 days after tumor inoculation. The number of tumor colonies in liver (A) or lung (B) was measured manually. *$p < 0.05$, **$p < 0.001$ compared with control.

specific organs for the expression, we next examined the effect of oral administration of Juzen-taiho-to and Ninjin-yoei-to on lung metastasis produced by intravenous injection of the same colon 26-L5 carcinoma cells into the same syngeneic mice. Although oral administration of Ninjin-yoei-to for 7 days before tumor inoculation was not effective at inhibiting liver metastasis, it significantly inhibited lung metastasis of the same tumor cells compared with the control (Figure 8.14). In contrast, Juzen-taiho-to showed no inhibitory effects against lung metastasis in contrast to its effect on inhibition of liver metastasis. Thus, Juzen-taiho-to and Ninjin-yoei-to clearly exhibited different inhibitory effects for liver and lung metastases of the same colon 26-L5 cells in syngeneic mice.

Although further research is needed to examine the underlying mechanism for organ-selective expression of the antimetastatic effects of the formulations, we have attempted to interpret our findings according to the theory of "Kei-Raku" (Jing and Lun in Chinese), in traditional Chinese medicine. "Kei" (Jing in Chinese) means "route" or "channels" and "Raku" (Lun in Chinese) means "collateral net" or "branch of channels." It is thought that Kei-Raku connect all parts of five viscera and six bowels to regulate their functions and keep them balanced. When dysfunction occurs in some organs, the relevant pathological changes take place. Based on the medicinal guides according to this theory (so-called "Inkei-hoshi" or "Ki-Kei" in Japanese), traditional Chinese medicines are classified for their respective therapeutic effect on the disease of a special Kei-Raku and its pertaining organs.

As illustrated in Figure 8.15, Cnidii Rhizoma (Senkyu) in Juzen-taiho-to and the Shimotsu-to formulation, are known to possess a selective effect for "Kan-Tan-Kei" (Gan-dan-Jing in Chinese). "Kan-Kei" (Gan-Jing in Chinese) is running from a point on the big toe just behind the nail to a point, "Ki-Mon" (Qimen in Chinese), which locates about 8 cm below the nipple on either side. Kan-Kei pertains to organ "Kan" (Gan in Chinese), which is considered to regulate mind and mood, digestion, and absorption function in traditional Chinese medicine, similar to the functions of the liver in Western medicine. The indications of the Kan-Tan-Kei are stuffiness in the chest and pain in the costal regions and in the top of the head. The most commonly used point, "Kei-Ketsu" (Jing-Xue in Chinese), is "Ki-Mon," whose indications are pains in chest and hypochondriac regions.

On the other hand, Aurantii Nobilis Pericarpium (Chimpi), Polygalae Radix (Onji), and Schisandrae Fructus (Gomishi) in Ninjin-yoei-to formulation are supposed to possess a selective effect for "Hai-Kei" (Fei-Jing in Chinese). Indications of Hai-Kei are cough, asthma, tightness in the chest,

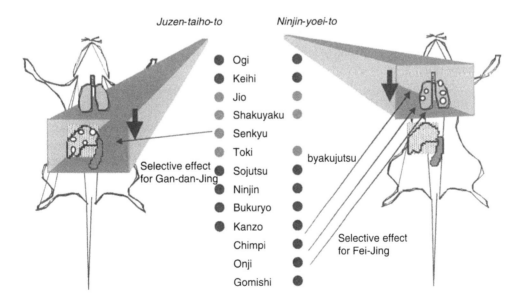

FIGURE 8.15 Possible interpretation for organ-selective effects of Kampo medicines on tumor metastasis according to medicinal guides based on the theory of Jing and Lun.

and a sore throat. Physiological functions of "Hai" (Fei in Chinese) are taking charge of Ki of respiration. Therefore, Hai in traditional Chinese medicine plays the same part as the respiratory system in Western medicine. The theory was advocated by LiGao, a physician in the Jin dynasty, in "Yo-yaku-ho-sho" (Yong-yao-fa-xiang in Chinese).

Although the concept of Kan and Hai in traditional Chinese medicine does not correspond to that of liver and lung in Western medicine, Juzen-taiho-to (containing Cnidii Rhizoma) and Ninjin-yoei-to (containing Aurantii Nobilis Pericarpium, Polygalae Radix, and Schisandrae Fructus without Cnidii Rhizoma) inhibited predominantly tumor metastases to liver and lung, respectively. Thus, it would be of prime interest that our pharmacological data be partly due to the theory of Kei-Raku formed in 13th century. Further study will be needed to examine antimetastatic effects using different types of tumors

8.6 ANTITUMOR EFFECTS OF CONSTITIUENT CRUDE DRUGS IN JUZEN-TAIHO-TO

Because Kampo formulations are generally prepared from the combination of many crude drugs, they may have effects that differ from the sum of the effects of the individual constituent crude drugs. However, some studies using individual crude drugs of Juzen-taiho-to and their components have been carried out in order to investigate the antitumor mechanism of Juzen-taiho-to *in vivo* and to discover active components, although it is doubtful that the results using crude drugs can completely reflect the efficacy of the whole formulation.

8.6.1 RHEMANNIAE RADIX, ATRACTYLODIS LANCEAE RHIZOMA, ANGELICAE RADIX, AND ASTRAGALI RADIX

Administration of Juzen-taiho-to in combination with cyclophosphamide after s.c. inoculation of EL-4 lymphoma cells resulted in a significant decrease of tumor size and the prolongation of the life span in mice (Yamaguchi et al., 1991). Combination of cyclophosphamide with the extracts of the individual composed crude drugs in Juzen-taiho-to, such as Rhemanniae Radix (Jio, in Japanese), Atractylodis Lanceae Rhizoma (Sojutsu), Angelicae Radix (Toki), and Astragali Radix (Ogi), also inhibited liver metastasis of EL-4 lymphoma cells and enhanced the survival rate of mice. Thus, some of the crude drugs in Juzen-taiho-to are effective at inhibiting metastasis indirectly through the augmentation of host immune responses without affecting direct cytotoxicity against tumor cells.

8.6.2 CINNAMOMI CORTEX

Haranaka et al. (1988) have shown that the combination of Juzen-taiho-to with mitomycin C (MMP) or OK432 inhibited the growth of Ehrlich ascites tumor in ddY mice or of Meth A fibrosarcoma in Balb/c mice, and enhanced the production of endogenous TNF. Similar effects by combination therapy were also observed using an extract of Cinnamomi Cortex in place of Juzen-taiho-to formulation.

8.6.3 GINSENG RADIX

Ginseng (the root of *Panax ginseng* C.A. Meyer, Araliaceae), a constituent crude drug of Juzen-taiho-to, has been used for traditional medicine in China, Korea, Japan, and other Asian countries for the treatment of various diseases, including psychiatric and neurologic diseases as well as diabetes mellitus. So far, ginseng saponins (ginsenosides) have been regarded as the principal components responsible for the pharmacological activities of ginseng. Ginsenosides are glycosides containing an aglycone (protopanaxadiol or protopanaxatriol) with a dammarane skeleton and have been shown to possess various biological activities, including the enhancement of cholesterol biosynthesis, stimulation of serum protein synthesis, immunomodulatory effects, and anti-inflammatory activity (Sakakibara et al., 1975; Shibata et al., 1976; Toda et al., 1990; Scaglione et al., 1990; Wu et al., 1992).

Several studies using ginsenosides have also reported antitumor effects, particularly the inhibition of tumor-induced angiogenesis (Sato et al., 1994), tumor invasion and metastasis (Mochizuki et al., 1995; Shinkai et al., 1996), and the control of phenotypic expression and differentiation of tumor cells (Odashima et al., 1985; Ota et al., 1987). Previously, it was reported that protopanaxadiol-type and protopanaxatriol-type ginsenosides are metabolized by intestinal bacteria after oral administration to their final derivative 20-O-β-D-glucopyranosyl-20(S)-protopanaxadiol (referred to as M1 [Hasegawa et al., 1996] or compound K [Kanaoka et al., 1994; Karikuma et al., 1991]) or 20(S)-protopanaxatriol [referred to as M4 (Hasegawa et al., 1996)] (Figure 8.16). This made it unclear whether or not the expression of antimetastatic effect by oral administration of ginsenosides can be induced by their metabolites.

We have recently reported that protopanxadiol- or protopanxatriol-type ginsenosides and their major metabolites M1 and M4 markedly inhibited lung metastasis of B16-BL6 melanoma cells when they were administered five times orally (Wakabayashi et al., 1997a, b). In contrast, three consecutive i.v. administrations of metabolite M4 after tumor inoculation resulted in a significant inhibition of lung metastasis, whereas ginsenosides Re and Rg_1 did not show any inhibitory effect. These findings suggest that the expression of the *in vivo* antimetastatic effect by oral administration of both types of ginsenosides was primarily based on their metabolites M1 and M4.

These results may be also supported by the finding that metabolites were detected in the serum from mice orally given ginsenosides, but ginsenosides were not detected by HPLC analysis. This pharmacokinetic study is in good agreement with previous reports on the low absorption rate of Rb1 from the intestines (Odani et al., 1983; Tanizawa et al., 1993) and high metabolic rate of Rb1 to M1 (Tanizawa et al., 1993) in rat and human by using HPLC and enzyme-immunoassay methods (Hasegawa et al., 1996; Kanaoka et al., 1994). Moreover, it has also been noted that ginsenosides are hardly decomposed by gastric juice with the exception of slight oxygenation (Karikuma et al., 1991). Therefore, our findings support the notion that ginsenosides may act as a natural pro-drug that can be transformed to M1 by intestinal anaerobe(s) after oral administration and consequently induce *in vivo* antimetastatic effect.

To investigate the incidence of intestinal bacteria possessing ginsenoside Rb1-hydrolyzing potential, the hydrolyzing potential of intestinal bacteria, expressed as the transformation rate of Rb1 to metabolite M1, was carried out by using fecal specimens of mice. There were some correlations of the transformation rate of Rb1 to M1 between mother mice and their litters, particularly statistically significant between the groups of litters born from mothers with different rates of hydrolyzing potential (Figure 8.17). This suggests that the intestinal microflora of a litter is primarily infected from the mother. On the other hand, consecutive administration of ginseng extract to the mice with transformation rate from Rb1 to M1 of $25 \pm 11\%$, resulted in a significant increase in the transformation rate, compared with untreated group (Figure 8.18). However, induction of Rb1-hydrolysing potential by the administration of ginseng extract was hardly effective for the mice with hydrolyzing potential of less than 10%. For such mice, the inoculation of fecal microflora from mice with high hydrolyzing potential was also ineffective. Therefore, the location of the bacteria capable of hydrolyzing Rb1 on intestinal epithelium cells may be associated with the genetic factors of hosts.

To examine the influence of Rb1-hydrolysing potential on antimetastatic efficacy of Rb1, Rb1 was orally administered to two sets of mice with low and high hydrolysing potential after s.c. inoculation of LLC tumor. Figure 8.19 shows a significant difference between active and inactive groups and also the tendency of a positive relationship between hydrolysing potential and inhibition of lung metastasis. These findings indicate that the transformation rate of Rb1 to its active metabolite M1 was dependent on the Rb1-hydrolysing potential of intestinal bacteria, which consequently

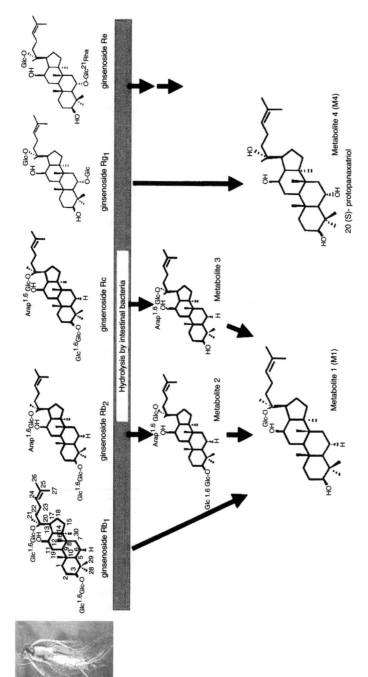

FIGURE 8.16 Chemical structure of ginsenosides and their metabolites. Protopanaxadiol-type and protopanaxatriol-type ginsenisides are metabolized by intestinal bacteria after oral administration to their final derivative 20-O-β-D-glucopyranosyl-20(S)-protopanaxadiol (referred to as M1 or compound K) or 20(S)-protopanaxatriol (referred to as M4). (Kanaoka, M. et al., 1994, *J. Traditional Med.*, 11, 241–245; Karikuma, M. et al., 1991, *Chem. Pharm. Bull.*, 39, 2357–2361; Hasegawa, H. et al., 1996, *Planta Medica*, 62, 453–457.)

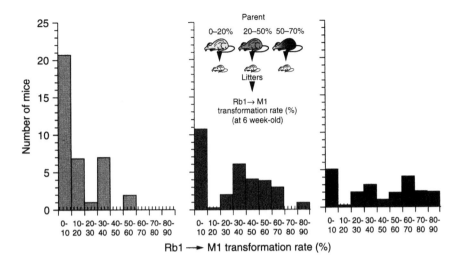

FIGURE 8.17 Relationship between parent and litters in Rb1-hydrolysing potential to M1. Rb1→M1 transformation rate of 90 fecal specimens of litter mice born from mothers with 0 to 20%, 20 to 50% or 50 to 70% of Rb1-hydrolysing potential of intestinal bacteria were determined by the HPLC method.

affected the expression of antimetastatic efficacy of orally administered Rb1. Thus, hydrolysing potential of intestinal bacteria for a crude drug or a formulation may be an important factor influencing the holistic pattern of symptoms and individual pathogenic alterations, so-called "Sho" (Zheng in Chinese), by which the diagnosis of a disease state and the ways of treatment in Kampo are determined.

On the other hand, these ginsenosides hardly inhibited the invasion, migration, and growth of tumor cells *in vitro*, whereas intestinal bacterial metabolites M1 and M4 showed the inhibitory

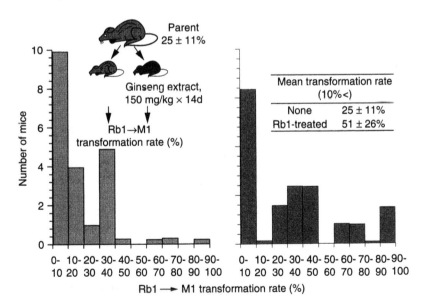

FIGURE 8.18 Effect of oral administration of Rb1 on Rb1→M1 transformation rate. Forty mice born from mothers with 25 ± 11% of Rb1-hydrolysing potential of intestinal bacteria were randomized into two groups at 4 weeks of age. One group was given the water dosed with ginseng extract (150 mg/kg). Rb1→M1 transformation rate was determined 2 weeks later by the HPLC method.

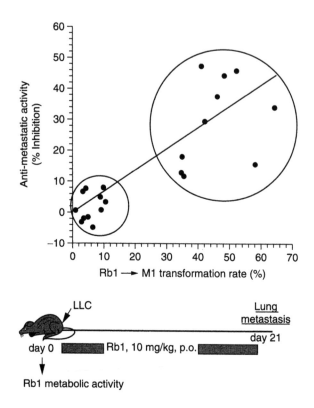

FIGURE 8.19 Influence of Rb1-hydrolysing potential on antimetastatic efficacy of Rb1. Mice with low or high Rb1-hydrolysing potential were given oral administration of Rb1 (25 mg/kg/day) for 2 weeks after s.c. inoculation of LLC cells. Mice were killed 3 weeks after tumor inoculation and tumor colonies in the lungs were examined.

effects dose dependently (Wakabayashi et al., 1997a, b). These findings indicate that M1 produced in the serum after the oral administration of ginsenoside Rb1 may induce the *in vivo* antimetastatic effect partly through the inhibition of tumor invasion, migration, and growth of tumor cells. Even if direct addition of ginsenosides into the culture *in vitro*, which is referred to as "Furikake" (in Japanese) assay, was effective at inhibiting tumor cell invasion and proliferation, these results should not be apparently accepted for the explanation of *in vivo* efficacy by oral administration of ginsenosides.

However, the detail of how M1 affects the growth of tumor cells remains unclear. Co-incubation of tumor cells with M1 at concentrations ranging from 5 to 40 μM resulted in a time- and concentration-dependent inhibition of tumor cell proliferation, with accompanying morphological changes (spindle shape) at the concentration of 20 μM (Wakabayashi et al., 1998). In addition, M1 at a concentration of 40 μM caused the cytotoxic response in tumor cells at an earlier time period (within 24 h) in the culture (Figure 8.20). Because the swelling-shape morphology of tumor cells is considered to be an apoptotic character, the cell death by treatment with 40 μM was due to the induction of apoptosis. The ladder fragmentation of the extracted DNA and swollen-round morphology indicated that the cell death was caused by apoptosis (Figure 8.20). In contrast, the incubation with ginsenoside-Rb$_1$ (40 μM) did not affect the morphology of tumor cells or cell proliferation (Wakabayashi et al., 1997a, 1998).

Although the molecular events that drive the apoptotic signaling pathway are not entirely clear, some apoptosis-related proteins such as cyclin D1, c-Myc, or cyclin-dependent kinase (CDK) inhibitors have been reported to be associated with cell division and proliferation (Fukumoto et al., 1997;

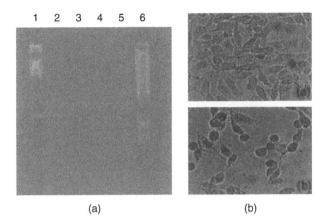

(a)   (b)

FIGURE 8.20 M1-induced DNA fragmentation of B16-BL6 melanoma cells and their morphology. B16-BL6 cells were cultured for 24 h in the presence or absence of 40 μM M1. The DNA was isolated, and electrophoresed on a 2% agarose gel. The fragmented DNA was detected by ethidium bromide staining. 100-bp ladder (lane 1), control (lane 2), and M1: 5 μM (lane 3), 10 μM (lane 4), 20 μM (lane 5), and 40 μM (lane 6), respectively. The apoptotic morphology of the cells untreated (a) or treated with 40 μM M1 (b). Magnification: ×200.

Rogatsky et al., 1997; Vlach et al., 1996). Therefore, in order to clarify the mechanism of M1-induced apoptosis, we investigated the effect of M1 on the expression of the apoptosis-related proteins, p21, p27^{Kip1}, c-Myc, and cyclin D1. M1 treatment (40 μM) markedly increased the expression of p27^{Kip1} compared with the untreated control (Figure 8.21). No expression of the other CDK inhibitor, p21, was detected in B16-BL6 cells in this experiment (data not shown). The up-regulation of p27^{Kip1},

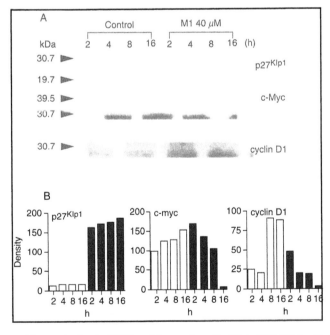

FIGURE 8.21 Western blot analysis of p27^{Kip1}, c-Myc, and cyclin D1 in B16-BL6 cells treated with M1. (A) Result of western blotting. B16-BL6 cells were cultured with or without 40 μM M1 for the time periods indicated. p27^{Kip1}, c-Myc, and cyclin D1 were detected using specific antibodies. (B) The density of the bands at each time point was densitometrically quantified. Control (□); M1 treated (■).

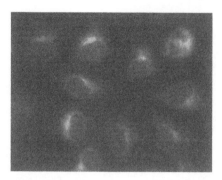

FIGURE 8.22 Fluorescent microscopy of B16-BL6 cells treated with dansyl M1. B16-BL6 cells were treated with dansyl M1 for 15 min. The cells were fixed with methanol and the photograph was taken under fluorescent microscopy. Magnification: ×400.

which is known to inhibit the CDK activity, was observed during the apoptotic process caused by anticancer agents, including etoposide and camptotecin.

On the other hand, a proto-oncogene product c-Myc, as well as cyclin D1, has been reported to be overexpressed in the proliferative phase of various types of tumor cells (Fukumoto et al., 1997; Rogatsky et al., 1997; Vlach et al., 1996). The expression of c-Myc and cyclin D1 was down-regulated by M1 treatment in a time-dependent manner. Thus, M1 might cause the cell-cycle arrest in tumor cells through the up- or down-regulation of these cell growth-related molecules and, consequently, induce apoptosis.

It has been reported that various molecules, such as Bcl-2 (an inhibitor of apoptotic cell death), Bax (promotion of apoptosis by antagonizing the function of Bcl-2), and caspases (interleukin-1β converting enzymes to trigger the execution of cell death), are involved in positively or negatively regulating apoptosis signaling (Story and Kodym, 1998; White, 1996; Vaux and Strasser, 1996). Recent studies have proposed some signaling pathways for apoptosis mediated by different regulatory molecules (Story and Kodym, 1998; Evans et al., 1995). Therefore, the possibility that M1 inhibits or promotes these apoptosis-related molecules will need to be examined.

We also examined the intracellular distribution of M1 after the incubation of tumor cells with dansyl M1. The fluorescent signal of dansyl M1 was detected in the cytosol and nuclei 15 min after incubation and was observed predominantly in the nuclei thereafter (Figure 8.22). These findings suggest that the apoptotic cell death is induced by intracellular M1 through the transcriptional regulation of several cell growth-associated proteins. Because M1 has a steroid-like chemical structure, it may interact with some intracellular receptors, including a steroid receptor, which are known to be involved in the rapid regulation of nuclear proto-oncogene transcription (Schuchard et al., 1993). The regulatory mechanisms of M1 at the transcriptional level will need to be investigated in detail.

REFERENCES

Aburada, M., Takeda, S., Ito, E., Nakamura, M., and Hosoya, E. (1983) Protective effects of Juzentaiho-to, dried decoctum of 10 Chinese herbs mixture upon the adverse effects of mitomycin C in mice. *J. Pharm. Dyn.,* 6, 1000–1004.

Bengmark, S. and Hafstrom, L. (1969) The natural history of primary and secondary malignant tumors of the liver. I. The prognosis for patients with hepatic metastases from colonic and rectal carcinoma verified by laparotomy. *Cancer,* 23, 198–202.

Cho, J.M., Sato, N., and Kikuchi, K., (1991) Prophylactic antitumor effect of Hochu-ekki-to (TJ41) by enhancing natural killer cell activity. *In Vivo,* 5, 389–391.

Doki, Y., Murakami K., Yamaura T., Sugiyama, S., Misaki, T., and Saiki, I. (1999) Mediastinal lymph node metastasis model by orthotopic intrapulmonary implantation of Lewis lung carcinoma cells in mice. *Br. J. Cancer,* 79, 1121–1126.

Ebisuno, S., Hirano, A., Okawa, T., Maruyama H., Kawamura, H., and Hosoya, E. (1990) Immunomodulating effects of Juzen-taiho-to in immunosuppressive mice. *Biotherapy,* 4, 112–116 (in Japanese).

Ebisuno, S., Hirano, A., Kyoku, I., Okawa, T., Iijima, O., Fujii, Y., and Hosoya, E. (1989) Basal studies on combination of Chinese medicine in cancer chemotherapy: preventive effects on the side effects of CDDP and antitumor effects with CDDP on murine bladder tumor (MBT-2). *J. Jpn. Soc. Cancer Ther.,* 24, 1305–1312 (in Japanese).

Eisenberg, B., DeCosse, J.J., Harford, F., and Michalek, J. (1982) Carcinoma of the colon and rectum: the national history reviewed in 1704 patients. *Cancer* (Phila), 49, 1131–1134.

Evans, G.I., Brown, L., Whyte, M., and Harrington, E. (1995) Apoptosis and the cell cycle. *Curr. Opin. Cell Biol.,* 7, 825–834.

Fidler, I.J. (1985) Macrophages and metastasis—a biological approach to cancer therapy (Presidential address). *Cancer Res.,* 45, 4714–4726.

Fidler, I.J. (1984) The evolution of biological heterogeneity in metastatic neoplasms. In *Cancer Invasion and Metastasis: Biologic and Therapeutic Aspects,* Nicolson, G.L. and Milas, L., Eds., 5–26, Raven Press, New York.

Fukumoto, S., Nishizawa, Y., Hosoi, M., Koyama, H., Yamakawa, K., Ohno, S., and Morii, H. (1997) Protein kinase C delta inhibits the proliferation of vascular smooth muscle cells by suppressing G1 cyclin expression. *J. Biol. Chem.,* 272, 13816–13822.

Galandivk, S., Wieand, H.S., Moertel, C.G., Cha, S.S., Filzgibbons, R.J., Jr., Premberton, J.H., and Wolff, B.G. (1992) Patterns of recurrence after curative resection of carcinoma of the colon and rectum. *Surg. Gynecol. Obstet.,* 174, 27–32.

Gastrointestinal Tumor Study Group (1985) Prolongation of the disease-free interval in surgically resected rectal cancer. *New Engl. J. Med.,* 312, 1465–1472.

Gavowski, T.J., Iwatsuki, S., Madariaga, J.R., Selby, R., Todo, S., Irish, W., and Startzl, T.S. (1994) Experience in hepatic resection for metastatic colorectal cancer: analysis of clinical and pathologic risk factors. *Surgery,* 116, 703–711.

Habu, S., Fukui, H., Shimamura, K., Kasai, M., Nagai, Y., Okumura, K., and Tamaoki, N. (1981) *In vivo* effects of anti-asialo GM1. *J. Immunol.,* 127, 34–38.

Hamada, M., Fujii, Y., Yamamoto, H., Miyazawa, Y., Shui, S. M., Tung, Y.C., and Yamaguchi, N. (1988) Effect of a Kampo medicine, Juzen-taiho-to, on the immune reactivity of tumor-bearing mice. *J. Ethnopharma.,* 24, 311–320.

Hanna, N. (1982) Role of natural killer cells and control of cancer metastasis. *Cancer Metast. Rev.,* 1, 45–65.

Haranaka, K., Sakurai, A., Satomi, N., Ono T., and Haranaka, R. (1995) Combination therapy with traditional Chinese medicines and *Streptococcus pyogenes* products (OK432) for endogenous tumor necrosis factor therapy. A preliminary report. *Cancer Biother.,* 10, 131–138.

Haranaka, R., Hasegawa, R., Nakagawa, S., Sakurai, A., Satomi, N., and Haranaka, K. (1988) Antitumor activity of combination therapy with traditional Chinese medicine and OK432 or MMC. *J. Biol. Response Mod.,* 7, 77–90.

Haranaka, K., Satomi, N., Sakurai, A., Haranaka, R., Okada, N., and Kobayashi, M. (1985) Antitumor activities and tumor necrosis factor producibility of traditional Chinese medicines and crude drugs. *Cancer Immunol. Immunother.,* 20, 1–5.

Hart, I.R. (1982) Seed and soil revisited: mechanisms of site specific metastasis. *Cancer Metastasis Rev.,* 1, 5–16.

Hasegawa, H., Sung, J., Matsumiya, S., and Uchiyama, M. (1996) Main ginseng saponin metabolites formed by intestinal bacteria. *Planta Medica,* 62, 453–457.

Hosokawa, Y. (1993). Modification of radioprotective effects by Chinese medicinal prescriptions in mice. *Jpn. J. Cancer Clin.,* 39, 1655–1659 (in Japanese).

Iijima, O., Fujii, Y., Funo, S., Hosoya, E., and Yamashita, M. (1988) Protective effects of Juzen-taiho-to on the adverse effects of mitomycin C. *J. Jpn. Soc. Cancer Ther.,* 23, 1277–1282 (in Japanese).

Ikekawa, T., Feng, W., Maruyama, H., and Kawamura, H. (1991) Synergistic effect of Kampo medicines and cis-dichlorodiammineplatinum (II) on Meth-A fibrosarcoma in BALB/c mice. *J. Med. Pharm. Soc. WAKAN-YAKU,* 8, 89–95.

Itoh, H. (1989) Study on antitumor effect of Japanese Kampo medicine. *Biotherapy,* 3, 760–766 (in Japanese).

Itoh, H. and Shimura, K. (1985a) Studies on antitumor activity of traditional Chinese medicines (I). *Jpn. J. Cancer Chemother.*, 12, 2145–2148 (in Japanese).

Itoh, H. and Shimura, K. (1985b) Studies on antitumor activity of traditional chinese medicines (II). Antitumor mechanism of traditional Chinese medicines. *Jpn. J. Cancer Chemother.*, 12, 2149–2154 (in Japanese).

Kanaoka, M., Akao, T., and Kobashi, K. (1994) Metabolism of ginseng saponins, ginsenosides, by human intestinal flora. *J. Traditional Med.*, 11, 241–245.

Karikuma, M., Miyase, T., Tanizawa, H., Taniyama, T., and Takino, Y. (1991) Studies on absorption, distribution, excretion and metabolism of ginseng saponins. VII. Comparison of the decomposition modes of ginsenoside-Rb1 and -Rb2 in the digestive tract of rats. *Chem. Pharm. Bull.*, 39, 2357–2361.

Kawamura, H., Maruyama, H., Takemoto, N., Komatsu, Y., Aburada, M., Ikehara, S., and Hosoya, E. (1989) Accelerating effect of Japanese Kampo medicine on recovery of murine haematopoietic stem cells after administration of mitomycin C. *Int. J. Immunother.*, 5, 35–42.

Kiyohara, H, Takemoto, N., Komatsu, Y., Kawamura, H., Hosoya, E., and Yamada, H. (1991) Characterization of mitogenic pectic polysaccharides from Kampo (Japanese herbal) medicine "Juzen-Taiho-To." *Planta Med.*, 57, 254–259.

Komiyama, K., Zhibo, Y., and Umezawa, I. (1989) Potentiation of the therapeutic effect of chemotherapy and hyperthermia on experimental tumor and reduction of immunotoxicity of mitomycin C by Juzen-taiho-to, a Chinese herbal medicine. *Jpn. J. Cancer Chemother.*, 16, 251–257 (in Japanese).

Komiyama, K., Hirokawa Y., Zhibo, Y., Umezawa, I., and Hata, T. (1988) Potentiation of chemotherapeutic activity by a Chinese herb medicine Juzen-taiho-to. *Jpn. J. Cancer Chemother.*, 15, 1715–1719 (in Japanese).

Kubota, A., Okamura, S., Shimoda, K., Harada, N., Omori, F., and Niho, Y. (1992) A traditional Chinese herbal medicine, Juzen-taiho-to, augments the production of granulocyte/macrophage colony-stimulating factor from human peripheral blood mononuclear cells *in vitro*. *Int. J. Immunother.*, 8, 191–195.

Liotta, L.A., Rao, C.V., and Barsky, S.H. (1983) Tumor invasion and the extracellular matrix. *Lab. Invest.*, 49, 636–649

Maruyama, H., Takemoto, N., Maruyama, N., Komatsu, Y., and Kawamura, H. (1993) Antitumour effect of Juzentaiho-to, a Kampo medicine, combined with surgical excision transplanted meth-A fibrosarcoma. *Int. J. Immunother.*, 9, 117–125.

Maruyama, H., Kawamura, H., Takemoto, N., Komatsu, Y., Aburada, M., and Hosoya, E. (1988) Effect of Juzentaihoto on phagocytes. *Jpn. J. Inflammation*, 8, 461–465 (in Japanese).

Mochizuki, M., Yoo, Y.C., Matsuzawa, K., Sato, K., Saiki, I., Tono-oka, S., Samukawa, K., and Azuma, I. (1995) Inhibitory effect of tumor metastasis in mice by saponins, ginsenoside-Rb2, 20(R)- and 20(S)-ginsenoside-Rg3, of red ginseng. *Biol. Pharm. Bull.*, 18, 1197–1202.

Nakamura, O., Okamoto, K., Kaneko, K., Nakamura, H., and Shitara, N., (1994) The combined effect of Juzen-taiho-to (TJ-48) and ACNU on mouse glioma, *Biotherapy*, 8, 1003–1006 (in Japanese).

Nicolson, G.L. (1987) Tumor cell instability, diversification, and progression to metastatic phenotype from oncogene to oncofetal expression. *Cancer Res.*, 47, 1473–1487

Nishizawa, Y., Fishiki, S., and Nishizawa, K. (1995) The mechanism on suppressive effect of Juzen-taiho-to on the proliferation of tumor in B-1F bearing mice. *J. Trad. Med.*, 12, 344–345 (in Japanese).

Odani, T., Tanizawa, H., and Takino, Y. (1983) Studies on absorption, distribution, excretion and metabolism of ginseng saponins. III. The absorption, distribution and excretion of ginsenoside Rb1 in the rat. *Chem. Pharm. Bull.*, 31, 1059–1066.

Odashima, S., Ohta, T., Kohno, H., Matsuda, T., Kitagawa, I., Abe, H., and Arichi, S. (1985) Control of phenotypic expression of cultured B16 melanoma cells by plant glycosides. *Cancer Res.*, 45, 2781–2784.

Ohnishi Y., Fujii, H, Hayakawa, Y., Sakukawa, R., Yamaura, T., Sakamoto, T., Tsukada, K., Fujimaki, M., Nunome, S., Komatsu, Y., and Saiki, I. (1998a) Oral administration of a Kampo (Japanese herbal) medicine Juzen-taiho-to inhibits liver metastasis of colon 26-L5 carcinoma cells. *Jpn. J. Cancer Res.*, 89, 206–213.

Ohnishi, Y., Yamaura, T., Tauchi, K., Sakamoto, T., Tsukada, K., Nunome, S., Komatsu, Y., and Saiki, I. (1998b) Expression of the anti-metastatic effect induced by Juzen-taiho-to is based on the content of Shimotsu-to constituents. *Biol. Pharm. Bull.*, 21, 761–765.

Ohnishi, Y., Sakamoto, T., Fujii, H., Kimura, F., Murata, J., Tazawa, K., Fujimaki, M., Sato, Y., Kondo, M., Une, Y., Uchino, J., and Saiki, I. (1997) Characterization of a liver metastatic variant of murine colon 26 carcinoma cells. *Tumor Biol.*, 18, 113–122.

Ohnishi, Y., Fujii, H, Kimura, F., Mishima, T., Murata, J., Tazawa, K., Fujimaki, M., Okada, F., Hosokawa, M., and Saiki, I. (1996) Inhibitory effect of a traditional Chinese medicine, Juzen-taiho-to, on progressive growth of weakly malignant clone cells derived from murine fibrosarcoma. *Jpn. J. Cancer Res.,* 87, 1039–1044.

Ohnishi, Y., Yasuzumi, R., Fan, H., Liu, L., Takao–Liu, F., Komatsu, Y., Hosoya, E., Good, R.A., and Ikehara, S. (1990) Effects of Juzen-taiho-toh (TJ-48), a traditional Oriental medicine, on hematopoietic recovery from radiation injury in mice. *Exp. Hematol.,* 18, 18–22.

Okada, F. and Hosokawa, M. (1995) Tumor progression accelerated by active oxygen species and its chemoprevention. *Int. Conf. Food Factors,* 71, Japan.

Okada, F., Hosokawa, M., Hasegawa, J., Kuramitsu, Y., Nakai, K., Yuan, L., Lao, H., Kobayashi, H., and Takeichi, N. (1994) Enhancement of *in vitro* prostaglandin E_2 production by mouse fibrosarcoma cells after co-culture with various anti-tumor effector cells. *Br. J. Cancer,* 70, 233–238.

Okada, F., Kobayashi, H., Hamada, J., Takeichi, N., and Hosokawa, M. (1993) Active radicals produced by host reactive cells in the malignant progression of murine tumor cells. *Proc. Am. Assoc. Cancer Res.,* 34, 1060.

Okada, F., Hosokawa, M., Hamada, J., Hasegawa, J., Kato, M., Mizutani, M., Ren, J., Takeichi, N., and Kobayashi, H. (1992) Malignant progression of a mouse fibrosarcoma by host cells reactive to a foreign body (gelatin sponge). *Br. J. Cancer,* 66, 635–639.

Okada, F., Hosokawa, M., Hasegawa, J., Ishikawa, M., Chiba, I., Nakamura, Y., and Kobayashi, H. (1990) Regression mechanisms of mouse fibrosarcoma cells after *in vitro* exposure to quercetin: dimunution of tumorigenicity with a corresponding decrease in the production of prostaglandin E_2. *Cancer Immunol. Immunother.,* 31, 358–364.

Ota, T., Fujikawa–yamamoto, K., Zong, Z.P., Yamazaki, M., Odashima, S., Kitagawa, I., Abe, H., and Arichi, S. (1987) Plant-glycoside modulation of cell surface related to control of differentiation in cultured B16 melanoma cells. *Cancer Res.,* 47, 3863–3867.

Rogatsky, I., Trowbridge, J.M., and Garabedian, M.J. (1997) Glucocorticoid receptor-mediated cell cycle arrest is achieved through distinct cell-specific transcriptional regulatory mechanism. *Mol. Cell. Biol.,* 17, 3181–3193.

Saiki, I., Yamaura, T., Ohnishi, Y., Hayakawa, Y., Komatsu, Y., and Nunome, S. (1999) HPLC analysis of Juzen-taiho-to and its variant formulations and their antimetastatic efficacies. *Chem. Pharm. Bull.,* 47, 1170–1174.

Saiki, I., Murata, J., Iida, J., Sakurai, T., Nishi, N., Matsuno, K., and Azuma, I. (1989) Antimetastatic effects of synthetic polypeptides containing repeated structures of the cell adhesive Arg-Gly-Asp (RGD) and Tyr-Ile-Gly-Ser-Arg (YIGSR) sequences. *Br. J. Cancer,* 60, 722–728.

Sairenji, M., Okamoto, T., Yanoma, S., Motohashi, H., Kabayashi, O., Okukawa, T., Takemiya, S., and Sugimasa, Y. (1992) Effect of Juzen-taiho-to in a murine tumor cachexia model, colon 26 adenocarcinoma. *Kampo Newest Ther.,* 1, 178–182 (in Japanese).

Saito, T. and Yamaguchi, J. (1981) 2-Chloroadenosine: a selective lethal effect to mouse macrophages and its mechanism. *J. Immunol.,* 134, 1815–1822.

Sakagami, Y., Mizoguchi, Y., Miyajima, K., Kuboi, H., Kobayashi, K., Kioka, K., Takeda, H., Shin, T., and Morisawa, S. (1988) Antitumor activity of Shin-Quan-Da-Bu-Tang and its effects on interferon-γ and interleukin 2 production. *Jpn. J. Allergol.,* 37, 57–60 (in Japanese).

Sakakibara, K., Shibata, Y., Higashi, T., Sanada, S., and Shoji, J. (1975) Effect of ginseng saponins on cholesterol metabolism. I. The level and the synthesis of serum and liver cholesterol in rats treated with ginsenosides. *Chem. Pharm. Bull.,* 23, 1009–1016.

Sakamoto, S., Furuichi, R., Matsuda, M., Kudo, H., Suzuki, S., Sugiura, Y., Kuwa, K., Tajima, M., Matsubara, M., Namiki, H., Mori, T., Kawashima, S., and Nagasawa, H. (1994) Effect of Chinese herbal medicines on DNA-synthesizing enzyme activities in mammary tumors of mice. *Am. J. Chin. Med.,* 22, 43–50.

Satoh, M., Miura, N., Naganuma, A., Matsuzaki, N., Kawamura, E., and Imura, N. (1989) Prevention of adverse effects of gamma-ray irradiation after metallothionein induction by bismuth subnitrate in mice. *Eur. J. Cancer Clin. Oncol.,* 25, 1727–1731.

Sato, K., Mochizuki, M., Saiki, I., Yoo, Y.C., Samukawa, K., and Azuma, I. (1994) Inhibition of tumor angiogenesis and metastasis by a saponin of *Panax ginseng*, ginsenoside-Rb2. *Biol. Pharm. Bull.,* 17, 635–639.

Scaglione, F., Ferrara, F., Dugnani, S., Falchi, M., Santoro, G., and Fraschini, F. (1990) Immunomodulatory effects of two extracts of *Panax ginseng* C.A. Meyer. *Drug Exp. Clin. Res.,* 16, 537–542.

Schuchard, M., Landers, J.P., Sandhu, N.P., and Spelsberg, T.C. (1993) Steroid hormone regulation of nuclear proto-oncogenes. *Endocr. Rev.,* 14, 659–669.

Shibata, Y., Nozaki, T., Higashi, T., Sanada, S., and Shoji, J. (1976) Stimulation of serum protein synthesis in ginsenoside treated rat. *Chem. Pharm. Bull.,* 24, 2818–2824.

Shinkai, K., Akedo, H., Mukai, M., Imamura, F., Isoai, A., Kobayashi, M., and Kitagawa, I. (1996) Inhibition of *in vivo* tumor cell invasion by ginsenoside Rg3. *Jpn. J. Cancer Res.,* 87, 357–362.

Story, M. and Kodym, R. (1998) Signal transduction during apoptosis; implications for cancer therapy. *Front. Biosci.,* 3, 365–375.

Sugiyama, K., Ueda, H., Ichio, Y., and Yokota, M. (1995a) Improvement of cisplatin toxicity and lethality by Juzen-taiho-to in mice. *Biol. Pharm. Bull.,* 18, 53–58.

Sugiyama, K., Ueda, H., and Ichio, Y. (1995b) Protective effect of Juzen-taiho-to against carboplatin-induced toxic side effects in mice. *Biol. Pharm. Bull.,* 18, 544–548.

Takahashi, H. and Nakazawa, S. (1995) Antitumor effect of Juzen-taiho-to, a Kampo medicine, for transplanted malignant glioma. *Int. J. Immunother.,* 11, 65–65.

Tanizawa, H., Karikuma, M., Miyase, T., and Takino, Y. (1993) Studies on the metabolism and/or decomposition and distribution of ginsenoside Rb2 in rats. *Proc. 6th Int. Ginseng Symp.,* 187–194, Seoul.

Tatsuta, M., Iishi, H., Baba, M., Nakaizumi, A., and Uehara, H. (1994) Inhbition of Shi-Quan-Da-Bu-Tang (TJ-48) of experimental hepatocarcinogenesis induced by *N*-nitrosomorpholine in Sprague–Dawley rats. *Eur. J. Cancer,* 30A, 74–78.

Toda, S., Kimura, M., and Ohnishi, M. (1990) Induction of neutrophil accumulation by red ginseng. *J. Ethnopharma.,* 30, 315–318.

Vaux, D.L. and Strasser, A. (1996) The molecular biology of apoptosis. *Proc. Natl. Acad. Sci.,* 93, 2239–2244.

Vlach, J., Hennecke, S., Alevizopoulos, K., Cinti, D., and Amati, B. (1996) Growth arrest by the cyclin-dependent kinase inhibitor p27Kip1 is abrogated by c-myc. *EMBO J.,* 15, 6595–6604.

Watanabe, T., Touge H., Koujimoto, Y., Ebisuno, S., and Ohkawa T. (1997) Study of inhibitory effects of Juzen-taiho-to on the bladder cancer model induced by *N*-butyl-*N*-(4-hydroxybutyl) nitrosamine. *Proc. 14th Kampo Seminar in Urol.,* 7–14. (in Japanese).

Wakabayashi, C., Murakami, K., Hasegawa, H., Murata, J., and Saiki, I. (1998) An intestinal bacterial metabolite of ginseng protopanaxadiol saponin has the ability to induce apoptosis in tumor cells. *Biochem. Biophys. Res. Commun.,* 246, 725–730.

Wakabayashi, C., Hasegawa, H., Murata, J., and Saiki, I. (1997a) *In vivo* antimetastatic action of ginseng protopanaxadiol saponins is based on their intestinal bacterial metabolites after oral administration. *Oncol. Res.,* 9, 411–417

Wakabayashi, C., Hasegawa, H., Murata, J., and Saiki, I. (1997b) The expression of *in vivo* anti-metastatic effect of ginseng protopanaxatriol saponins is mediated by their intestinal bacterial metabolites after oral administration. *J. Traditional Med.,* 14, 180–185.

White, E. (1996) Life, death, and the pursuit of apoptosis. *Genes Dev.,* 10, 1–15.

Wu, J.Y., Gardner, B.H., Murphy, C.I., Seals, J.R., Kensil, C.R., Recchia, J., Beltz, G.A., Newman, G.W., and Newman, M.J. (1992) Saponin adjuvant enhancement of antigen-specific immune responses to an experimental HIV-1 vaccine. *J. Immunol.,* 148, 1519–1525.

Yamaguchi, N., Yamada, T., and Sugiyama, K. (1991) Effect of herbal medicine and its component herbs on antitumor immunity and nonspecific immune reactivity in lymphoma host. *Biotherapy,* 5, 1840–1849 (in Japanese).

9 Juzen-taiho-to in Combination with Chemotherapy and Radiotherapy

Kiyoshi Sugiyama

CONTENTS

9.1   INTRODUCTION

For more than half a century, it has been expected that the problem of the treatment of cancer would be overcome sometime in the early 21st century. However, there is no definite prospect of this as yet. At present, the treatment of cancer is mostly concentrated on surgical therapy, chemotherapy, and radiotherapy, and each of these therapeutic approaches has its own defects and limits.

In the chemotherapy and radiotherapy described here, limiting toxicity is an important factor. The influence of chemical therapeutic agents or radioactivity on normal cells and their influence on cancer cells are not substantially different. As a result, normal cells are also subject to the negative effects of these methods and severe side effects develop. In this respect, the dosage of chemotherapeutic agents and the radiation dose for the treatment of cancer are limited, thus hindering satisfactory treatment. The side effects most frequently observed include: myelopathy; digestive organ disorder; renal disorder; vomiting; peripheral nerve disorder; diarrhea; alopecia; hearing disorder; systemic malaise; loss of appetite; etc. Alleviating the side effects of chemotherapy and radiotherapy would be good news for cancer patients.

9.2 USE OF JUZEN-TAIHO-TO TO REDUCE AND ALLEVIATE SIDE EFFECTS OF RADIOTHERAPY

The therapeutic methods used to treat cancer are immunotherapy, endocrinotherapy, surgical therapy, chemotherapy, and radiotherapy. However, the anticancer effect of each of these therapeutic methods is limited by the extent to which the disease has progressed. Accordingly, in clinical practice, attempts are made to treat the disease by adequately combining these therapeutic methods in order to obtain more therapeutic effects. For example, radiotherapy and chemotherapy are combined together to treat small cell carcinoma of the lung. To treat maxillary cancer, radiotherapy and chemotherapy are combined with surgical therapy. It is believed that free radicals are generated in the human body as the result of radiation, and these free radicals may cause disorder in the cells. In the sensitivity of each individual cell to the radioactivity, normal cells and cancer cells do not differ significantly; this is why side effects such as a high degree of myelopathy or inflammation of skin are induced.

In an earlier chapter, Hisha and Ikehara have given a detailed description of the effectiveness of Juzen-taiho-to for the treatment of myelopathy caused by the side effects of radiotherapy, so that aspect is not described here again. The energetic efforts of Ikehara et al. have made clear for the first time that Juzen-taiho-to provides the mechanism to improve hematopoietic function that has decreased as a result of radioactive radiation.

Apart from the study by Ikehara et al., Hosokawa et al. (Hosokawa, 1985) gave Juzen-taiho-to to mice in order to evaluate its effectiveness in preventing radioactive disorders. These researchers found that, when Juzen-taiho-to was given to a mouse in advance of radiation, the survival rate was significantly improved. Platelet count had significantly increased 14 days after radiation and thereafter, and internal bleeding had reduced remarkably 11 days after radiation and thereafter.

It has been demonstrated in clinical cases that Juzen-taiho-to is effective against the side effects of radiotherapy. Miyamoto et al. (1985) performed a study on 22 patients who had been receiving radiotherapy at the Department of Radiology, Keio University, and evaluated the effects of Juzen-taiho-to. Major side effects found in these patients were loss of appetite, systemic malaise, diarrhea, nausea, and vomiting. When Juzen-taiho-to (TJ-48) was given to these patients for an average period of 95 days after radiotherapy, indefinite complaints were subdued. Diarrhea and bloody stool healed without resort to surgery, and loss of appetite and malaise were reduced. Taniguchi et al. (1984) also assessed the effects of Juzen-taiho-to on 59 patients with cervical cancer who had received radiotherapy. Juzen-taiho-to was effective in treating leukopenia and also decreased the number of cases for which radiotherapy would have had to be abandoned.

At present, no definite therapeutic method to eliminate side effects of radiotherapy exists. Much expectation is now placed on the effects of Juzen-taiho-to and the application of Kampo medicines in this field.

9.3 USE OF JUZEN-TAIHO-TO TO ALLEVIATE SIDE EFFECTS OF MITOMYCIN C (MMC)

Mitomycin (MMC) has a wide anticancer spectrum and gives a strong carcinostatic effect; it is still widely used in Japan. MMC is effective for the treatment of digestive organ cancer such as gastric, colic, liver, pancreatic, lung, breast, uterine, and bladder cancers, as well as malignant lymphoma and chronic myelocytic leukemia. Because of its strong anticancer properties, MMC also has extensive side effects. In particular, it has high myelotoxicity, which often causes problems during treatment. In general, with the increase of total dosage, symptoms such as leukopenia, thrombocytopenia, and bleeding appear to a great extent. In addition, digestive organ disorder such as loss of appetite, nausea, vomiting, and diarrhea, or systemic malaise, alopecia, liver disorder, and renal disorder occur from time to time.

Aburada et al. (1983) scientifically demonstrated for the first time that Juzen-taiho-to was effective against the side effects of MMC. They used BDF1 mouse implanted with P-388 and gave

MMC once or twice per day and then evaluated the effects of Juzen-taiho-to against antitumor effects and the side effects of MMC. As a result, it was found that Juzen-taiho-to reinforces the antitumor effect of MMC and alleviates the side effects of MMC (which were determined using only the decrease of WBC count and body weight decrease as indices). With this study as a turning point, attempts and studies to use Kampo medicines as the agents in alleviating the side effects of anticancer drugs have increased extensively.

Study of this type was then energetically performed by the research group of Tsumura & Co. with the Ikehara and Yamada groups. As a result (as described in detail by Ikehara), effective components of Juzen-taiho-to were elucidated and justified its use in overcoming the problems of myelotoxicity in the field of cancer chemotherapy.

9.4 USE OF JUZEN-TAIHO-TO TO ALLEVIATE SIDE EFFECTS OF CISPLATIN

Cisplatin (CDDP, Figure 9.1) is a platinum (Pt) complex discovered by Rosenberg et al. in 1965. It has a broad-spectrum antitumor effect and has often been used to treat various solid type cancers such as testicular, ovarian, bladder and uterine cancers. This drug has a serious disadvantage, however, in that it is associated with strong side effects such as renal toxicity, vomiting, and nausea. These side effects limit continuous treatment with this agent. Our study was undertaken to search for Kampo medicines that would reduce the side effects of CDDP through systematic study of the efficacy, active components, mechanisms of action, and roles of crude drug combinations.

9.4.1 ESTABLISHMENT OF ANIMAL MODEL

In the clinical application of CDDP, patients were given from 10 to 50 mg/m^2/day of CDDP two or more times. Acute toxicity of this drug can be reduced considerably by using large amounts of intravenous fluids and/or diuretics. However, its subacute toxicity has not been reduced satisfactorily and has become a serious problem. Thus, we attempted to establish a mouse model for the study of the subacute toxicity of CDDP.

Figure 9.2 shows the model that was established. The ddY mice with implanted sarcoma 180 (S-180) were given a daily dose of CDDP (3 mg/kg/day) for 9 consecutive days. Biochemical and histopathological studies of these animals revealed symptoms resembling those clinically known as the toxic effects of CDDP (Gonzalez–Vitale et al., 1977). As markers of nephrotoxicity, the blood urea nitrogen (BUN) and creatinine levels showed fourfold and twofold increases over the normal levels, respectively, and urinary volume decreased by 42%. As the markers of marrow toxicity, the white blood cell count (WBC), platelet count (PLT), relative spleen weight, and relative thymus weight decreased to 29, 27, 60, and 24% of their normal levels, respectively. As the markers of hepatic toxicity, sGOT and sGPT rose to approximately three times and five times the normal level, respectively. In addition, body weight and food consumption decreased to 65 and 31% of

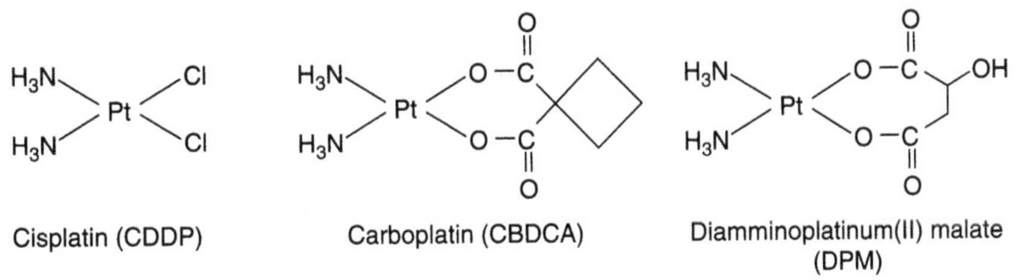

Cisplatin (CDDP)   Carboplatin (CBDCA)   Diamminoplatinum(II) malate (DPM)

FIGURE 9.1 Structures of antitumor platinum compounds. (From Sugiyama, K., 1996, *J. Trad. Med.*, 13, 27–41. With permission.)

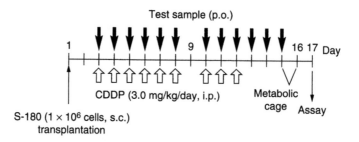

FIGURE 9.2 Experimental design for examining the effect of Kampo medicines on CDDP-induced toxicity. Five-week-old, male, ddY mice (average weight, 30 g) were used ($n = 10$). Mice were inoculated with sarcoma 180 (S-180) cells in the left thigh s.c. on day 1. On day 17, blood was collected from the inferior vena cava, and biochemical data were measured. CDDP: *cis*-diamminedichloroplatinum (II). (From Sugiyama, K., 1996, *J. Trad. Med.*, 13, 27–41. With permission.)

the normal level, respectively. The stomach weight (including the weight of gastric contents) increased to four times that of the normal level (Sugiyama et al., 1993).

9.4.2 ASSESSMENT OF KAMPO FORMULATIONS' EFFICACY

Because serious toxic effects of CDDP occur in the kidneys, marrow, and gastrointestinal organs, Kampo formulations known to be effective on the urinary system or to have tonifying effects were selected and the effects of these formulations assessed (Table 9.1).

Figure 9.3 shows the results of oral treatment with each formulation at ten times the usual daily dose. Each formulation was prepared by extraction in hot water and subsequent freeze-drying. The BUN, a marker of nephrotoxicity, showed an approximately fourfold increase after treatment with CDDP alone, compared to the control level. This indicates the onset of severe renal disorder. When CDDP was administered in combination with Juzen-taiho-to; Toki-shakuy-aku-san (Dang-Gui-Shao-Yao-San in Chinese); Hochu-ekki-to (Bu-Zhong-Yi-Qi-Tang); Hachimi-jio-gan (Ba-Wei-Di-Huang-Wan); Chorei-to (Zhu-Ling-Tang); Gorei-san (Wu-Ling-San); Ryokei-jutsukan-to (Ling-Gui-Shu-Gan-Tang); or Boi-ogi-to (Fang-Yi-Huang-Qi-Tang),

---

TABLE 9.1
Prescription and Experimental Doses of Kampo Formulations

Kampo Formulation	Dose[a]
Juzen-taiho-to	1.7
Toki-shakuyaku-san	1.4
Hochu-ekki-to	1.4
Hachimi-jio-gan	1.1
Rikkunshi-to	0.8
Chorei-to	0.6
Gorei-san	0.5
Ryo-kei-jutsu-kan-to	0.3
Boi-bukuryo-to	0.4
Boi-ogi-to	1.0
Shimbu-to	0.4

[a] Ten times the usual clinical dose (g/kg/day).
Note: Each experimental dose for an animal was calculated on the basis of each yield of the water extract of Kampo formulations.

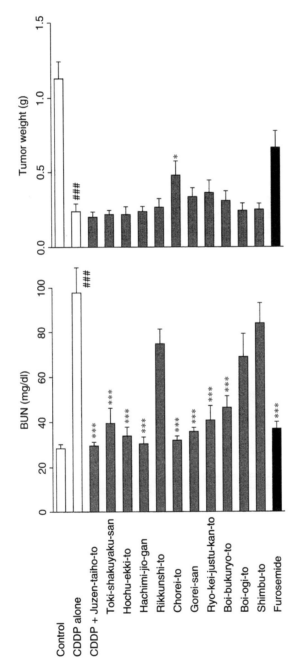

FIGURE 9.3 Effects of Kampo formulations on toxicity and antitumor effect of CDDP. Kampo formulations (ten times the usual daily dose) or furosemide (20 mg/kg/day) was administered p.o. to mice once a day 30 min before CDDP i.p. injection. Each value is the mean ± S.E. ($n = 10$). Significant difference from the control group, ###: $p < 0.001$. Significant difference from the CDDP alone group, *: $p < 0.05$; **: $p < 0.01$; ***: $p < 0.001$. (From Sugiyama, K., 1996, *J. Trad. Med.*, 13, 27–41. With permission.)

the increase in BUN over the control level was prevented almost completely (Sugiyama et al., 1993). The effect of these Kampo formulations in reducing the nephrotoxicity of CDDP was comparable to that of furosemide, a frequently used diuretic (Pera et al., 1979).

The antitumor effect of CDDP was slightly increased by the combined use of tonifying Kampo formulations, but was reduced by the combined use of Kampo formulations acting on the urinary system. Furosemide reduced the antitumor effect of CDDP more markedly. These results suggest that the mechanism of reducing the toxicity of CDDP differs between a tonifying Kampo formulation and a Kampo formulation acting on the urinary system. The latter types of Kampo formulations seem to reduce the toxicity of CDDP by promoting the urinary elimination of CDDP through its diuretic action, similar to the mechanism of the action of furosemide (Vogl et al., 1980). Tonifying Kampo formulations seemed to reduce the toxicity of CDDP by some other mechanisms (Sugiyama et al., 1996).

9.4.3 Use of Juzen-taiho-to to Reduce CDDP Toxicity

Figure 9.4 shows the dose–response curve of Juzen-taiho-to. Oral treatment with Juzen-taiho-to (freeze-dried extracts) at a dose level of 1.7 g/kg/day (ten times the usual daily dose level) resulted in approximately normal BUN, WBC, and body weight. The other markers of toxicity were also improved markedly by treatment with Juzen-taiho-to at a daily dose more than ten times the usual dose. The antitumor effect of CDDP was not affected by this Kampo formulation at any dose.

Table 9.2 shows histopathological changes of the kidneys. Treatment with CDDP alone resulted in moderate to severe degeneration and necrosis of the renal cortex and tubules, accompanied by ureteral dilation and cast formation in a relatively narrow area. These changes are identical to those reported clinically (Chopra et al., 1982). When CDDP was combined with Juzen-taiho-to, these changes were almost completely prevented (Sugiyama et al., 1996).

Figure 9.5 shows the effects of Juzen-taiho-to in reducing the toxicity of CDDP administered in higher doses. When mice were treated daily with 3.0 mg/kg of CDDP, their death was first noted on the 20th day, and the survival rate on the 27th day was 60%. When the same dose of CDDP

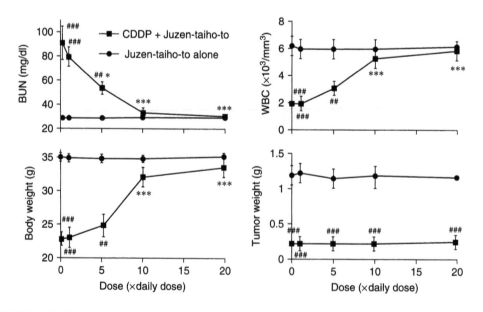

FIGURE 9.4 Relationship between the dose of Juzen-taiho-to and its effect in reducing the toxicity of CDDP. Juzen-taiho-to (0.17, 0.85, 1.7, and 3.4 g/kg/day) was administered p.o. to mice once a day 12 times. Each result is the mean ± S.E. ($n = 10$). Significant difference from the Juzen-taiho-to alone group, ##: $p < 0.01$; ###: $p < 0.001$. Significant difference from the CDDP alone group, *: $p < 0.05$; ***: $p < 0.001$. (From Sugiyama, K., 1996, *J. Trad. Med.*, 13, 27–41. With permission.)

TABLE 9.2
Pathological Changes in Kidneys (HE Staining)

	Normal (n = 9)				CDDP (n = 8)				CDDP + JTT (n = 9)				JTT (n = 9)			
	Grade															
Pathological Changes	−	±	+	++	−	±	+	++	−	±	+	++	−	±	+	++
Cortex																
Tubular degeneration/necrosis	2	7	0	0	0	1	4	3	7	2	0	0	9	0	0	0
Tubular dilation/cast formation	3	5	1	0	0	0	5	3	6	0	3	0	8	1	0	0
Changes of glomerulus	9	0	0	0	8	0	0	0	9	0	0	0	9	0	0	0
Medulla																
Degeneration of Henle's loop	9	0	0	0	7	0	1	0	9	0	0	0	9	0	0	0
Tubular dilation/cast formation	7	2	0	0	4	2	1	1	8	1	0	0	9	0	0	0
Interstitium																
Edema	9	0	0	0	6	0	2	0	9	0	0	0	9	0	0	0
Cell increase	9	0	0	0	8	0	0	0	3	0	6	0	4	2	3	0

Notes: −: negative or within the borderline of normal variation; ±: slightly positive; +: positive; ++: strongly positive. Mice were sacrificed on day 17. Kidneys were removed, weighed, and processed for light microscopy by routine histology.

was combined with Juzen-taiho-to at a daily dose level 20 times that of the usual dose, all mice were alive on the 27th day. When the dose of CDDP was increased to 4.5 mg/kg/day while keeping the Juzen-taiho-to dose unchanged, all mice were alive on the 27th day. When the CDDP dose level was further increased to 6.0 mg/kg/day, 50% of mice were alive on the 27th day. At a CDDP dose level over 7.5 mg/kg/day, the combined use of Juzen-taiho-to did not prolong the survival period of animals any further. The daily dose of 3.0 mg/kg is close to the maximum clinical dose used for continuous CDDP therapy. Juzen-taiho-to was found to be capable of reducing the toxicity of CDDP even when the CDDP dose was twice the maximum clinical dose (Sugiyama et al., 1995a).

Figure 9.6 shows the effects of Juzen-taiho-to on different days after the start of treatment. Nephrotoxicity of CDDP appeared on the eighth day. Marrow toxicity began to be noted immediately after the start of treatment. Weight loss began on the eighth day. When CDDP was combined with Juzen-taiho-to, no change in BUN, WBC, or body weight was seen at any point of assessment during the treatment period, suggesting that Juzen-taiho-to prevents the toxic effects of CDDP (Sugiyama et al., 1995a).

Figure 9.7 shows the relationship between the timing of Juzen-taiho-to treatment and its effect in reducing the toxicity of CDDP. When this Kampo formulation was administered simultaneously with or prior to CDDP treatment, the toxicity of CDDP was reduced. However, when the Kampo formulation was administered after CDDP treatment, the toxicity remained almost unchanged. The antitumor effect of CDDP was not affected by Juzen-taiho-to, irrespective of the timing of its administration (Sugiyama et al., 1996).

Figure 9.8 shows the effects of Juzen-taiho-to in reducing the toxicity of carboplatin (Figure 9.1), a derivative of CDDP with reduced nephrotoxicity. Marrow toxicity, which limits the dose levels of carboplatin (Los et al., 1991; Guchelaar et al., 1994; Markman et al., 1993), was reduced by the combined use of Juzen-taiho-to at a daily dose ten times the usual dose amount; the antitumor effect of carboplatin was not affected by this Kampo formulation. The other toxic effects of carboplatin were also reduced markedly by Juzen-taiho-to (Sugiyama et al., 1995b).

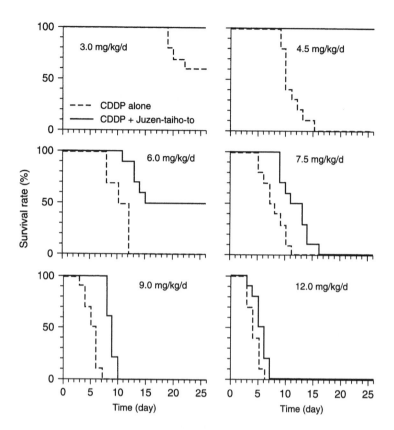

FIGURE 9.5 Effect of Juzen-taiho-to on lethal toxicity of CDDP. Juzen-taiho-to (3.4 g/kg/day) was administered p.o. to mice inoculated with S-180 cells 30 min before CDDP (3.0 to 12.0 mg/kg/day, i.p.) injection once a day during the experimental period. All mice that received CDDP alone at doses of 4.5, 6.0, 7.5, 9.0, and 12.0 mg/kg/day died within 15, 12, 11, 7, and 6 days after the initial CDDP injection, respectively. All mice that received Juzen-taiho-to and 3.0 and 4.5 mg/kg/day of CDDP were alive 25 days after the initial CDDP injection. (From Sugiyama, K., 1996, *J. Trad. Med.*, 13, 27–41. With permission.)

A comparative study of the effect of Juzen-taiho-to and some other drugs on the CDDP toxicity has been undertaken. Furosemide (a diuretic) and sodium thiosulfate (a neutralizing agent) (Kobayashi et al., 1991) reduced the toxicity and the antitumor effect of CDDP. Metallothionein inducers zinc chloride (Satoh et al., 1993) and bismuth subnitrate (Satoh et al., 1988), an antioxidant vitamin E (Sugihara and Gemba, 1986), and lysosome membrane stabilizers fosfomycin (Umeki et al., 1988) and urinastain (Umeki et al., 1989) reduced the toxicity of CDDP less markedly than Juzen-taiho-to (data not shown).

Figure 9.9 shows the effects of Juzen-taiho-to on P388 cells. The average survival period for the untreated controls (8 days) was not prolonged by treatment with Juzen-taiho-to alone (at a daily dose ten times the usual dose), but was prolonged up to 22 days by treatment with CDDP alone. Treatment with a combination of CDDP and Juzen-taiho-to further prolonged the survival period to 33 days. Because the survival period was not prolonged by Juzen-taiho-to alone, the extended survival period following the combined use of CDDP and Juzen-taiho-to seems to be attributable to the reduction of the toxic effects of CDDP by use of Juzen-taiho-to (Sugiyama et al., 1996).

These results suggest that Juzen-taiho-to prevents the toxic effects of CDDP and that this Kampo formulation is expected to exert the most favorable effects when it is administered at a daily dose of more than ten times the level of the usual dose simultaneously with or prior to CDDP treatment.

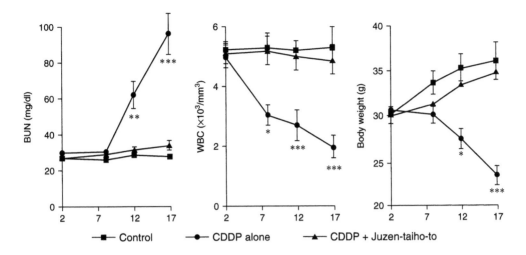

FIGURE 9.6 Time course of effect of Juzen-taiho-to on CDDP-induced toxicity. Juzen-taiho-to (1.7 g/kg/day, p.o.) and CDDP (3.0 mg/kg/day, i.p.) were administered to mice with S-180 cells once a day. Each value is the mean ± S.E. ($n = 10$). Significant difference from the control group, *: $p < 0.05$; **: $p < 0.01$; ***: $p < 0.001$. (From Sugiyama, K., 1996, *J. Trad. Med.*, 13, 27–41. With permission.)

9.4.4 ISOLATION OF EFFECTIVE COMPONENTS

It has been already explained that Juzen-taiho-to reduces nephrotoxicity and myelotoxicity of cisplatin. After elaborate examination of the effective components of Juzen-taiho-to, it was found that the components to reduce nephrotoxicity and the component to reduce myelotoxicity are different from each other.

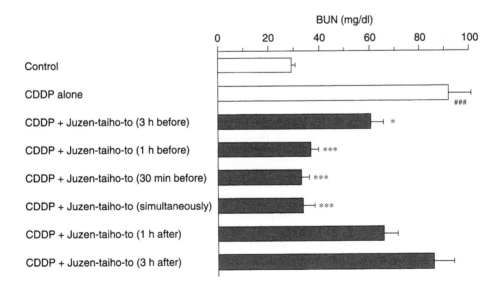

FIGURE 9.7 Relationship between the timing of Juzen-taiho-to treatment and its effectiveness in reducing the toxicity of CDDP. Juzen-taiho-to (1.7 g/kg/day, p.o.) was administered to mice with S-180 cells after (1 h or 3 h), simultaneously with, or prior to (3h, 1h, or 30 min) CDDP treatment. Each value is the mean ± S.E. ($n = 10$). Significant difference from the CDDP alone group, *: $p < 0.05$; **: $p < 0.01$; ***: $p < 0.001$. Significant difference from the control group, ###: $p < 0.001$. (From Sugiyama, K., 1996, *J. Trad. Med.*, 13, 27–41. With permission.)

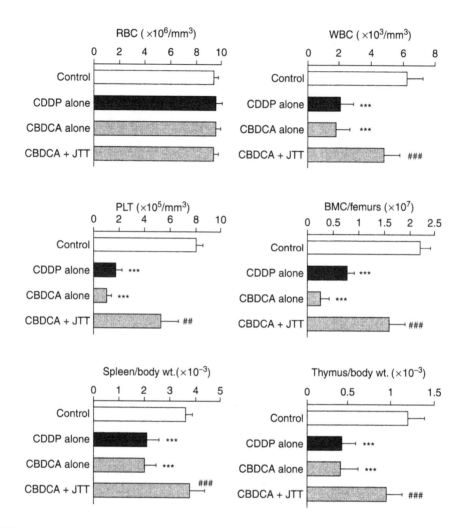

FIGURE 9.8 Effect of Juzen-taiho-to (JTT) on CBDCA-induced myelosuppression. JTT (1.7 g/kg/day × 12) was administered p.o. to mice inoculated with S-180 cells 30 min before CBDCA (15 mg/kg/day) i.p. injection nine times. All parameters were determined on day 17. Each value is the mean ± S.E. ($n = 10$). Significant difference from the control group, ***: $p < 0.001$. Significant difference from the CBDCA alone group, ##: $p < 0.01$; ###: $p < 0.001$. BMC: bone marrow cell. (From Sugiyama, K., 1996, *J. Trad. Med.*, 13, 27–41. With permission.)

Juzen-taiho-to is composed of ten crude drugs. Figure 9.10 shows the effects of each of these drugs (freeze-dried extracts) in reducing the nephrotoxicity of CDDP. Angelicae Radix (Toki), Ginseng Radix (Ninjin), and Poria (Bukuryo) reduced the nephrotoxicity, and the effect was comparable to that of Juzen-taiho-to when these crude drugs were administered in amounts equal to those contained in the Juzen-taiho-to. A general assessment was made of the effects of individual crude drugs in reducing marrow, hepatic, and gastrointestinal toxicity and of their antitumor effects. This assessment revealed that Toki plays a more significant role than any other ingredient of Juzen-taiho-to in reducing the nephrotoxicity, hepatotoxicity, and gastrointestinal toxicity of CDDP (data not shown).

Figure 9.11 shows the procedure to isolate an active constituent from Toki. Using various methods of isolation, pure sodium malate (5.2 mg, Figure 9.12) was obtained from 2 kg of Toki, with a recovery of 0.0003% (Sugiyama et al., 1994). When sodium malate (p.o.) and CDDP (i.p.) with a molar ratio over 0.5:1 were used, the renal, hepatic, and gastrointestinal toxicity of CDDP

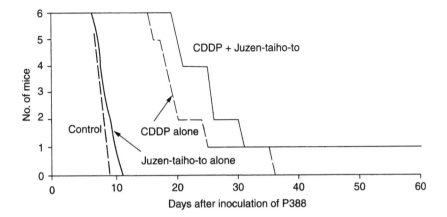

FIGURE 9.9 Effect of Juzen-taiho-to on lethal toxicity of CDDP in BDF1 mice inoculated with P388 cells. P388 cells (10^6 cells/mouse) were inoculated s.c. on day 1. Juzen-taiho-to (1.7 g/kg/day, p.o.) and/or CDDP (3.0 mg/kg/day, i.p.) was administered to BDF1 mice once a day. (From Sugiyama, K., 1996, *J. Trad. Med.*, 13, 27–41. With permission.)

were suppressed almost completely, without affecting the antitumor effect of CDDP. The effect of sodium malate in reducing this toxicity was comparable to that of Juzen-taiho-to, but its effect in reducing the marrow toxicity of CDDP was weaker (Figure 9.13). These results, combined with the results of quantitative analysis of sodium malate contained in Juzen-taiho-to (0.98%) as determined using high-performance liquid chromatography (HPLC), indicate that sodium malate plays an important role in Juzen-taiho-to in reducing the nephrotoxicity of CDDP (Ueda et al., 1998a).

It was also made clear that sodium malate is contained in all of the crude drugs blended in Juzen-taiho-to with 0.98% of the salt in the usual daily dose of this Kampo medicine (Table 9.3). Furthermore, it was found that Juzen-taiho-to contains a number of carboxylic acids (such as fatty

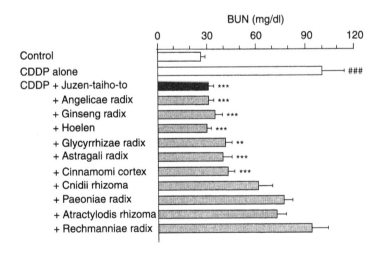

FIGURE 9.10 Effects of ingredients of Juzen-taiho-to on CDDP-induced nephrotoxicity. All samples tested were administered p.o. to mice at a dose of ten times the usual daily dose 30 min before CDDP (3 mg/kg/day, i.p.) injection. The control group was treated with water (p.o.) and saline (i.p.). S-180 cells were inoculated s.c. on day 1 and nephrotoxicity was determined on day 17. Each value is the mean ± S.E. ($n = 10$). Significant difference from the control group, ###: $p < 0.001$. Significant difference from the CDDP alone group, **: $p < 0.01$; ***: $p < 0.001$. (From Sugiyama, K., 1996, *J. Trad. Med.*, 13, 27–41. With permission.)

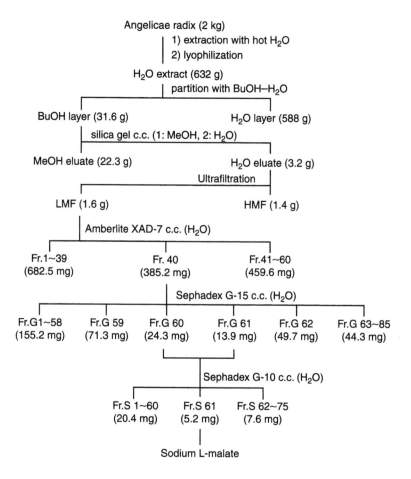

FIGURE 9.11 Isolation procedure for sodium L-malate from Angelicae Radix. Isolation was performed guided by BUN measurements in mice. The yield of each fraction obtained from 2 kg of Angelicae Radix is designated in parentheses. LMF and HMF indicate low molecular weight fraction (MW < 10,000) and high molecular weight fraction (MW > 10,000), respectively. (From Sugiyama, K., 1996, *J. Trad. Med.*, 13, 27–41. With permission.)

acids, aromatic carboxylic acids, etc.) in addition to sodium malate, and that these carboxylic acids also have the effect of reducing the nephrotoxicity of cisplatin (data not shown).

On the other hand, it was identified that the component to alleviate myelotoxicity is a polysaccharide with molecular weight of more than 5000. It was made clear that this component contains a large quantity of Atractylodis Rhizoma (Byakujutsu). In some types of Juzen-taiho-to, Atractylodis Lanceae Rhizoma (Sojutsu) is blended instead of Byakujutsu. It was found that Sojutsu contains a larger amount of the polysaccharide with molecular weight of more than 5000 and that this component increases platelets to a great extent.

$$
\begin{array}{c}
COONa \\
| \\
HO-C-H \\
| \\
H-C-H \\
| \\
COONa
\end{array}
$$

FIGURE 9.12 Structure of sodium L-malate.

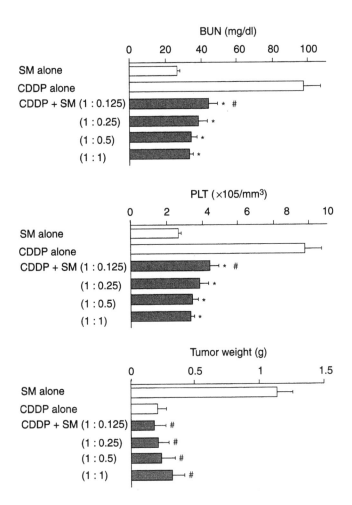

FIGURE 9.13 Effect of sodium malate on CDDP-induced toxicity and antitumor effect. Sodium malate (SM, p.o.) and CDDP (i.p.) with a molar ratio over 0.125 to 1 : 1 were administered to mice with inoculated S-180 cells. Each value is the mean ± S.E. ($n = 10$). Significant difference from the SM alone group, #: $p < 0.05$. Significant difference from the CDDP alone group,*: $p < 0.05$. (From Sugiyama, K., 1996, *J. Trad. Med.*, 13, 27–41. With permission.)

Hisha et al. (1997, 2002) found that the unsaturated fatty acids contained in Juzen-taiho-to (TJ-48) stimulate the proliferation of marrow stem cells. We also demonstrated that the polysaccharide with molecular weight over 5000 contained in Juzen-taiho-to promoted the proliferation of marrow stem cells (CFU-meg and CFU-GM). Therefore, the effect of Juzen-taiho-to in reducing the marrow toxicity of CDDP involves these substances rather than sodium malate.

9.4.5 Mechanism of Suppression of Nephrotoxicity by Juzen-taiho-to

It is thought that the toxic actions of CDDP on cells undergoing rapid mitosis (i.e., the antitumor effects, marrow toxicity, etc. of CDDP) are induced by the binding of CDDP to DNA through cross-linking to the N-7 positions of the adjacent guanines (Takahara et al., 1995; Petsko, 1994; Treiber et al., 1994; Brown et al., 1993). The toxic actions of CDDP on the cells of the kidney and some other organs, however, cannot be fully explained by its binding to DNA and involve many unresolved questions. CDDP in blood is eliminated via the glomerulus and the proximal

TABLE 9.3
Contents of Sodium Malate in Juzen-taiho-to and Its Ingredients

	Clinical Dose (herbs)	Content of Sodium Malate	
	(g)	(mg)	(%)
Juzen-taiho-to	29	79.5	0.98
Angelicae Radix	3	23.1	2.33
Ginseng Radix	3	14.8	1.34
Rehmanniae Radix	3	10.7	0.74
Astragali Radix	3	9.2	1.20
Glycyrrhizae Radix	2	6.0	0.77
Atractylodis Rhizome	3	6.0	0.81
Cnidii Rhizome	3	4.8	0.46
Paeoniae Radix	3	4.5	0.81
Cinnamomi Cortex	3	1.1	0.67
Poria	3	0.3	0.79

Notes: Contents of sodium malate in Kampo medicines were determined by HPLC; column: TSK ODS-80 Ts (4.6 × 250 mm); column temp.: 40°C; eluent: 0.05 M KH2PO4/0.05 M H3PO4 (1 : 1); flow rate: 0.5 ml/min; detection: UV 210 nm, injection: 10 μl (50 mg/ml). Retention time of sodium malate was 8.2 min.

tubule. At the same time, CDDP can be reabsorbed into blood from the proximal tubule (Daley–Yates and McBrien, 1983). Enzymes in the proximal tubule, exposed to high concentrations of CDDP, lose their activity through binding to CDDP. Inactivation of ATPase (Daley–Yates and McBrien, 1982), glutathione and its related enzymatic systems (Feinfeld and Fuh, 1986), cytochrome P-450 (Bompart, 1989), and Na+-dependent transport systems (Gautier et al., 1995) can result in the necrosis of cells, leading to severe renal damage (Courjault et al., 1993).

Some derivatives of CDDP with reduced nephrotoxicity have been used clinically, including carboplatin. These derivatives have a lower potential to bind to protein compared to CDDP (Micetich et al., 1985; DeNeve et al., 1990). Taking note of the finding that sodium malate, an effective component of Juzen-taiho-to, is a dicarboxylic acid resembling the ligand of carboplatin, we assumed that sodium malate binds to CDDP *in vivo* to yield a less toxic derivative of CDDP.

To test the validity of this assumption, we conducted two experiments. First, we analyzed the pharmacokinetics of Pt following the combined administration of CDDP and sodium malate. This experiment revealed that the administration of CDDP in combination with sodium malate selectively reduced renal accumulation of Pt (Figure 9.14) and prolonged the half-life of non-protein-bound Pt (Figure 9.15). These kinetics of Pt were similar to those of carboplatin (data not shown). We subsequently synthesized a derivative of CDDP, diamminoplatinum (II) malate (DPM, Figure 9.1) (Dus and Jaworska, 1982; Janina and Boguslawa, 1981), which seems to be formed in the body following the administration of CDDP and sodium malate, and we assessed its effects. DPM had an antitumor effect and marrow toxicity comparable to those of CDDP but had no nephrotoxicity (Figure 9.16). DPM was also detected in blood, using HPLC (Ueda et al., 1998b).

In conclusion, because it was found that the effective component of Juzen-taiho-to to alleviate nephrotoxicity is sodium malate, the mechanism of sodium malate to alleviate nephrotoxicity was assessed. As a result, it was identified that sodium malate is combined with cisplatin in the human body and is converted to DPM. This newly generated DPM in the body is a compound with the

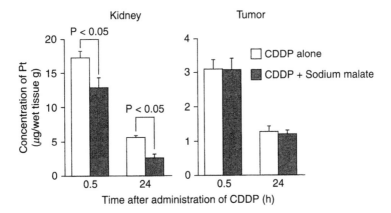

FIGURE 9.14 Accumulation of Pt in kidneys and tumor 30 min or 24 h after administration of CDDP. S-180 cells (10^6 cells/mouse) were inoculated s.c. to ddY mice on day 1. On day 14, sodium malate (14.8 mg/kg) was administered to the mice 30 min prior to CDDP (12.5 mg/kg) treatment. The concentration of Pt was detected by use of an inductively coupled plasma spectrometer (ICP) (Seiko SPS 1200A). Each value is the mean ± S.E. ($n = 10$). (From Sugiyama, K., 1996, *J. Trad. Med.*, 13, 27–41. With permission.)

same degree of antitumor effect as cisplatin, but it has almost no possibility of causing nephrotoxicity. It is considered that this is the mechanism by which sodium malate (and, thus, Juzen-taiho-to) alleviates nephrotoxicity.

9.4.6 MECHANISM OF MYELOTOXICITY SUPPRESSION BY JUZEN-TAIHO-TO

Because it has been demonstrated that the effective component of Juzen-taiho-to producing the effect to alleviate myelotoxicity is a polysaccharide with molecular weight of more than 5000, the mechanism of this component to alleviate myelotoxicity was evaluated. First, it was found that this polysaccharide has the effect of increasing the platelet count. In this respect, a study was performed *in vitro* to determine the influence of this polysaccharide on the proliferation of CFU-meg, i.e., a

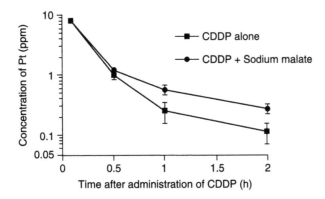

FIGURE 9.15 Filterable Pt levels in plasma of mice after treatment with CDDP and/or sodium malate. S-180 cells (10^6 cells/mouse) were inoculated s.c. to ddY mice on day 1. On day 14, sodium malate (14.8 mg/kg) was administered to the mice 30 min prior to CDDP (12.5 mg/kg) treatment. After 5, 30, 60, and 120 min, blood was collected and plasma was ultrafiltered through Centricon 10 (Amicon Co., Ltd.). Pt was measured by an SPS 1200A. Significant difference from the CDDP alone group,*: $p < 0.05$. (From Sugiyama, K., 1996, *J. Trad. Med.*, 13, 27–41. With permission.)

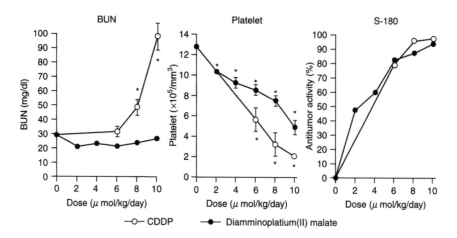

FIGURE 9.16 Toxicity and antitumor effect of diamminoplatinum (II) malate (DPM). S-180 cells (10^6 cells/mouse) were inoculated s.c. in the left thigh on day 1. Diamminoplatinum (II) malate (2, 4, 6, 8, and 10 μmol/kg/day) and CDDP (6, 8 and 10 μmol/kg/day) were administered i.p. on days 3, 4, 5, 6, 7, 8, 10, 11, and 12. BUN, platelet count, and tumor weight were measured on day 17. Each value is the mean ± S.E. ($n = 10$). Significant difference from the control group, *: $p < 0.05$. (From Sugiyama, K., 1996, *J. Trad. Med.*, 13, 27–41. With permission.)

precursor cell of platelet in bone marrow. As a result, it became clear that this polysaccharide has almost no influence *in vitro* on the proliferation of CFU-meg (data not shown).

Then, the serum was collected from the rat to which this polysaccharide had been administered, and evaluation was made on its influence on the proliferation of CFU-meg. When the polysaccharide was given to rat, the serum of the rat significantly promoted CFU-meg (Figure 9.17). Thus, it was found that the polysaccharide with molecular weight of more than 5000 is effective only when it enters the human body.

A further study was performed on the mechanism using the ELISA kit for quantitative determination of cytokines, and it was found that GM-CSF is remarkably increased in the serum of the rat to which the polysaccharide was given (production of IL-3 or IL-6 was not noted) (Table 9.4).

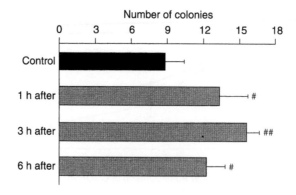

FIGURE 9.17 Effect of plasma treated with the polysaccharide on the growth of CFU-meg. Plasma was prepared from the BDF1 mice 1, 3, or 6 h after treatment with the polysaccharide. Bone marrow cells were incubated at 37°C for 7 days. CFU-meg colonies were measured under a microscope of 100 magnifications (mean ± S.D.; $n = 3$). Significant differences from the control group, #: $p < 0.05$; ##: $p < 0.01$. (From Sugiyama, K., 1996, *J. Trad. Med.*, 13, 27–41. With permission.)

TABLE 9.4
Detection of Cytokines in Serum Treated with MW > 5000

	Cytokine (pg/ml)		
	GM-CSF	IL-3	IL-6
Water	1.62	ND	ND
MW > 5000	3.50	ND	ND

Notes: Cytokines in serum were measured by ELISA kits. ND: not detected.

These results suggest that the polysaccharide promotes production of GM-CSF and then enhances the proliferation of CFU-meg in bone marrow. As a result, the platelet count in peripheral blood increases.

9.4.7 ROLE PLAYED BY THE CRUDE DRUGS CONSTITUTING JUZEN-TAIHO-TO

Table 9.3 and Figure 9.18 show the roles played by individual herbs in the expression of the effects of Juzen-taiho-to, as assessed on the basis of the amount of sodium malate contained and the degree of reduction in nephrotoxicity. The actual measurement of the effect of Juzen-taiho-to in reducing renal toxicity was about one fourth of its theoretical magnitude and did not correlate with the amount of sodium malate contained. On the other hand, the actual magnitude of this effect of Toki was approximately equal to its theoretical magnitude and correlated well with the amount of sodium malate contained.

Rehmanniae Radix (Jukujio) reduced renal toxicity only very slightly, although it contained large amounts of sodium malate. When the activity of Juzen-taiho-to assessed after Rehmanniae Radix with the application of steaming (Jukujio) was removed from this formulation, the actual activity agreed with the theoretical activity. The other herbs showed a good correlation between activity and the amount of sodium malate contained, resembling the relationship seen for Toki. These results indicate that sodium malate is the most important component involved in the activity of Juzen-taiho-to and that Jukujio markedly suppresses the action of sodium malate (Ueda et al., 1997).

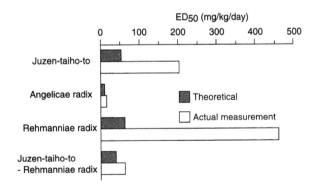

FIGURE 9.18 Comparison of the experimental ED50 with calculated ED50 of Kampo formulation on CDDP-induced nephrotoxicity. Theoretical ED50 was calculated on the basis of the amount of sodium malate contained and the degree of reduction in nephrotoxicity (BUN). (From Sugiyama, K., 1996, *J. Trad. Med.*, 13, 27–41. With permission.)

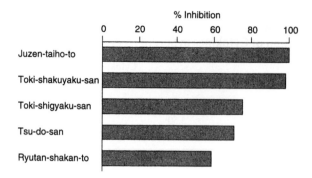

FIGURE 9.19 Effects of Kampo formulations contained a similar amount of Angelicae Radix (Toki) on CDDP-induced nephrotoxicity. Kampo formulations were administered p.o. to mice at a dose of ten times the usual daily dose. BUN was measured on day 17. (From Sugiyama, K., 1996, *J. Trad. Med.*, 13, 27–41. With permission.)

The addition of the actions of the crude drugs to the actions of Toki and Jukujio leads to the manifestation of the overall effects of Juzen-taiho-to. These elaborate interactions among different crude drugs are a characteristic of Kampo formulations. Figure 9.19 shows the data yielded by our recent study, in which the effect in reducing nephrotoxicity was compared among five different Kampo formulations, each of which contained a similar amount of Toki. The effect differed among different formulations; that is, nephrotoxicity was reduced less by formulations indicated for "excess syndrome." This suggests that the toxicity of CDDP pertains to "deficient syndrome" and provides a key to clarifying why the same crude drug is used in different Kampo formulations with different indications (Ueda et al., 1997).

9.5 CONCLUSION

In Kampo medicine, "Ki," "Ketsu," and "Sui" are understood to be essential components of the body, indispensable for life. Individuals are healthy when these three components perform their functions well; illness causes their dysfunction. Kampo medicine treats diseases by causing these three components to resume their functions, thus eliminating the cause of diseases. This is called the "Fusei-Kyoja" (Fu Zheng-Qu Xie in Chinese) rule.

"Kyoja" (Qu Xie in Chinese) means elimination of pathogenic factors. This concept is close to the concept of treatment used in Western medicine. "Fusei" (Fu Zheng in Chinese), the most important concept in Kampo medicine, means resumption of the proper functioning of Ki, Ketsu, and Sui. Western medicine also has concepts corresponding to Fusei, but they are not so clearly defined and are not supported by the actual methods used. The author believes that if the concept of Fusei is incorporated into Western medicine, we may be able to make up for the weak points that exist in the present forms of medicine and achieve satisfactory therapeutic effects.

We studied the Kampo formulations used for the purpose of Fusei (i.e., tonifying Kampo formulations) to provide scientific evidence to the concept of Fusei and to translate the concept into scientific terms. To these ends, we have conducted systematic studies, including the establishment of animal models; evaluations of efficacy; analyses of effective components; clarification of the mechanisms of action; and analyses of the roles played by constituent crude drugs (Sugiyama et al., 1993, Sugiyama et al., 1995).

Conducted within this framework, the present study revealed that the effect of Juzen-taiho-to in reducing the toxicity of CDDP is not attributable to any single component or herb, but rather is due to composite actions of nephrotoxicity-reducing substances (carboxylic acids such as sodium malate), substances controlling the actions of these substances, and substances promoting hemopoiesis (polysaccharides). In recent years, attempts have been made to confirm the effects of Juzen-taiho-to in clinical applications based on the results of our study. Up to now, only fragmentary data have been obtained; however, a number of reports have been produced on cases in which the effectiveness of Juzen-taiho-to against side effects of cisplatin was confirmed. Much expectation is placed now on the application of Juzen-taiho-to in this field.

At present, Kampo medicines are used with increasing frequency in Japan for medical treatment. This may be mainly attributed to the fact that drug preparation technique for Kampo medicines has shown rapid progress and also that Kampo medicines can be used now under the application of health insurance. In addition, distrust of the strong side effects of the drugs used in chemotherapy may be present. It is not that Kampo medicines have no side effects, but rather that, compared with the side effects of the drugs used in chemotherapy, those of Kampo medicines have a very low incidence and are mostly mild when the medicines are used correctly.

Here, description has been concentrated on the fact that Juzen-taiho-to is effective against the side effects caused by chemotherapy drugs or side effects caused by radiotherapy radiation. Many other reports describe the effects of Kampo medicines in alleviating the side effects of the drugs used in chemotherapy. Sakata et al. (1994) and Mori et al. (1998) elucidated that the side effect (diarrhea) of irinotecan hydrochloride is reduced by Hange-shashin-to (Ban-Xia-Xie-Xin-Tang in Chinese). A number of reports confirm that side effects of anticancer drugs such as 5-fluorouracil (5-FU), cyclophosphamide, etc. can be improved by tonics such as Juzen-taiho-to, Ninjin-yoei-to, Hochu-ekki-to, Shikunshi-to (Si-Jun-Zi-Tang in Chinese), etc. Side effects of various types of the drugs used in chemotherapy, such as anticancer drugs, cause a serious social problem. Great expectation is now put on the new application of Kampo medicines when coprescribed with the drugs used in chemotherapy.

Currently, Western medicine is the mainstream of medical treatment. The principle of Western medicine holds that almost all types of diseases can be overcome if the cause of the disease is identified and treatment is performed in accordance with scientific theory and practice. In reality, however, a great number of diseases cannot be completely healed. If it became possible to present highly evaluated data from Kampo medicine using the viewpoint of Western medicine, the latter would become a very useful supplement for overcoming the defective aspects of Western medicine.

ACKNOWLEDGMENTS

The author is indebted to Dr. Harumi Ueda, Dr. Yoshimasa Ichio, and Dr. Yasuhiro Komatsu (Tsumura & Co.) for their support of this study; to Professor Masami Yokota (University of Shizuoka) for his advice and guidance during the pursuit of this study; to Dr. Taichi Ezaki and Kenjiro Matsuno (Kumamoto University) for their cooperation with histopathological examination; and to all members of the Institute of Traditional Chinese Medicines (University of Shizuoka) for their technical assistance.

REFERENCES

Aburada, M., Takeda, S., Ito, E., Nakamura, M. and Hosoya, E. (1983) Protective effects of Juzentaihoto, dried decoctum of ten Chinese herb mixtures, upon the adverse effects of mitomycin c in mice. *J. Pharm. Dyn.*, 6, 1000–1004.

Bompart, G. (1989) Cisplatin-induced changes in cytochrome P-450, lipid peroxidation and drug-metabolizing enzyme activities in rat's kidney cortex. *Toxicol. Lett.*, 48, 193–199.

Brown, S.J., Kellett, P.J. and Lippard, S.J. (1993) Ixr1, a yeast protein that binds to platinated DNA and confers sensitivity to cisplatin. *Science*, 261, 603–605.

Chopra, S., Kaufman, J.S., Jones, T. W., Hong, W. K., Gehr, M.K., Hamburger, R.J., Flamenbaum, W. and Trump, B.F. (1982) *Cis*-diamminedichloroplatinum-induced acute renal failure in the rat. *Kidney Int.*, 21, 54–64.

Courjault, F., Leroy, D., Coquery, I. and Toutain, H. (1993) Platinum complex-induced dysfunction of cultured renal proximal tubule cells. *Arch.Toxicol.*, 67, 338–346.

Daley–Yates, P.T. and McBrien, D.C.H. (1982) The inhibition of renal ATPase by cisplatin and some biotransformation products. *Chem. Biol. Interactions*, 40, 325–334.

Daley–Yates, P.T. and McBrien, D.C.H. (1983) Cisplatin metabolites: a method for their separation and for measurement of their renal clearance *in vivo*. *Biochem. Pharmacol.*, 32, 191–184.

DeNeve, W., Valeriote, F., Tapazoglou, E., Everett, C., Khatana, A. and Corbett, T. (1990) Discrepancy between cytotoxicity and DNA interstrand crosslinking of carboplatin and cisplatin *in vivo*. *Invest. New Drugs*, 8, 17–24.

Dus, D. and Jaworska, J.K. (1982) Cytostatic activity *in vitro* of new *cis*-dichlorodiammine-platinum (II) analogs. *Arch. Immunol. Therap. Exp.*, 30, 357–361.

Feinfeld, D.A. and Fuh, V.L. (1986) Urinary glutathione-S-transferase in cisplatin nephrotoxicity in the rat. *Clin. Chem. Clin. Biochem.*, 24, 529–832.

Gautier, F.C., Grimellec, C.L., Giocondi, M.C. and Toutain, H.J. (1995) Modulation of sodium-coupled uptake and membrane fluidity by cisplatin in renal proximal tubular cells in primary culture and brush-border membrane vesicles. *Kidney Int.*, 47, 1048–1056.

Gonzalez–Vitale, J.C., Hayes, D.M., Cvitkovic, E. and Sternberg, S.S. (1977) The renal pathology in clinical trials of *cis*-platinum (II) diamminedichloride. *Cancer*, 39, 1362–1371.

Guchelaar, H.J., Vries, E.G., E.de, Meijer, C., Esselink, M.T., Vellenga, E., Uges, D.R.A. and Mulder, N.H. (1994) Effect of ultrafilterable platinum concentration on cisplatin and carboplatin cytotoxicity in human tumor and bone marrow cells *in vitro*. *Pharm. Res.*, 11, 1265–1269.

Hisha, H., Yamada, H., Sakurai, M.H., Kiyohara, H., Li, Y., Yu, C., Takemoto, N., Kawamura, H., Yamaura, K., Shinohara, S., Komatsu, Y., Aburada, M. and Ikehara, S. (1997) Isolation and identification of hematopoietic stem cell-stimulating substances from Kampo (Japanese herbal) medicine, Juzen-taiho-to. *Blood*, 90, 1022–1030.

Hisha, H., Kohdera, U., Hirayama, M., Yamada, H., IguchiñUehira, T., Fan, T.X., Cui, Y.Z., Yang, G.X., Li, Y., Sugiura, K., Inaba, M., Kobayashi, Y. and Ikehara, S. (2002) Treatment of Shwachman syndrome by Japanese herbal medicine (Juzen-taiho-to): stimulatory effects of its fatty acids on hemopoiesis in patients. *Stem Cells*, 20, 311–319.

Hosokawa, Y. (1985) Inhibitory effect of Juzen-taiho-to and Hochu-ekki-to against radiation injury in mice. *Kampo Igaku*, 9, 13–17.

Janina, K.J. and Boguslawa, J.T. (1981) New platinum complexes with expected antineoplastic activity. *Polish J. Chem.*, 55, 1143–1149.

Kobayashi, H., Hasuda, K., Taniguchi, S. and Baba, T. (1991) Therapeutic efficacy of two-route chemotherapy using *cis*-diamminedichloroplatinum(II) and its antidote, sodium thiosulfate, combined with the angiotensin-II-induced hypertension method in a rat uterine tumor. *Int. J. Cancer*, 47, 893–898.

Los, G., Verdegaal, E., Noteborn, H.P.J.M., Ruevekamp, M., Graeff, A.D., Meesters, E.W., Huinink, D.T.B. and McVie, J.G. (1991) Cellular pharmacokinetics of carboplatin and cisplatin in relation to their cytotoxic action. *Biochem. Pharmacol.*, 42, 357–363.

Markman, M., Reichman, B., Hakes, T., Rubin, S., Jones, W., Lewis, J.L., Jr., Barakat, R., Curtin, J., Almadrones, L. and Hoskins, W. (1993) The use of recombinant human erythropoietin to prevent carboplatin-induced anemia. *Gynecol. Oncol.*, 49, 172–176.

Micetich, K.C., Barnes, D. and Erickson, L.C. (1985) A comparative study of the cytotoxicity and DNA-damaging effects of *cis*-(diammino) (1,1-cyclobutanedicarboxylato)-platinum (II) and *cis*-diamminedichloroplatinum (II) on L1210 cells. *Cancer Res.*, 45, 4043–4047.

Miyamoto, H., Shigematsu, N., Yamashita, S., Tominaga, S., Kondo, M. and Hashimoto, S. (1985) Effect of Juzen-taiho-to and Hochu-ekki-to against radiation injury. *Shindann To Chiryo*, 7, 153–159.

Mori, K., Hirose, T., Machida, S. and Tominaga, K. (1998) Kampo medicines for the prevention of irinotecan-induced diarrhea in advanced non-small cell lung cancer. *Gan To Kagaku Ryoho*, 25, 1159–1163.

Pera, M.F., Jr., Zook, B.C. and Harder, H.C. (1979) Effect of mannitol or furosemide diuresis on the nephrotoxicity and physiological disposition of *cis*-dichlorodiammineplatinum-(II) in rats. *Cancer Res.*, 39, 1269–1278.

Petsko, G.A. (1995) Heavy metal revival. *Nature*, 377, 580–581.

Rosenberg, B., Van Camp, L. and Krigas, T. (1965) Inhibition of cell division in *Escherichia coli* by electrolysis products from a platinum electrode. *Nature*, 205, 698–699.

Sakata, Y., Suzuki, H. and Kamataki, T. (1994) Preventive effect of TJ-14, a Kampo (Chinese herb) medicine, on diarrhea induced by irinotecan hydrochloride (CPT-11). *Gan To Kagaku Ryoho,* 21, 1241–1244.

Satoh, M., Naganuma, A. and Imura, N. (1988) Metallothionein induction prevents toxic side effects of cisplatin and adriamycin used in combination. *Cancer Chemother. Pharmacol.,* 21, 176–178.

Satoh, M., Kloth, D.M., Kadhim, S.A., Chin, J.L., Naganuma, A., Imura, N. and Cherian, M.G. (1993) Modulation of both cisplatin nephrotoxicity and drug resistance in murine bladder tumor by controlling metallothionein synthesis. *Cancer Res.,* 53, 1829–1832.

Sugihara, K. and Gemba, M. (1986) Modification of cisplatin toxicity by antioxidants. *Jpn J. Pharmacol.,* 40, 353–355.

Sugiyama, K., Yokota, M., Ueda, H. and Ichio, Y. (1993) Protective effects of Kampo medicines against *cis*-diamminedichloroplatinum (II)-induced nephrotoxicity and bone marrow toxicity in mice. *J. Med. Pharm Soc. WAKAN-YAKU,* 10, 76–85.

Sugiyama, K., Ueda, H., Suhara, Y., Kajima, Y., Ichio, Y.and Yokota, M. (1994) Protective effect of sodium L-malate, an active constituent isolated from Angelicae Radix, on *cis*-diamminedichloroplatinum (II)-induced toxic side effect. *Chem. Pharm. Bull.,* 42, 2565–2568.

Sugiyama, K., Ueda, H., Ichio, Y. and Yokota, M. (1995a) Improvement of cisplatin toxicity and lethality by Juzen-taiho-to in mice. *Biol. Pharm. Bull.,* 18, 53–58.

Sugiyama, K., Ueda, H. and Ichio, Y. (1995b) Protective effect of Juzen-taiho-to against carboplatin-induced toxic side effects in mice. *Biol.Pharm.Bull.,* 18, 544–548.

Sugiyama, K. (1996) Effects of Juzen-taiho-to in reducing the side effect of cisplatin. *J. Trad. Med.,* 13, 27–41.

Takahara, P.M., Rosenzweig, A.C., Frederik, C.A. and Lippard, S.J. (1995) Crystal structure of double-stranded DNA containing the major adduct of the anticancer drug cisplatin. *Nature,* 377, 649–652.

Taniguchi, I., Iwasato, K., Sato, M., Terawaki, S., Tomonari, M., Abe, A., Hadano, K. and Hida, K. (1984) Clinical effect of Juzen-taiho-to in radiation therapy. *Kampo Igaku,* 9, 21–23.

Treiber, D.K., Zhai, X., Jantzen, H.M. and Essigmann, J.M. (1994) Cisplatin-DNA adducts are molecular decoys for the ribosomal RNA transcription factor hUBF (human upstream binding factor). *Proc. Natl. Acad. Sci. USA,* 91, 5672–5676.

Ueda, H., Sugiyama, K., Kajima, Y. and Yokota, M. (1997) Role of sodium malate in the inhibitory effect of of Juzen-taiho-to against cisplatin-induced toxic side effect. *J. Trad. Med.,*14,199–203.

Ueda, H., Sugiyama, K., Tashiro S. and Yokota, M. (1998a) Mechanism of the protective effect of sodium malate on cisplatin-induced toxicity in mice. *Biol. Pharm. Bull.,* 21, 121–128.

Ueda, H., Sugiyama, K., Yokota, M., Matsuno, K. and Ezaki, T. (1998b) Reduction of cisplatin toxicity and lethality by sodium malate in mice. *Biol. Pharm. Bull.,* 21, 34–43.

Umeki, S., Watanabe, M., Yagi, S. and Soejima, R. (1988) Supplemental fosfomycin and/or steroids that reduced cisplatin-induced nephrotoxicity. *Am. J. Med. Sci.,* 295, 6–10.

Umeki, S., Tsukiyama, K., Okimoto, N. and Soejima, R. (1989) Urinastatin reducing cisplatin nephrotoxicity. *Am. J. Med. Sci.,* 298, 221–226.

Vogl, S.E., Zaravinos, T. and Kaplan, B.H. (1980) Toxicity of *cis*-diamminedichloroplatinum II given in a 2-hour outpatient regimen of diuresis and hydration. *Cancer,* 45, 11–15.

10 Antitumor Effects in Combination with Chemotherapy

Kazuo Tarao and Takashi Okamoto

CONTENTS

10.1 INTRODUCTION

Juzen-taiho-to, Hochu-ekki-to, and Ninjin-yoei-to are the main agents of the so-called "Hozai" in Kampo medicines, as well as the drugs to activate digestion and absorption in the alimentary tract, to improve the nutritional state, to restore the defensive immune system in the whole body, and to accelerate the healing process as a result. The chemotherapy of anticancer drugs has many side effects, such as gastrointestinal discomfort, disturbances of bone marrow, liver damage, renal damage, and immunologic disturbances. Among the herbal drugs, Juzen-taiho-to and Hochu-ekki-to have been known to alleviate these side effects of anticancer drugs (Okamoto et al., 1993, 1995; Kurokawa and Tamakuma, 1995).

Okamoto et al. (1995) and Kurokawa and Tamakuma (1995) have demonstrated the beneficial effects of Juzen-taiho-to in the alleviation of gastrointestinal side effects of anticancer drugs such as anorexia, epigastric fullness, nausea, and vomiting, in the postoperative periods of gastrointestinal cancer. Miura et al. (1995), Adachi (1987), and Okimoto et al. (1993) have demonstrated the beneficial effects of Juzen-taiho-to in alleviation of side effects of anticancer drugs on the bone marrow and the improvements of anemia and leukopenia as a result. Furthermore, Kurokawa and Tamakuma (1995) demonstrated the inhibitory effects of Juzen-taiho-to on the nonspecific immune suppressive factor, and Konno et al. (1997) demonstrated significant decrease in the ratio of suppressor T-lymphocytes, together with an increase of cytotoxic T-lymphocytes, in patients treated with tegafur + uracil (UFT) and Juzen-taiho-to.

10.2 CANCER OF GASTROINTESTINAL TRACT IN GENERAL

Konno et al. (1997) and Konno (1998) observed the beneficial effects of the administration of Juzen-taiho-to (TJ-48) in postoperative adjuvant chemotherapy for patients with gastric cancer and found improvement in performance status, anorexia, and fatigability (feeling of exhaustion). They compared the performance status (PS) and percentages of patients claiming anorexia and a feeling of exhaustion between the postoperative patients for gastric cancer administered by UFT and Juzen-taiho-to (group A), and those administered by UFT alone (group B). PS was improved and percentages of patients who claimed anorexia and a feeling of exhaustion were decreased in group A. These researchers stressed that the feeling of exhaustion disappeared in all the patients in group A. They also demonstrated a significant decrease in the ratio of suppressor T-lymphocytes, together with an increase in cytotoxic T-lymphocytes in group A patients.

Furthermore, Kurokawa and Tamakuma (1995) examined the benefical effects of Juzen-taiho-to on the improvement of appetite (as an indication of quality of life) in patients operated on for cancer of the gastrointestinal tract (39 cases of stomach cancer, 15 cases of colon cancer, and 11 cases of rectum cancer) treated with anticancer drugs such as tegafur (FT); 5-fluorouracil (5-FU); carmofur (HCFU); UFT; cisplatin (CDDP); adriamycin (ADM); mitomycin C (MMC); and cyclophoshamide (CPA) thereafter. The result was examined after 4 weeks' administration of Juzen-taiho-to, which was shown to be markedly effective in 36.4%, effective in 46.6%, and not effective in 16.0% of patients.

Moreover, Yamada et al. (1991a, b) and Yamada (1992) conducted a clinical study on Juzen-taiho-to (JTT, TJ-48) in a combination use with anticancer drug administration to digestive organ cancers: esophageal carcinoma (46 cases); colorectal carcinoma (35 cases); gastric carcinoma, parcial gastrectomy (53 cases); and gastric carcinoma, total gastrectomy (40 cases). Blood tests and immunological examinations were made pre- and postoperation.

Changes were seen in the red blood cell (RBC) count and hemoglobin content in the gastric carcinoma, total gastrectomy, group at 2 and 3 months, and significant increases were seen in the Juzen-taiho-to group in comparison with the control group (Figure 10.1). Changes were seen in

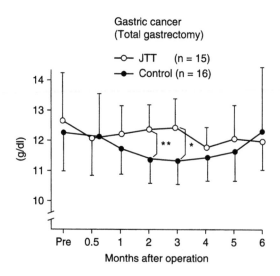

FIGURE 10.1 Changes in hemoglobin content in patients with gastric carcinoma who underwent total gastrectomy and were administered anticancer drugs and Juzen-taiho-to (O) and those receiving anticancer drug only (●). *$p < 0.05$. (From Yamada, T., 1992, *J. Med. Pharm. Soc. WAKAN-YAKU.* 9, 157–164. With permission.)

the white blood cell (WBC) count in the gastric carcinoma, total gastrectomy, group and a distinct decrease was seen in the control group compared to the Juzen-taiho-to group. The WBC count in the Juzen-taiho-to group did not decrease. In phytohemaglutinin (PHA)-induced lymphocyte blastgenesis, differences were seen between the Juzen-taiho-to group and the control group; esophageal carcinoma at postoperative 1 and 3 months, and gastric carcinoma, total gastrectomy, at postoperative 3 and 4 months. The Juzen-taiho-to group was restored to preoperative condition more rapidly than the control group. In natural killer (NK) cell activity, an increase was seen in the Juzen-taiho-to group 1 month postoperative in esophageal carcinoma and gastric carcinoma, total gastrectomy; both were treated with a combination of anticancer drug therapy. A difference was seen between the Juzen-taiho-to group and the control group.

From these observations, and due to the effectiveness of postoperative Juzen-taiho-to administration with anticancer drugs, the recommendation was for treatment to proceed.

10.3 GASTRIC CANCER

Combined treatment of Juzen-taiho-to was administered to patients undergoing curative resection of gastric cancer in which oncolytic chemotherapy was performed within 1 month postoperatively. The administration method of anticancer drugs for gastric cancer is centered on MMC and fluorinated pyrimidine analogues. The side effect of these anticancer drugs is bone marrow suppressive action. The cases assigned to that study were: Juzen-taiho-to administered group, 26 cases; nonadministered group, 17 cases. The changes in hemoglobin (Hb) concentration, WBC, serum albumin level, and cellular immunity were evaluated preoperatively, before administration, and at 1, 2, 3, 4, 5, and 6 months after operation.

At 2 months postoperatively, a significant increase was seen in Hb concentration and WBC counts in the group administered Juzen-taiho-to compared with the nonadministered group. In cellular immunity changes at 2 months postoperatively, the OKT8 increased significantly in the nonadministered group, and OKT4/8 ratio, as well as the helper-T/suppressor T ratio, tended to show a high value in the administered group receiving Juzen-taiho-to.

Moreover, Miura et al. (1985) studied the beneficial effects of Juzen-taiho-to in patients with gastric cancer who underwent chemoimmune therapy after gastrectomy; the study was performed on a total of 59 patients. The 29 group A patients received Juzen-taiho-to in addition to the chemoimmune therapy (MFO therapy; MMC + 5-FU + picibanil [OK-432]), and in the 30 group B patients, only the chemoimmune therapy was performed. Results indicated that leukopenia was less prominent (Figure 10.2) and recovery from weight loss was more rapid in the group A patients (group administered Juzen-taiho-to).

Kuboki et al. (1991) studied the effectiveness of Juzen-taiho-to (7.5 g/day) for patients with inoperable advanced gastric cancer or with advanced pancreatic cancer who were treated with regimen of EAP (etoposide, ADM, CDDP) administration. These researchers surveyed the improvement of side effects such as leukopenia, nausea, or vomiting in the controlled study comprising 23 patients with inoperable advanced gastric cancer and 9 patients with advanced pancreatic cancer (Juzen-taiho-to was administered to 12 cases of gastric cancer and 5 cases of pancreatic cancer in addition to EAP administration). They found improvement in side effects of EAP regimen such as leukopenia, nausea, vomiting, and cancer pain in the group administered Juzen-taiho-to.

Additionally, Kaneuchi et al. (1987) surveyed the ability of Kampo medicines to promote cellular immunity in 25 patients who underwent gastrectomy due to gastric cancer, using blast-formation of lymphocytes by mitogens such as phytohaemagglutinin (PHA), concanavalin A (Con A), and pork wheat mitogen (PWM) and found that Juzen-taiho-to promoted cellular immunity.

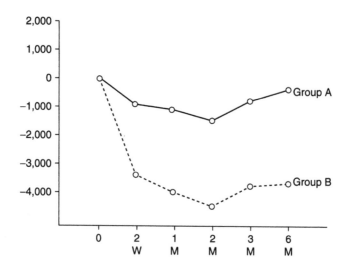

FIGURE 10.2 Changes in WBC counts in group A patients (chemoimmune therapy + Juzen-taiho-to) and in group B patients (chemoimmune therapy only). (From Miura F. et al., 1985, *Geka-shinryo*, 27, 127–130 [in Japanese]. With permission.)

10.4 COLORECTAL CANCER

Toda et al. (1998) surveyed the effectiveness of preoperative and postoperative combination therapy with slow-release tegafur capules and Juzen-taiho-to in patients with colorectal cancer, with special reference to tissue concentrations and thymidine phosphorylase activity. Slow-release tegafur capsules (SF-SP) and Juzen-taiho-to (TJ-48) were administered preoperatively and postoperatively to 51 patients with colorectal cancer. SF-SP alone was also administered in the same manner. Blood and tissue concentrations of tegafur, 5-FU, and tissue thymidine phosphorylase (TP) activity were examined.

In the combination group, the concentration of 5-FU in normal tissue was significantly lower and the tumor–normal tissue ratio of 5-FU tended to be higher. The TP-mediated conversion of tegafur to 5-FU in normal tissue tended to be lower in the combination group. The conversion ratio of tegafur to 5-FU in tumor–normal tissue was significantly higher in the combination group (Figure 10.3). The incidence of hepatotoxicity was lower in the combination group. Juzen-taiho-to (JTT) inhibited the conversion of tegafur to 5-FU in normal tissue and decreased the incidence of hepatotoxicity. The researchers also mentioned that the duration until the first appearance of side effects such as anorexia and nausea tended to be longer in the combination group (263.8 ± 66.3 days) than that in the control group (123.1 ± 28.7 days). They concluded that Juzen-taiho-to can be considered useful as a biochemical modulator in the treatment of colorectal cancer.

10.5 HEPATOCELLULAR CARCINOMA

Yamauchi et al. (1986a, b) examined the effectiveness of Juzen-taiho-to in the treatment of greatly advanced hepatocellular carcinoma (HCC); they compared the clinical course of the unresectable cases of HCC who were treated with transcatheter arterial infusion (TAI) of mitomycin C and Juzen-taiho-to (7.5 g/day), with those treated with TAI alone. They found that clinical symptoms such as anorexia and fatigability are less common in patients treated with TAI and Juzen-taiho-to than in those treated with Juzen-taiho-to alone. They also found that weight loss was not observed in patients treated with TAI and Juzen-taiho-to, compared with significant decrease (about 6 kg)

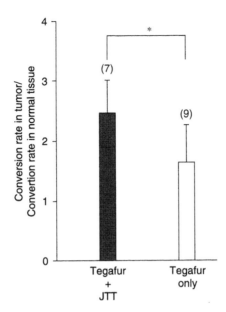

FIGURE 10.3 Difference in the conversion rate of tegafur to 5-FU in tumor and normal tissue between patients with colorectal cancer treated with combination of tegafur and Juzen-taiho-to and those treated with tegafur only. (From Toda, T. et al., 1998, *Jpn. J. Cancer Clin.*, 44, 317–323 [*Gann no Rinsho* in Japanese] With permission.)

in body weight in patients treated with TAI alone (Figure 10.4). Furthermore, they found significant ($p < 0.05$) elongation of the survival periods in the group treated with TAI and Juzen-taiho-to compared with those in the group treated with TAI alone (10.2 months vs. 3.9 months, in mean).

With regard to the antitumor effects of Juzen-taiho-to against HCC, Hirose (1998) demonstrated a marked reduction in size of HCC tumor with marked decrease in the serum α-fetoprotein (AFP) level in a case report (Figure 10.5). He administered Juzen-taiho-to (main)

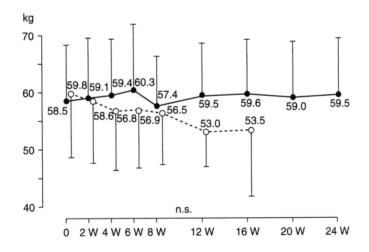

FIGURE 10.4 Changes in body weight in advanced and unresectable HCC patients treated with TAI of MMC and Juzen-taiho-to (—) and those treated with TAI of MMC only (---). (From Yamauchi, H. et al., 1986, *Kampo-igaku*, 10, 17–26 [in Japanese]. With permission.)

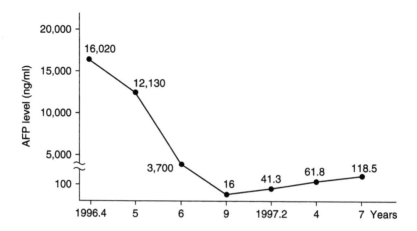

FIGURE 10.5 Time course of AFP in a patient with advanced hepatocelluar carcinoma treated with Juzen-taiho-to. (From Hirose, S., 1998, *Igaku-no-Ayumi*, additional vol., 141–142 [in Japanese]. With permission.)

and Hanshi-ren-to (Ban-Zhi-Lian-Tang in Chinese) to 69-year-old male patients who had greatly advanced HCC tumors and whose expected life span was thought to be about 3 months. He found a marked reduction in the size of HCC tumors with marked decrease in the serum AFP levels as a result.

Moreover, with regard to the direct beneficial effects of Juzen-taiho-to combined with chemotherapy on hepatocellular carcinoma (HCC), we experienced a case of advanced HCC in which combination therapy of UFT and Kampo medicines (Juzen-taiho-to + Ginseng Radix with the application of steaming/Kojin powder) demonstrated a remarkable effectiveness (Figure 10.6). The patient was a 59-year-old male, HBs-antigen-negative, cirrhotic patient confirmed by liver biopsy. On the July 29, 1988, a lesion of HCC was confirmed by US in the right lobe of his liver (S7-S8), and this lesion was treated by transcatheter arterial embolization (TAE) on August 17. After that, on January 11, 1989, a new lesion of HCC was again confirmed by dynamic CT in his left hepatic lobe (S4), and this lesion was also treated by TAE.

After his discharge, the serum level of AFP was increasingly elevated from April of 1989, reaching 57,500 ng/ml on August 7. By this time, the lesions of HCC were scattered in all the bilateral lobes and in the left portal trunk and only transcatheter arterial infusion (TAI) of anticancer drugs and lipiodol was performed on August 14. Because the serum AFP level was still elevated after that regimen, the combination therapy of UFT and Kampo medicines (Juzen-taiho-to [JTT, TJ-48] 7.5 g/day + Kojin-powder 7.5 g/day) was carried out thereafter. Surprisingly, the AFP level decreased sharply to 27.5 ng/ml on December 2, 1989 (Figure 10.6). Regrettably, the patient died from rupture of esophageal varies on the December 28.

Although the combined administration of glycyrrhizin (SNMC) and ursodeoxycholic acid (UDCA) is usually used for intractable patients with active HCV-associated chronic hepatitis (HCV-CH) or cirrhosis (LC), many cases do not respond to this combination therapy. Recently, Tarao et al. (2003) reported the effects of adding Juzen-taiho-to (TJ-48) to lower the levels of serum alanine aminotransferase (s-ALT = s-GPT) in such cases. When the average s-ALT levels for 6 months were compared before and after Juzen-taiho-to was added to the combined therapy of SNMC and UDCA, in the HCV-CH cases ($n = 9$) s-ALT levels were significantly decreased in three of nine (33%) cases in 6 months (about 30 INU on average). In the HCV-LC cases ($n = 12$), s-ALT levels were also significantly decreased in 5 of 12 (42%) cases in 6 months (more than 40 INU on average). Some patients' s-ALT levels decreased significantly after 6 months. As to the improvement of clinical symptoms, general fatigability improved in 12 out of 20 cases (60%) and anorexia improved in 10 out of 19 cases (53%). From these results, the researchers concluded that, added to the

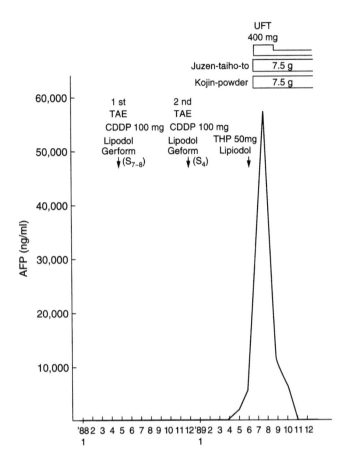

FIGURE 10.6 Time course of 59-year-old males with hepatocelluar carcinoma and non-B liver cirrhosis treated with Kampo medicines (Juzen-taiho-to + Kojin powder) and UFT.

combined therapy of SNMC and UDCA, Juzen-taiho-to (TJ-48, 7.5g daily) may be an effective therapy for intractable cases of active HCV-CH or LC.

10.6 BREAST CANCER

Adachi (1987) and Adachi and Watanabe (1989) studied the effectiveness of Juzen-taiho-to (TJ-48) as an adjuvant therapy to chemoendocrine therapy in patients with advanced breast cancer and found its ability to lengthen survival periods. They divided the 74 patients into two groups at random (37 of group A and 37 of group B). Juzen-taiho-to was administered to the group A patients in addition to the chemoendocrine therapy (chemotherapy: ADM + CPA, or CPA + methotrexate (MTX) + FT; endocrine therapy: tamoxifen (TAM) or fluoxymesterone), but only chemoendocrine therapy was provided to the group B patients.

No significant difference was observed in the background characteristics of the two groups. The researchers found that in the observation periods of 14 and 13 months, 4 of 37 (11%) group A patients died, in contrast to 10 of 37 (27%) group B patients. From the analysis of the cumulative survival curve of Kaplan–Meier, it was determined that the group A patients survived significantly longer than the group B patients ($p < 0.05$; Figure 10.7). Moreover, the side effects of chemoendocrine therapy, such as general malaise and anorexia, improved significantly in the group A patients.

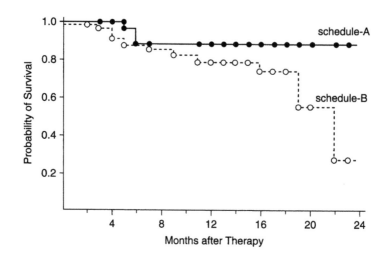

FIGURE 10.7 Cumulative survival curves (A vs. B; $p < 0.05$ with Cox–Mantel test). (From Adachi, I. and Watanabe, T., 1989, *Jpn. J. Cancer Chemother.*, 16, 1538–1543 [in Japanese]. With permission.)

Adachi and Watanabe (1989) and Adachi (1989) also surveyed the effectiveness on survival periods of supporting therapy of Juzen-taiho-to in advanced breast cancer patients. The patients were randomized with the envelope method and divided into two groups (group A and B). Group A was treated with Juzen-taiho-to combined with chemoendocrine therapy; group B was treated with chemoendocrine therapy only. Group A had 58 patients and 61 patients belonged to group B. The patients' characteristics were well balanced in both groups. ADM, CPA, MTX, and FT were administered as chemotherapy. As endocrine therapy, TAM and fluoxynesterone was used.

In the group A patients, 5.0 to 7.5 g of Juzen-taiho-to was administered in addition to the chemoendocrine therapy. Results showed that the cumulative survival rate (Kaplan–Meir method) showed no significant difference in generalized Wilcoxon method up to 38 months. However, in the Greenwood method, there was a tendency of higher survival rate in the group A patients (Juzen-taiho-to added group, $p < 0.1$) in the period of 24 to 27 months from the beginning of the therapy. Moreover, in the Juzen-taiho-to-Sho (Si-Quan-Da-Bu-Tang-Zheng in Chinese; Kampo diagnosis/the set holistic pattern of a patients' symptoms that indicates the appropriate Kampo prescription) group, the survival rate was significantly higher in group A patients ($p < 0.05$) in the Greenwood test, see Figure 10.7). In addition, the quality of life (QOL) was expressed as self-assessment scores for physical condition, appetite, and the coldness of extremities; QOL improved significantly in group A patients. The researchers concluded that the supporting therapy of Juzen-taiho-to was more effective on patients with advanced breast cancer.

10.7 GYNECOLOGICAL CANCERS

Hasegawa et al. (1994) studied the usefulness of Kampo medicines (Ninjin-yoei-to and Juzen-taiho-to) for side effects in chemotherapy of ovarian and uterine cancer. In that study, the 32 patients (19 cases treated with Kampo medicines in addition to chemotherapy and 13 cases treated with chemotherapy alone) received at least one cycle of CAP therapy (CDDP: 50 to 75 mg/m²; ADR: 30 to 40 mg/m²; CPM: 150 to 400 mg/m²). Administration of Kampo medicines (Ninjin-yoei-to [NIN, TJ-108 by Tsumura & Co.]: 7.5 g/day or Juzen-taiho-to [JTT, TJ-48]: 7.5 g/day) was started 1 week before chemotherapy and continued for 5 weeks.

In the group treated with Kampo medicines, the leucocyte count recovered to the pretreatment level at 4 weeks after chemotherapy, and the platelet count recovered at 3 weeks after chemotherapy

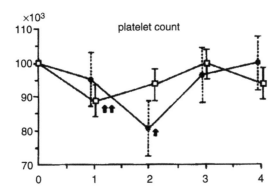

FIGURE 10.8 Effect of Kompo medicines (Ninjin-yoei-to + Juzen-taiho-to) on plantelet cout changes in CAP therapy. A group: CAP therapy + Kampo medicines (Ninjin-yoei-to + Juzen-taiho-to) (□—□) (n = 25); B group: CAP therapy only (●—●) (n = 17); ↑: p < 0.05, ↑↑: p < 0.01 (from Hasegawa K et al., 1994. J. Trad. Met., 11, 181–187 [in Japanese] with permission.)

(Figure 10.8). Blood urea nitrogen levels were significantly reduced in the Kampo medicine groups compared with the nontreated group. Clinical usefulness of Kampo medicines for side effects of anticancer agents such as general fatigue, anorexia, and vomiting was shown in 36.8% of the cases as response and in 89.5% of the cases as slight response. These results indicate that Kampo medicines are clinically useful to reduce side effects of bone marrow suppression such as leucopenia, thrombocytopenia, and nephrotoxicity in gynecologic cancer chemotherapy (CAP therapy).

Furthermore, Fujiwara (1986) studied the effectiveness of Juzen-taiho-to in preventing the side effects of chemotherapy after operation in 18 patients with cancer of gynecological origin and found that general malaise was improved in 86.7% of patients; anorexia was improved in 83.3% of patients; and coldness of extremities was improved in 73.3% of patients. Also, leukopenia in the continued chemotherapy group was prevented in almost all patients. Yoshimura et al. (1984) studied the beneficial effects of Juzen-taiho-to to the clinical discomfort in 30 patients with gynecological cancer (chiefly uterine cancer) who underwent operation and chemotherapy or chemoradiation therapy thereafter. They found that general malaise was improved in 19 of 30 (63.3%); anorexia was improved in 12 (40.0%); coldness in extremities was improved in 5 (33.3%); and that no significant leukopenia was observed in almost all patients.

Additionally, Kure et al. (1985) found an improvement in side effects such as anorexia, nausea, and general fatigue in patients with gynecologic cancer (chiefly uterine cancer) treated with chemotherapy (oral 5-FU administration) after operation (chiefly hysterectomy). Chin (1992) reported a case of advanced ovarian carcinoma (IIIc stage) that showed resistance to CDDP therapy and became cachectic but was finally improved by administration of Kampo medicines (Juzen-taiho-to and Aconiti Tuber with the application of steaming/Kakoo-bushi). The patient was a 66-year-old female whose chief complaint was abdominal fullness in the lower abdomen. Physical examination revealed a huge mass extending to the Douglas cavity and many metastasis, including mesenterium, peritoneum, and omentum. The chemotherapy consisted of 100 mg CDDP, 50 mg ADR, 500 mg CPA, and 500 mg 5-FU 500 and was repeated every 4 weeks.

After the sixth course of this regimen, the result was no change. At the eighth course, the jaundice due to drug injury appeared and the patient's serum AST level reached 1368U./I; and the tenth course, severe anemia appeared and the result was still PD. For improvement of performance status (PS), the chemotherapy was discontinued and Kampo medicines (Juzen-taiho-to and Kakoo-bushi) were administered. Surprisingly, anorexia disappeared after 2 weeks of Kampo therapy; abdominal pain disappeared and anemia was also improved (Hb 10.5 g/dl). The size of the tumor ceased to increase and cachectic tendency stopped. The patient was discharged from hospital but, unfortunately, she slipped down the stairs and injured her cervical vertebra and died.

10.8 UROLOGICAL CANCER

In the urological fields, Sasakawa et al. (1996) demonstrated the beneficial effects of Juzen-taiho-to in preventing side effects of chemotherapy such as CDDP, ADM, MMC, CPA, UFT, and MTX in the treatment of urological malignancy. With regard to the side effects to gastrointestinal tracts such as anorexia, nausea, vomiting, and abdominal fullness, 9 out of 13 (69.2%) patients improved. They mentioned that anorexia was markedly improved. As to the preventie effects of Juzen-taiho-to administration on the side effects of chemotherapy in the hematopoietic system, the researchers mentioned that the amount of granulocyte-colony stimulating factor (G-CSF) required to improve the leukopenia, thrombocytopenia, was reduced by administration of Juzen-taiho-to. Ebisuno et al. (1992) also recognized the same beneficial effects in patients with urological cancer.

Furthermore, Fukui et al. (1992) studied the effectiveness of Juzen-taiho-to on the elongation of survival periods in patients with invasive bladder cancer who underwent total cystectomy followed by adjuvant combination chemotherapy. They found, however, that administration of Juzen-taiho-to had no favorable effects on survival.

10.9 LUNG CANCER

With regard to antitumor effects or immune activation effects of Juzen-taiho-to against lung cancer, Sato et al. (1997) demonstrated the effectiveness in a case study; they treated a case of lung cancer (88-year-old male, probably adenocarcinoma of 7 cm in diameter) with Juzen-taiho-to (7.5 g/day) alone and found very slight enlargement of the tumor during more than 2 years. Sasaki and Fukuda (1993) administered Juzen-taiho-to (7.5 g/day) and Kojin powder (3.0 g/day) to patients with laryngial or pharyngial cancer who were treated by chemoradiation therapy. These researchers found an improvement in the side effects of chemotherapy such as general fatigue, nausea, weight loss, and a loss of vitality.

10.10 CONCLUSION

Juzen-taiho-to has been demonstrated to have beneficial effects in the alleviation of the side effects of anticancer agents such as a feeling of exhaustion, general fatigue, anorexia, vomiting, and bone marrow suppression in general. Therefore, the administration of Juzen-taiho-to is recommended in various kinds of chemotherapy regimens of cancer in almost all clinical fields in order to improve QOL. With regard to the elongation of the survival periods of patients with cancer, Juzen-taiho-to has demonstrated improvement of the cumulative survival rate in combination with chemotherapy only in patients with breast cancer or advanced unresectable hepatocellular carcinoma. In a few cases, Juzen-taiho-to has, by itself, had a direct beneficial effect in eradicating a cancer, especially in patients with lung cancer, hepatocellular carcinoma, and ovarian cancer. Although the mechanism of this phenomenon is unknown, activation of the immune mechanism against cancer by Juzen-taiho-to is the most probable cause.

REFERENCES

Adachi, I. (1989) Juzen-taiho-to as a supporting therapy in advanced breast cancer. *Biotherapy*, 3, 782–788 (in Japanese).

Adachi, I. and Watanabe, T. (1989) Role of supporting therapy of Juzen-taiho-to (JTT) in advanced breast cancer patients. *Jpn. J. Cancer Chemother.*, 16, 1538–1543 (in Japanese).

Adachi, I. (1987) The effectiveness of Juzen-taiho-to as a adjuvant therapy to chemo-endocrine therapy in the patients with advanced breast cancer. *J. Med. Pharm. Soc. for WAKAN-YAKU*, 4, 338–339 (in Japanese).

Chin, Z. (1992) A case of advanced ovarian cancer (IIIc stage) who showed resistance to CDDP therapy and become cachectic but was finally improved by the administration of Kampo medicines (Juzen-taiho-to and Kako-bushi). *Gendai-Toyo-Igaku,* 13, 361–363 (in Japanese).

Ebisuno, S., Watanabe, S., Hirano, A., et al. (1992) Clinical benefits of JTT on the side effects of chemotherapy in patients with urological cancer. *Biotherapy,* 6, 222–226 (in Japanese).

Fujiwara, A.(1986) Beneficial effects of JTT on the improvements of indefinite complaints after the treatments of gynecological malignancy. *Nippon Ishikai Zasshi,* 95, extra pages (in Japanese).

Fukui, I., Gotoh, S., Kihara, K., et al. (1992) Adjuvant chemotherapy for invasive bladder cancer. *Jpn. J. Urol.,* 83, 1633–1639 (in Japanese).

Hasegawa, K., Fukunishi, H., Kiyoshige K., et al. (1994) Clinical usefulness of Kampo medicines (Juzen-taiho-to, Ninjin-yoei-to) for side effects in gynecologic cancer chemotherapy. Effects on reducing side effects of CDDP in CAP therapy. *J. Trad. Med.,* 11, 181–187 (in Japanese).

Hirose, S. (1998). A case of far advanced hepatocellular carcinoma who showed marked decrease in size of the tumor and serum α-fetoprotein level by the administration of Juzen-taiho-to and Hanshi-ren-to. *Igaku-no-Ayumi,* additional vol., 141–142 (in Japanese).

Kaneuchi, S., Nagahama, M., Ogawa, Y., et al. (1987) Activative effects of Kampo medinine in cellular immunity; with special reference to cancer of digestive organ. *JAMA* (Japanese ed.), 22–23.

Konno, H. (1998). Improvement of host immunity by Juzen-taiho-to in the postoperative adjuvant chemotherapy for patients with gastric cancer. *Therapeutic Res.,* 19, 711–713 (in Japanese).

Konno, H., Maruo, Y., Baba, S., et al. (1997) Improvement of host immunity by Juzen-taiho-to in the postop-erative adjuvant chemotherapy for patients with gastric cancer. *Biotherapy,* 11, 193–199 (in Japanese).

Kuboki, M., Shinzawa, Y., and Takahashi, T. (1991). The effectiveness of Juzen-taiho-to in the alleviation of side effects of EAP chemotherapy. *Biotherapy,* 5, 1862–1866 (in Japanese).

Kure, A., Ichikawa, Y., Ohsawa, M., et al. (1985). The effectiveness of Juzen-taiho-to in patients with gynecologic cancer after operation. *Sanka-to-Fujinka,* 5, 539–545 (in Japanese).

Kurokawa T. and Tamakuma S. (1995) The role of Kampo medicines as the combination therapy for cancer. *Prog. Med.,* 15, 155–160 (in Japanese).

Miura F., Saito J., Nakamura K., et al. (1985) The effectiveness of the administration of Juzen-taiho-to in the postoperative adjuvant chemotherapy in patients with gastric cancer. *Geka-shinryo,* 27, 127–130 (in Japanese).

Okamoto, T., Sairenji, M., Motohashi, H., et al.(1995) Effects of Japanese Kampo medicines on restoration after surgery of the patients with digestive tract cancer. *Jpn. J. Gastroenterol. Surg.,* 28, 971–975.

Okimoto, J., Yoshida, K., Tamada, S., et al. (1993) The effectiveness of Juzen-taiho-to on the suppression of bone marrow from the administration of anticancer drugs. *Shindan-to-Chiryo,* 2040–2043 (in Japanese).

Sasakawa, I., Hashimoto, T., Yaguchi, H., et al. (1996). The clinical effects of JTT administration on the anorexia in the chemotherapy of patients with urological malingnancy. *JAMA,* 17, 32–33 (in Japanese).

Sasaki, S. and Fukuda, H. (1993) The effects of combination therapy of Juzen-taiho-to and Kojin powder after the therapy of cervical malignancy. *Prog. Med.,* 13, 1737–1742 (in Japanese).

Sato, H., Ishikawa, H., Ebara, K. et al. (1997). A case of lung cancer which showed extremely slow development under the treatment with Juzen-taiho-to. *Kampo-Shinryo,* 16, 15–17 (in Japanese).

Tarao, K., Okamoto, T., Miyakawa, K., Endo O., Tarao, N., and Masaki T. (2003) Effect of Juzen-taiho-to on the serum alanine aminotransferase levels in patients with HCV-associated chronic hepatitis or liver cirrhosis. *Kampo Med.,* 54, 191–198 (in Japanese).

Toda, T., Matsuzaki, K., and Kawano, T., et al. (1998). Preoperative and postoperative combination therapy with slow-release tegafur capsules and Juzen-taiho-to in patients with colorectal cancer. Tissue con-centrations and thymidine phosphorylase activity. *Jpn. J. Cancer Clin.,* 44, 317–323 (*Gann no Rinsho* in Japanese).

Yamada, T. (1992) Clinical study of Juzen-taiho-to administration for postoperative esohageal carcinoma, gastric carcinoma and colorectal carcinoma. Influence of surgical intervention and postoperative chemotherapy to cell-mediated immunity *J. Med. Pharm. Soc. WAKAN-YAKU.,* 9, 157–164.

Yamada, T., Nabeya, K., and Ri, S. (1991a) Efficacy of Kampo medicines in patients after surgery of alimentary tract. *J. Med. Pharm. Soc. WAKAN-YAKU.,* 8, 207–210 (in Japanese).

Yamada, T., Nabeya, K., Ri, S. (1991b) Postoperative chemotherapy for gastric carcinoma with combined treatment of Juzen-taiho-to. *Biotherapy,* 5, 1850–1856 (in Japanese).

Yamauchi, H., Nakayama, S., Komatsuzaki O., et al. (1986a) The clinical effects of combination therapy of anticancer drugs (chemotherapy) and herbal drugs in the treatment of hepatocellular carcinoma. *Kampo-igaku,* 10, 17–26 (in Japanese).

Yamauchi, H., Nakayama, S., Sato, S., et al. (1986b) Combination therapy of anticancer drugs and herbal drugs in the treatment of hepatocellular carcinoma. *Curr. Ther.,* 4, 79–87.

Yoshimura, O., Aihara, M., Utsunomiya, M., et al. (1984) Beneficial effects of Juzen-taiho-to to the clinical discomfort in patients with gynecological cancer. *Sanka-to-Fujinka,* 51, 963–972 (in Japanese).

11 Toxicology and Side Effects of Juzen-taiho-to

Osamu T. Iijima, Toshihiko Yanagisawa, Hiroshi Takeda, and Teruhiko Matsumiya

CONTENTS

11.1 INTRODUCTION

Traditional remedies such as Japanese herbal medicines (Kampo medicines) have been widely used throughout the world. One reason for their popularity is the belief that herbal medicines are safe. Although this belief is generally correct, some epidemiological surveys have revealed that Chinese herbal medicines do cause severe adverse reactions in some cases (Chan et al., 1992; Shaw et al., 1997). Usually, adverse reactions occur as a result of misuses such as (Chan and Critchley, 1996):

- Long-term use of drugs known to be highly toxic (e.g., preparations containing aconitine or anticholinergics)
- Misidentification of a drug (e.g., herbs containing podophyllin mistaken for the intended harmless herbs)
- Use of doses larger than recommended

In addition to these mistakes, some herbs are known to have potential side effects. For example, Glycyrrhizae Radix and Rehmanniae Radix, which are constitutional crude drugs of Juzen-taiho-to, induce pseudoaldosteronism and gastrointestinal problems respectively in some patients.

Although a scientific approach to assessing toxicity of Kampo medicines in man is very important, only a few formulas have been evaluated in clinical reevaluation systems in man. Unfortunately, Juzen-taiho-to is not one of these formulas. In this chapter, we have reviewed the toxicity of Juzen-taiho-to mostly with data from experimental animals and clinical case reports.

11.2 GENERAL TOXICITY

Toxicological data of the constitutional crude drugs of Juzen-taiho-to may help us to understand the toxicological characteristics of the formula. Hitherto, oral toxicological data of seven of the drugs (with the exception of Astragali Radix, Cinnamomi Cortex, and Rehmanniae Radix) have been reported (Shoji and Kisara, 1975; Tsukui et al., 1975; Komiyama et al., 1977; Hayashi, 1977; Tanaka et al., 1983, 1986; Takase et al., 1983). Most of these data are from single-dose (acute) toxicity studies in mice or rats. In these studies, the investigators administered bolus doses of test substances to obtain toxicological profiles of the crude drugs. Only symptoms or signs difficult to distinguish from bulk toxicity, such as sedation or decrease in motor activity, were found. Oral LD50s of these crude drugs were >6 to 18 g/kg, which is commonly thought to be "relatively harmless" (Eaton and Klaassen, 1995).

11.2.1 SINGLE-DOSE TOXICITY

A single-dose toxicity study in mice and rats has been reported by Minematsu et al. (1989). Both sexes of mice and rats received 3.75, 7.5, or 15 g/kg of Juzen-taiho-to extract (TJ-48) orally. After the 14-day observation period, they were euthanized for gross pathological examination. No deaths or abnormal symptoms occurred throughout the experimental period. The oral LD50s of Juzen-taiho-to were estimated to be more than 15 g/kg in both sexes of rats and mice.

Safety pharmacological studies sometimes may be more practical in estimating acute response to chemical substances. According to one such study, oral administration of Juzen-taiho-to up to 2 g/kg did not show detrimental effects on the function of the central nervous and digestive systems in mice, water and electrolyte metabolism in rats, or respiratory/cardiovascular systems in dogs (Tsumura & Co., internal data).

11.2.2 REPEATED-DOSE TOXICITY

Two 13-week repeated-dose toxicity studies of Juzen-taiho-to in rats were reported (Minematsu et al., 1989; Fujii et al., 1995). Results of the studies are summarized in Table 11.1. There is no item that changed in both studies, except for slight changes in some hematological values.

In the study done by Minematsu et al. (1989), slight decreases in hemoglobin (Hb) and red blood cells (RBCs) and a slight increase of reticulocyte in female rats were observed dose dependently at the end of the administration period. Fujii et al. (1995) also reported similar findings in male rats. In the constitutional crude drugs of Juzen-taiho-to, effects of decreasing hemoglobin (Hb) or RBCs were reported in Glycyrrhizae Radix (Komiyama et al., 1977, Tanaka et al., 1986), Paeoniae Radix (Tanaka et al., 1983), and Cnidii Rhizoma (Tanaka et al., 1983). Although the findings observed in the study of Juzen-taiho-to might be due to the effects of these constitutional crude drugs, the effects were very small and no abnormality was found in histological examination of bone marrow cells.

Based on these data alone, it would be unreasonable to conclude that Juzen-taiho-to is toxic to the hematopoietic system. In other studies, improvement in reduction of peripheral white blood cell, RBC, and hematcrit values (Iijima et al., 1989) and enhancement of colony-forming units in spleen (Kawamura et al., 1989; Hisha et al., 1997) were reported in experimental anemia induced by mitomycin C. As far as we could discover, no study has reported that Juzen-taiho-to induced hematological toxicity in clinical use.

The preceding studies lead us to the conclusion that the safety of Juzen-taiho-to has been demonstrated in repeated dose toxicity studies.

The exploratory behavior in mice after repeated dosing of Juzen-taiho-to was measured by Tanaka and Ichikawa (1993). Juzen-taiho-to was administered to male mice (Crj: CD-1) between 8 to 9 weeks of age at levels of 0, 0.0625, 0.25, and 1.0% in drinking water. The highest dose

TABLE 11.1
Summary of Repeated-Dose Toxicity Studies of Juzen-taiho-to

Study Design	Sex	Dose	Stage	CS	BW	FC	Hem	BC	Oph	Urin	Path
Strain of rats used: Crj-CD(SD); method of administration: by gavage; administration period: 13 weeks; recovery period: 4 weeks; (Minematsu et al., 1989)	Male	0 mg/kg	EDA	—[b]	—	—	—	—	—	—	—
			ER	—	—	—	—	—	—	—	—
		300 mg/kg	EDA	—	—	—	—	IP ↑[c]	—	—	—
			ER	—	—	—	—	—	—	—	—
		1000 mg/kg	EDA	—	—	—	—	IP, TBIL ↑[c]	—	—	—
			ER	—	—	—	—	—	—	—	—
		3000 mg/kg	EDA	—	—	—	—	IP, TBIL ↑[c]	—	—	—
			ER	—	—	—	—	—	—	—	—
	Female	0 mg/kg	EDA	—	—	—	—	—	—	—	—
			ER	—	—	—	—	—	—	—	—
		300 mg/kg	EDA	—	—	—	—	—	—	—	—
			ER	—	—	—	—	—	—	—	—
		1000 mg/kg	EDA	—	—	—	RBC ↓[c]	—	—	—	—
			ER	—	—	—	—	—	—	—	—
		3000 mg/kg	EDA	—	—	—	RBC, Hb ↓[c]	—	—	—	—
			ER	—	—	—	—	—	—	—	—

(continued)

**TABLE 11.1
(Continued)**

Study Design	Sex	Dose	Stage	CS	BW	FC	Hem	BC	Oph	Urin	Path
Strain of rats used: F344/DuCrj; method of administration: mixed in diet;[a] administration period: 13 weeks; recovery period: none; ref. (Fujii et al., 1995)	Male	0%	EDA	—	—	—	—	—	ND	ND	—
		0.2%	EDA	—	—	—	WBC↓c	—	ND	ND	—
		0.8%	EDA	—	—	—	WBC↓c	TG↓c	ND	ND	—
		3.2%	EDA	—	—	—	WBC, RBC, Hb↓c	TG↓c	ND	ND	—
	Female	0%	EDA	—	—	—	—	—	ND	ND	—
		0.2%	EDA	—	—	—	—	—	ND	ND	—
		0.8%	EDA	—	—	—	—	CPK, AST, LDH ↑c	ND	ND	—
		3.2%	EDA	—	—	—	—	CPK, AST, LDH ↑c	ND	ND	—

[a] The mean intakes of the 0.2, 0.8, or 3.2% groups were 142, 561, and 2272 mg/kg in males and 150, 610, and 2458 mg/kg in females, respectively.

[b] No remarkable changes were observed.

[c] These changes were very slight and considered toxicologically insignificant by the authors.

Note: CS: clinical sign, BW: body weight, FC: food consumption, Hem: hematology, BC: blood chemistry, Oph: ophthalmology, Urin: urinalysis, Path: pathology, EDA: end of drug administration period, ER: end of recovery period, RBC: red blood cell, Hb: hemoglobin, WBC: white blood cell, IP: inorganic phosphorus, TBIL: total bilirubin, TG: triglyceride, CPK: creatine phosphokinase, AST: aspartate aminotranspherase, LDH: lactic dehydrogenase, ND: not done corresponds to about 2490 mg/kg/day. Juzen-taiho-to had no adverse effect on the exploratory behavior of mice.

11.3 GENOTOXICITY

11.3.1 MUTAGENICITY

The mutagenic potential of extracts of Juzen-taiho-to and its ten constitutional crude drugs were tested with the Ames test by Fujita et al. (1994). According to them, Juzen-taiho-to extract was mutagenic to *Salmonella typhimurium* TA97, TA98, TA100, and TA102 without mammalian microsomal enzymes (S9 mix). Because of mutagenicity of Cinnamomi Cortex and Paeoniae Radix extract to TA97, TA98, and TA100 without S9 mix, they speculated that the mutagenicity of Juzen-taiho-to might be attributed to these constitutional herbal drugs.

Other studies dealing with the mutagenic potential of extracts of the constitutional drugs have observed mutagenicity against TA98 without S9 mix in Angelicae Radix (Yamamoto et al., 1982), TA 100 with S9 mix in Paeoniae Radix (Czeczot et al., 1990), and TA98 with S9 mix in Astragali Radix (Xue-jun et al., 1991). The latter study also reported that intraperitoneal administration of Astragali Radix and Cinnamomi Cortex extracts was positive in micronucleus assay in TAI strain mice. However, some reports have indicated that Juzen-taiho-to extract did not show any mutagenicity in the Ames test (Tsumura & Co.). The contradicting negative findings were also reported in Cinnamomi Cortex, Paeoniae Radix, and Angelicae Radix (Watanabe et al., 1983).

The conflicting results described here point to the complexity of investigating the pharmacology of a Kampo medicine, especially in *in vitro* assays. One reason for such findings may be variations in the origin of crude drugs, which have many subspecies and are altered by climatic and geographical conditions. There are many variations of crude drugs, from different origins and in different grades, as well as drugs with different pharmacological activity in the market (Kashiwada et al., 1986). These variations can easily affect the results of *in vitro* assays such as the Ames test. Pharmacological research of Kampo medicines should be considered in light of these facts. Contradictory results and complicating factors described point to the need for more information when dealing with the problem.

11.3.2 TERATOGENICITY

A teratogenicity study of Juzen-taiho-to in mice has been reported by Ogata et al. (1993). Pregnant mice (Crj: CD-1) received 500, 1000, 2000, or 4000 mg/kg of Juzen-taiho-to extract (TJ 48) orally from day 7 through day 15 of gestation. All fetuses were removed from the uterus on day 18 of gestation and were examined for external and skeletal anomalies. Body weight gain, organ weights, and reproductive parameters of pregnant mice and skeletal variations and ossification of the fetuses in the groups treated with Juzen-taiho-to extract were not significantly different from those of the control group.

11.4 SIDE EFFECTS AND CONTRAINDICATIONS

It has been thought that Juzen-taiho-to does not have serious side effects or contraindications in clinical use. In fact, severe adverse reaction associated with the prescription has not been reported. However, Juzen-taiho-to includes constitutional crude drugs known to have the potential to cause adverse reactions. A few cases of adverse reactions thought to be related to these drugs are reported next.

11.4.1 PSEUDOALDOSTERONISM

It is well known that Glycyrrhizae Radix (licorice), which is one of the constitutional crude drugs of Juzen-taiho-to, induces pseudoaldosteronism (Conn et al., 1968), the symptoms of which include hypokalemia, hypertension, and edema. In particular, hypokalemia can be a cause of myopathy (potassium deficiency myopathy) (Knochel and Schlein, 1972).

Many studies designed to investigate the pathogenic mechanism of pseudoaldosteronism from licorice have been conducted. It was demonstrated in 1975 that glycyrrhetinic acid, the active principle of licorice might displace aldosterone from its renal receptors (Ulmann et al., 1975). Recently, it has been found that glycyrrhizin derivatives block 11β-hydroxysteroid-dehydrogenase (Stewart et al., 1987; Farese et al., 1991)—the enzyme that converts cortisol into cortisone and therefore protects mineralocorticoid receptors from binding to cortisol (Edwards et al., 1988; Funder et al., 1988). This latter effect has been considered the only pathogenic mechanism of pseudoaldosteronism from Glycyrrhizae Radix because the affinity of glycyrrhetinic acid for its receptors seemed too low for binding *in vivo* to the receptor (Stewart et al., 1987).

Morimoto and Nakajima (1991) reviewed all case reports of pseudoaldosteronism induced by glycyrrhizin derivatives in Japan (1974–1990). According to them, about 60% of all cases were females aged 50 to 79 years. Although pseudoaldosteronism was thought to occur when taking a large amount of Glycyrrhizae Radix (Hanasaki et al., 1976), now it is clear that pseudoaldosteronism does occur in those taking small doses of 150 mg or less of glycyrrhizin, or 2 g or less of Glycyrrhizae Radix extract (Morimoto and Nakajima, 1991; Iwamoto et al., 1991). There was no relationship between the period of Glycyrrhizae Radix intake and the occurrence of pseudoaldosteronism (Morimoto and Nakajima, 1991). Because of a wide variation in sensitivity to glycyrrhizin (Epstein et al., 1977; Cumming et al., 1980), it is reasonable to anticipate that, in small amounts, it can cause severe symptoms in people with hypersensitivity to licorice. However, Juzen-taiho-to has not been reported to cause pseudoaldosteronism. Nonetheless, we should pay attention to the possibility of pseudoaldosteronism because the daily amount of Juzen-taiho-to includes 1.5 g Glycyrrhizae Radix, which may be enough to induce pseudoaldosteronism.

Pseudoaldosteronism, caused by the intake of a large amount of Glycyrrhizae Radix, is thought to be ameliorated after withdrawal of intake. Then, 2- to 3-week therapeutic period is required for normalization of hypokalemia even if potassium is used as the therapeutic agent (Morimoto and Nakajima, 1991).

11.4.2 Gastrointestinal Disorder

Prescriptions that contain Astragali Radix, Cnidii Rhizoma, or Angelicae Radix had been thought to have the potential to cause anorexia, stomach discomfort, nausea, emesis, or diarrhea when administered to patients with infirmity of gastrointestinal tract, or to aggravate symptoms of patients with these conditions. It was thought that two groups of prescriptions met the conditions described here; one is predisposed to cause those disorders, and the other is not so susceptible. Juzen-taiho-to was thought to belong to the latter group (Kikutani, 1982). The only case of gastrointestinal disorder has been reported by Kurokawa et al. (1989). They used Juzen-taiho-to for the purpose of alleviating the negative side effects of anticancer agents and observed only three cases of soft feces in 88 patients. They speculated that the side effect was associated with Rehmanniae Radix. However, no study describing the disorder is available except for clinical descriptions in old Chinese literature and modern clinical care studies.

11.4.3 Rash

It is thought that prescriptions that contain Cinnamomi Cortex or Ginseng Radix may cause a rash. Kanasashi et al. (1995) reported a case of rash that was thought to be caused by the administration of Juzen-taiho-to. The case was inferred to be an allergic reaction against the cinnamic aldehyde that is a major component of Cinnamomi Cortex. The rash was ameliorated after withdrawal of Juzen-taiho-to.

The Committee for National Health Insurance of the Japan Society for Oriental Medicine (1988) used a mailed questionnaire to collect data about side effects that were suspected to be caused by administration of Kampo medicines. According to the 1246 doctors who answered the questionnaire, side effects caused by administration of Juzen-taiho-to were rare; only a case of a sense of incompatibility, a case of edema, and a case of nettle rash were recorded.

Because of a lack of data concerning administration to pregnant women, it is recommended that the prescription be administered to pregnant or lactating women only when the therapy is judged to be more beneficial than the associated risks. Safety to children has not been established.

11.5 INTERACTION WITH MODERN MEDICINES

A few studies concerning interaction of Juzen-taiho-to with modern medicines have been reported. Whether given singly or repeatedly, Juzen-taiho-to had no effect on morphine-induced analgesia in rats (Yamamoto et al., 1990). Juzen-taiho-to also had no effect on the rate and extent of bioavailability or renal excretion of levofloxacin (Hasegawa et al., 1995). Although no scientific study has dealt with this problem, it is easy to speculate that combined use of Juzen-taiho-to with agents that include licorice is likely to cause pseudoaldosteronism, hypokalemia, or myopathy. Because it is suspected that more than 1000 medical preparations, nutrients, and dietary supplements contain Glycyrrhizae Radix, attention should be paid to the possible combined use of Juzen-taiho-to with these preparations (Morimoto and Nakajima, 1991).

With regard to the toxicology and side effects of Juzen-taiho-to, only a little information is available. Hitherto, Juzen-taiho-to seems to be a relatively safe drug. On the other hand, side effects of Kampo medicines sometimes appear to be dose independent and the reactions may be idiosyncratic (Shaw et al., 1997). Care must be taken during very long-term administration; use with old people or infants; or people in pregnant or critical conditions. Efforts to collect toxicological data and other safety information regarding Kampo medicines, including Juzen-taiho-to, should continue.

REFERENCES

Chan, T.Y.K., Chan, A.Y.W., and Critchley, J.A.J.H. (1992) Hospital admissions due to adverse reactions to Chinese herbal medicines. *J. Tropical Med. Hyg.*, 95, 296–298.

Chan, T.Y.K. and Critchley, J.A.J.H. (1996) Usage and adverse effects of Chinese herbal medicines. *Hum. Exp. Toxicol.*, 15, 5–12.

Committee for National Health Insurance, the Japan Society for Oriental Medicine. (1988) A questionnaire relating to the problems between national health insurance and Kampo medicine. *Jpn. J. Orient. Med.*, 38, 191–217 (in Japanese).

Conn, J.W., David, R.R., and Edwin, L.C. (1968) Licorice-induced pseudoaldosteronism. *JAMA*, 205, 492.

Cumming, A.M., Boddy, K., Brown, J.J., Fraser, R., Lever, A.F., Padfield, P.L., and Robertson, J.I. (1980) Severe hypokalaemia with paralysis induced by small doses of licorice. *Postgrad. Med. J.*, 56, 526–529.

Czeczot, H., Tudek, B., Kusztelak, J., Syzmczyk, T., Dobrowolska, B., Glinkowska, G., Malinowski, J., and Strzelecka, H. (1990) Isolation and studies of the mutagenic activity in the Ames test of flavonoids naturally occurring in medical herbs. *Mutation Res.*, 240, 209–216.

Eaton, D.L. and Klaassen, C.D. (1995) Principle of toxicology. In C.D. Klaassen, M.O. Amdur, and J. Doull (Eds.), *Casarett and Doull's Toxicology* (5th ed.), McGraw-Hill, New York, 13–33.

Edwards, C.R., Stewart, P.M., Burt, D., Brett, L., McIntyre, M.A., Sutanto, W.S., de Kloet, E.R., and Monder, C. (1988) Localization of 11 beta-hydroxysteroid dehydrogenase—tissue specific protector of the mineralocorticoid receptor. *Lancet*, 2, 986–989.

Epstein, M.T., Espiner, E.A., Donald, R.A., and Hughes, H. (1977) Effect of eating licorice on the renin–angiotensin aldosterone axis in normal subjects. *Br. Med. J.*, 19, 488–490.

Farese, R.V., Jr., Biglieri, E.G., Shackleton, C.H., Irony, I., and Gomez–Fontes, R. (1991) Licorice-induced hypermineralocorticoidism. *N. Engl. J. Med.*, 325, 1223–1227.

Fujii, T., Mikuriya, H., Yano, N., Yuzawa, K., Nagasawa, A., Tada, Y., Fukumori, N., Ikeda, T., Takahashi, H., Sakamoto, Y., and Sasaki, M. (1995) Toxicity test of Juzen-taiho-to in rats by dietary administration for 13 weeks. *Annu. Rep. Tokyo Metr. Res. Lab. P. H.*, 46, 243–253 (in Japanese).

Fujita, H., Aoki, N., and Sasaki, M. (1994) Mutagenicity of Juzen-taiho-to with *Salmonella typhimurium*. *Annu. Rep. Tokyo Metr. Res. Lab. P. H.*, 45, 185–187 (in Japanese).

Funder, J.W., Pearce, P.T., Smith, R., and Smith, A.I. (1988) Mineralocorticoid action: target tissue specificity is enzyme, not receptor, mediated. *Science*, 242, 583–585.

Hanasaki, N., Katoo, H., Shinomura, Y., Nakao, K., and Morimoto, Y. (1976) Pseudoaldosteronism induced by administration of a massive dose of glycyrrhizin. *Jap. J. Clin. Med.*, 34, 390–394.

Hasegawa, T., Yamaki, K., Muraoka, I., Nadai, M., Takagi, K., and Nabeshima, T. (1995) Effects of traditional Chinese medicines on pharmacokinetics of levofloxacin. *Antimicrob. Agents Chemother.*, 39, 2135–2137.

Hayashi, M. (1977) Pharmacological studies of Shikon and Toki. (1) Pharmacological effects of the water and ether extracts. *Yakugaku Zasshi*, 73, 177–191.

Hisha, H., Yamada, H., Sakurai, M.H., Kiyohara, H., Li, Y., Yu, C., Takemoto, N., Kawamura, H., Yamaura, K., Shinohara, S., Komatsu, Y., Aburada, M., and Ikehara, S. (1997) Isolation and identification of hematopoietic stem cell-stimulating substances from Kampo (Japanese herbal) medicine, Juzen-taiho-to. *Blood*, 90, 1022–1030.

Iijima, O.T., Fujii, Y., Kobayashi, Y., Kuboniwa, H., Murakami, C., Sudo, K., Aburada, M., Hosoya, E., and Yamashita, M. (1988) Protective effects of Juzen-taiho-to on the adverse effects of mitomycin C. *Nippon Gan Chiryo Gakkai Shi*, 20, 1277–1282 (in Japanese).

Iwamoto, H., Adachi, S., Hinoshita, F., and Shiigai, T. (1991) Two cases of Sho-saiko-to induced pseudoaldosteronism confirmed by loading test. *Jin to to-seki*, 30, 647–650 (in Japanese).

Kanasashi, M., Kitamura, K., Osawa, J., and Ikezawa, Z. (1995) Cinnamic aldehyde gan'yu kampo-yaku ni yoru yakushin no 4 syourei. *Hifuka no Rinsyou*, 37, 715–719 (in Japanese)

Kashiwada, Y., Nonaka, G., and Nishioka, I. (1986) Tannins and related compounds. XLVIII. Rhubarb. (7). Isolation and characterization of new dimeric and trimeric procyanidins. *Chem. Pharm. Bull.*, 34, 4083–4091.

Kawamura, H., Maruyama, H., Takemoto, N., Komatsu, Y., Aburada, M., Ikehara, S., and Hosoya, E. (1989) Accelerating effect of Japanese Kampo medicine on recovery of murine haematopoietic stem cells after administration of mitomycin C. *Int. J. Immunother.*, 5, 35–42.

Kikutani, T. (1982) Kampo-yaku no fukusayou (II). *Nihon Yakuzaishi Gakkaishi*, 34, 727–731 (in Japanese).

Knochel, J.P. and Schlein, E.M. (1972) On the mechanism of rhabdomyolysis in potassium depletion. *J. Clin. Invest.*, 51, 1750–1758.

Komiyama, K., Kawakubo, Y., Fukushima, T., Sugimoto, K., Takeshima, H., Ko, Y., Sato, T., Okamoto, M., Umezawa, I., and Nishiyama, Y. (1977) Acute and subacute toxicity test on the extract from Grycyrrhiza. *Pharmacometrics (Oyo Yakuri)*, 14, 535–578 (in Japanese).

Kurokawa, K., Imai, J., and Tamakuma, M. (1989) Kougan-zai fukusayou ni taisuru Juzen-taiho-to (TJ-48) no rinsyou-teki, men'eki-teki kentou. *Prog. Med.*, 9, 819–825 (in Japanese).

Minematsu, S., Sudo, K., Suzuki, W., Iijima, O., Taki, M., Kanitani, M., Uchiyama, T., Kobayashi, Y., Kuboniwa, H., Kobayashi, N., Yoshida, C., Izumi, K., Kato, C., Hirakawa, Y., Funo, S., Fujii, Y., Aburada, M., and Hosoya, E. (1989) Safety evaluation of Juzentaiho-to. Acute toxicity study in mice and rats and 13-week subacute toxicity study in rats with 4-week recovery period. *Pharmacometrics (Oyo Yakuri)*, 38, 215–229 (in Japanese).

Morimoto, Y. and Nakajima, C. (1991) Pseudoaldosteronism induced by licorice derivatives in Japan. *J. Med. Pharm. Soc. WAKAN-YAKU*, 8, 1–22.

Ogata, A., Kubo, Y., Ando, H., Ysuda, I., and Sasaki, M. (1993) Teratological tests of Juzen-taiho-to extract in ICR mice. *Annu. Rep. Tokyo Metr. Res. Lab. P. H.*, 44, 268–273 (in Japanese).

Shaw, D., Leon, C., Kolev, S., and Murray, V. (1997) Traditional remedies and food supplements. A 5-year toxicological study (1991–1995). *Drag Safety*, 17, 342–356.

Shoji, T. and Kisara, K. (1975) Pharmacological studies of crude drugs showing antitussive and expectorant activity (report 1). The combined effects of some crude drugs in antitussive activity and acute toxicity. *Pharmacometrics (Oyo Yakuri)*, 10, 407–415 (in Japanese).

Stewart, P.M., Wallace, A.M., Valentino, R., Burt, D., Shackleton, C.H., and Edwards, C.R. (1987) Mineralocorticoid activity of liquorice: 11-beta-hydroxysteroid dehydrogenase deficiency comes of age. *Lancet*, 1, 821–824.

Takase, M., Narita, T., Yoshioka, K., Azuma, H., Kiriki, N., Fukuda, S., Shinohara, K., Tadano, K., Okada, T., Arano, R., and Muramoto, A. (1983) A comparative study of acute toxicity and pharmacological activity between Ninjin (*Panax ginseng*) and Den-shichi (*Panax notoginseng*). *Jpn. Pharmacol. Ther.,* 11, 1173–1191 (in Japanese).

Tanaka, S., Takahashi, A., Onoda, K., Kawashima, K., Nakaura, S., Nagao, S., Endo, T., Ohno, Y., Kawanishi, T., Takanaka, A., Kasuya, Y., Yoshihira, K., Fukuoka, M., Sekita, S., Suzuki, H., Harada, M., Natori, N., Nakaji, Y., Kobayashi, K., Yasuhara, K., Saito, M., Ishii, Y., Tobe, M., and Shoji, J. (1983) Toxicological studies on biological effects of the herbal drug extracts in rats and mice—peony root, peach kernel, Japanese angelica root and Cnidium rhizome. *Yakugaku Zasshi*, 103, 937–955 (in Japanese).

Tanaka, S., Takahashi, A., Onoda, K., Kawashima, K., Nakaura, S., Nagao, S., Ohno, Y., Kawanishi, T., Nakaji, Y., Kobayashi, K., Suzuki, S., Naito, K., Uchida, O., Yasuhara, K., Takada, K., Saito, M., Sekita, S., Ozaki, Y., Suzuki, H., Takanaka, A., Tobe, M., and Harada, M. (1986) Toxicological studies on biological effects of the herbal drug extracts in rats and mice. II. Mountain bark, Glycyrrhiza and Bupleurum root. *Yakugaku Zasshi*, 106, 671–686 (in Japanese).

Tanaka, T. and Ichikawa, H. (1993) Effects of Kampo extract medicine "Juzen-taiho-to" on male mice behavior. *Annu. Rep. Tokyo Metr. Res. Lab. P. H.,* 44, 265–267 (in Japanese).

Tsukui, M., Otsuka, H., Suji, K., and Matsuoka, T. (1975) Effects of "Toki-shakuyaku-san" (one of prescriptions in Chinese medicine) on mice. *Jpn. J. Orient. Med.,* 25, 186–190 (in Japanese).

Ulmann, A., Menard, J., and Corvol, P. (1975) Binding of glycyrrhetinic acid to kidney mineralocorticoid and glucocorticoid receptors. *Endocrinology,* 97, 46–51.

Watanabe, F., Morimoto, I., Nozaka, T., Koyama, M., and Okitsu, T. (1983) Mutagenicity screening of hot water extracts from crude drugs. *Shoyakugaku Zasshi,* 37, 237–240.

Xue–jun, Y., De–xiang, L., Hechuan, W., and Yu, Z. (1991) A study on the mutagenicity of 102 raw pharmaceuticals used in Chinese traditional medicine. *Mutation Res.,* 260, 73–82.

Yamamoto, H., Mizutani, T., and Nomura, H. (1982) Studies on the mutagenicity of crude drug extracts, I. *Yakugaku Zasshi,* 102, 596–601.

Yamamoto, H., Kishioka, S., Osaki, M., Miyamoto, Y., Kitabata, Y., Morita, N., and Yamanishi, T. (1990) Effects of Juzen-taiho-to on morphine-induced analgesia in rats. *Pharmacometrics (Oyo Yakuri),* 40, 1–6 (in Japanese).

12 Formulations Related to Juzen-taiho-to: Hochu-ekki-to and Ninjin-yoei-to

Yasuhiro Komatsu

CONTENTS

12.1 INTRODUCTION

In Kampo medicines, a group of formulae called "Hozai" in Japanese, "Buji" in Chinese, and "tonics" in English typically has three formulations—namely, Juzen-taiho-to, Hochu-ekki-to, and Ninjin-yoei-to—in clinical use in the Japanese medical care system (see Appendix 1 and Appendix 2). Juzen-taiho-to and Ninjin-yoei-to have the same clinical indications and are applied to patients with symptoms of fatigue, anemia, or loss of appetite; recovering from surgery; suffering from chronic disease; and needing to recover from a general condition. Hochu-ekki-to also has been applied to patients with loss of appetite and deteriorated body conditions during chronic disorders.

Hochu-ekki-to, however, has more complicated indications in clinical applications than Juzen-taiho-to and Ninjin-yoei-to, e.g., common cold, tuberculosis, persistent malaria, chronic gonorrhea, hemorrhoids, gastroptosis, rectal prolapse, uterine prolapse, hernia, partial paralysis, etc. Hochu-ekki-to is the only formulation that has such wide clinical indications among the three Hozai. This is really Kampo medicine's theory for clinical treatments. All these indications, however, have not yet been fully elucidated by scientific evidence in the fields of clinical and basic biomedicine. In this chapter, the differences among the biomedical activities of the three Hozai Kampo medicines will be discussed.

12.2 NINJIN-YOEI-TO

Ninjin-yoei-to is described in the classic book of traditional Chinese medicine (TCM), *Taihei Keimin Wazai-Kyokuho* (*Da ping hui ming he ji ju fang* in Chinese), published the 1400s. Originally,

Ninjin-yoei-to was prescribed to patients with the following symptoms:

- Muscle pain
- Rapid and shallow respiration
- Asthmatic respiratory sound
- Severe pain in back and loin
- Dry lips and throat with palpitation
- Loss of sense of taste
- Loss of appetite
- Sweat
- Habitual chills in foot and hand
- Anemia

In ancient China, people used Ninjin-yoei-to for elderly people to prevent or cure amnesia, to strengthen will and intellectual power, and to prolong longevity. Ninjin-yoei-to was called a tonic agent because it helped man to live long without aging. In general, Ninjin-yoei-to helps to normalize deteriorated body conditions. In modern Japan, Ninjin-yoei-to is used for anemia of pregnant women and for prevention or retardation of adverse reactions of antitumor chemotherapeutic agents.

12.2.1 CLINICAL APPLICATION OF NINJIN-YOEI-TO

Ninjin-yoei-to was traditionally applied to patients with the following symptoms: irregular menstruation with scanty pale discharge; sallow complexion; palpitation; dizziness; shortness of breath; fatigue; poor appetite; pale tongue and anemia that are attributed to deficiency of liver-blood, controlling blood, spleen energy and vital energy.

12.2.2 CONSTITUENTS OF THE CRUDE DRUGS IN NINJIN-YOEI-TO

Each crude drug constitution of the three "Hozai" Kampo medicines is show in Appendices 1 and 2. In Ninjin-yoei-to crude drug constituents, 9 out of 12 medicinal plants are the same as those for Juzen-taiho-to. Therefore, the biological activities of Ninjin-yoei-to are very similar to those of Juzen-taiho-to. Examples of such biological activities are augmentation of immune responses, effects on the bone marrow stem cells, and increase of appetite. Ninjin-yoei-to contains four different medicinal plants: Polygalae Radix, Schisandrae Fructus, Aurantii Nobilis Pericarpium, and Atractilodi Rhizoma.

In ancient China, plants such as Polygalae Radix, Schisandrae Fructus, and Aurantii Nobilis Pericarpium were well-known drugs believed to stimulate increase of the intellect or mental power in amnesia patients. The antiamnesia activity of Ninjin-yoei-to has a special effect different from those of Juzen-taiho-to and Hochu-ekki-to. At present, there are very few scientific data to elucidate their actions.

12.2.3 DIFFERENT BIOLOGICAL ACTIVITIES BETWEEN NINJIN-YOEI-TO
AND JUZEN-TAIHO-TO

Ninjin-yoei-to has almost the same biological activities and clinical indications as Juzen-taiho-to. However, Ninjin-yoei-to also shows a different effect on the brain function. Diagnostic tests distinguish Ninjin-yoei-to clinical use from that of Juzen-taiho-to by the existence of an amnesic symptom. From ancient times, Polygalae Radix, Schisandrae Fructus, and Aurantii Nobilis Pericarpium have been thought to enforce "Ki" ("Qi" in Chinese; biological function and bioenergetics). From a point of the stimulated Ki, they could improve brain functions. However, under modern biomedical science, the mechanism of these actions cannot be depicted.

Yamamoto (1995) reported that Ninjin-yoei-to treatments improved symptoms of Alzheimer's disease in patients who showed incontinence and an unstable state. Experimentally, Egashira et al. (1996) reported that Ninjin-yoei-to administration improved amnesia in rats induced by scopolamine injection in terms of the methods of passive avoidance tasks. In this study, rats were subjected to electric shock in a step-through type apparatus for the passive avoidance task. After the training, the rats received the test samples and were injected with 0.5 mg/kg of scopolamine intraperitoneally and their amnesia was assessed. The experimental results indicated that the treatment of 1.0 g/kg of Ninjin-yoei-to was able to bring about recovery from amnesia to the normal response level. However Juzen-taiho-to and Hochu-ekki-to could not effect this recovery response.

These researchers also tested the effect of the three Hozai on the cholinergic nervous system. Ninjin-yoei-to (1 g/kg) reduced the tremor induced by an intraperitoneal injection of 0.3 mg/kg oxotremorine to the same degree as that caused by the action of tetrahydroaminoacridine hydrochloride. The preceding indicates that Ninjin-yoei-to stimulates cerebral function.

Yabe et al. (1995, 1996) and Yabe and Yamada (1996) have studied the effects of Ninjin-yoei-to on the function of the central nervous systems in laboratory animals. They have studied the inducing activity of Ninjin-yoei-to for a neurotrophic factor and expression of its receptors in cultured astrocytes of rats. They cultured isolated basal forebrain cells from 17 to 19 days' gestation of rats with or without Ninjin-yoei-to and Juzen-taiho-to. Then, NGF (nerve growth factor) and NGF mRNA in the cultured cells were estimated. Ninjin-yoei-to stimulated induction of NGF in the culture fluid in a dose-dependent manner and mRNA of NGF increased in the cultured cells. Juzen-taiho-to, however, did not induce any NGF.

All 12 medicinal plants in Ninjin-yoei-to were tested for NGF-inducing activity by the same *in vitro* experimental system. As a consequence of the study, Polygalae Radix and Ginseng Radix were found to be the active medicinal plants. Further *in vivo* tests revealed that the content of NGF was increased in cortical cortex of aged rats (16 months old) given orally 400 to 500 mg/kg of Ninjin-yoei-to for 1 month, but not in young rats. This suggests that Ninjin-yoei-to may be a useful agent for prophylaxis of senile dementia in aged people and contributes to understanding of why the ancient Chinese used Ninjin-yoei-to for patients with amnesic symptoms.

On the basis of this phenomenon, the question concerns which medicinal plant is most active. Yabe et al. (2003) reported that Polygalae Radix and Ginseng Radix are active drugs in Ninjin-yoei-to for NGF induction of cultured cerebral cells. Polygalae Radix, in particular, has a very potent activity and onjisaponins were identified as its active ingredient for NGF induction in astorocytes (Yabe et al. 2003). However, more precise studies are needed in order to use Ninjin-yoei-to in clinical practice for the prophylaxis of senile dementia.

Yabe and Yamada (1996) and Yabe et al. (1997) reported that the other formulation, Kami-untan-to (Jia-Wei-Wen-Dan-Tang in Chinese) (Appendix 2), which contains 13 dried crude drugs including Polygalae Radix, but not related to Hozai, showed almost the same biological activities as Ninjin-yoei-to. When 2-year-old rats were given Kami-untan-to orally, their passive avoidance behavior was significantly improved. They also showed an increase in choline acetyltransferase (ChAT) activity and induction of mRNA of NGF in the frontparietal cortex. This meant that Ninjin-yoei-to and Kami-untan-to containing Polygalae Radix could be useful drugs for prophylaxis of developing senile dementia.

It is generally believed that herbal products have a lot of antioxidative agents in them. Egashira et al. (1999a, b) investigated antioxidative activity by examining the scavenging action of Ninjin-yoei-to on superoxide, hydroxyl radical, DPPH radical, and alkyl radical stimulated by t-BuOH using electron spin resonance (ESR) spectrometry. In *in vitro* experiments, Ninjin-yoei-to scavenged superoxide, hydroxyl radical, and DPPH radical in a dose-dependent manner at concentrations of 5 to 0.002 mg/ml from hepatic cells and brain homogenate. Commonly, herbal medicines contain flavonoids and other active compounds for scavenging superoxides. When Ninjin-yoei-to was given orally to rats for 5 weeks, superoxide induction from leukocytes stimulated with PMA was suppressed, but the Kampo medicines significantly increased the lipid peroxides in plasma, brain, and liver.

This discrepancy cannot be explained at the moment. Ninjin-yoei-to, however, has been previously reported to induce many cytokines, augment immune responses, and stimulate phagocytic activities of macrophages and leucocytes (Nakai et al., 1996). Therefore, the elevation of superoxide in the plasma was explained by reason of causing immunostimulating activity and induction of interleukines and other cytokines. Then, the superoxide scavenging activity in Ninjin-yoei-to-treated leucocytes could be explained also by the flavonoids and the polyphenolic compounds in it, when the leucocytes release or induce such active oxygen-reactive compounds to eliminate foreign substances as non-self. This antioxidative action will benefit aging and elderly people.

Wang et al. (1994) reported biological activity of Polygalae Radix to the central nervous system (CNS). They revealed that the active compounds inducing prolongation of sleeping time by hexobarbital were 3,4,5-trimethoxy cinnamic acid, metyl 3,4,5-trimethoxy cinnamate, and *p*-methoxy cinnamic acid in rats. Probably these compounds' effects on the sedative and other unknown chemicals will act on some other function of the CNS such as improving the memory system. More precise studies are required in this area.

12.3 HOCHU-EKKI-TO

Hochu-ekki-to formulation is one of the very popular Kampo medicines in Japanese medical care systems. The formulation is described in the classic medical book, *Nai-Gai-Sho-Ben-Waku-Ron*, in Japanese (*Nei-wai-shang-bian-huo-lun* in Chinese) written by Li Dong Yuan, in the 13th century. This formulation has a wide range of clinical indications for diseases. Originally, it was said that Hochu-ekki-to could give vital energy ("Bu-Qi" in Chinese) by stimulating the central nervous system and by the absorption of nutrients from the gut to patients with symptoms of severe chronic fatigue, feelings of malaise, and coldness in the toe, foot, and leg.

12.3.1 CLINICAL APPLICATION OF HOCHU-EKKI-TO

Hochu-ekki-to also has been prescribed to patients with symptoms of poor appetite, gastrointestinal track disorders (gastritis and/or enteritis), prolapse of uterus, apareunia (impotentia coeundi), tuberculosis, and the common cold. These complicated indications, which are typical in Kampo medicine, are not yet fully understood and therefore cannot be elucidated by scientific evidence. The explanation of clinical use from the viewpoint of Kampo medicine is to improve absorption of nutrients from the small intestine and supply them to all organs and tissues to restore their functions to original levels.

12.3.2 CONSTITUENTS OF CRUDE DRUGS IN HOCHU-EKKI-TO

The crude drugs of Hochu-ekki-to are quite different from Juzen-taiho-to and Ninjin-yoei-to (Table 12.1). Hochu-ekki-to has Cimifugae Rhizoma, Zingiberis Rhizoma, and Bupleuri Radix. These have anti-inflammatory activities. Hochu-ekki-to, therefore, shows potent antiallergic and anti-inflammatory effects, such as activity against hepatitis. Hochu-ekki-to has immunopotentiating activities like Juzen-taiho-to and Ninjin-yoei-to. It has been prescribed to patients in the opportunistic condition who are infected with some bacteria, such as MRSA, *Salmonella enteritidis*, and influenza A (Yamaoka et al., 1998; Matsui et al., 1997; Mori et al., 1999). Cancerous patients also are treated with Hochu-ekki-to for the purpose of recovering from the deteriorated general body conditions.

12.3.3 HOCHU-EKKI-TO FOR TREATMENT OF MALE IMPOTENCE

In this section, the effects of Hochu-ekki-to on impotent patients will be discussed. Juzen-taiho-to has no such effect. In Japan today, male sterility is a serious clinical problem. Hormonal agents, testosterone derivatives, are available for therapy but these agents are not always effective. Patients with some defects in their genital organs take Kampo medicines as alternative medicines. The treatment was probably often used in ancient times. In Japan, urologists now are using Hochu-ekki-to for male sterility.

Kazama (1998) reported that the efficacy of Hochu-ekki-to on male sterility was more potent than that of Kallikrein by comparative clinical studies. In this study, 32 patients with idiopathic oligospermia were enrolled and divided into two groups: one (16 patients) treated with Hochu-ekki-to and the other (16 patients) with Kallikrein. Efficacy rate on the increasing sperm concentration was 56.3% in the patients treated with Hochu-ekki-to and 25.0% in patients treated with Kallikrein. Improvement of the sperm motility rate (25.0%) by Hochu-ekki-to was better than that (18.8%) of Kallikrein. These results clearly indicated that Hochu-ekki-to treatment gave the patients beneficial effects for spermatogenesis. Tamaya et al. (1987) also reported that Hochu-ekki-to treatment was effective for improving the low number of sperm in semen of the oligospermia patients.

Inoue et al. (1997) conducted a study to confirm the effectiveness of Hochu-ekki-to on patients with oligospermia and asthenospermia. It is said that these patients usually must take the Kampo medicine for a rather long period to improve their function. Nevertheless, evidence that Hochu-ekki-to directly influenced motility of sperm *in vitro* has already accumulated. Therefore, one would consider it unnecessary to take Hochu-ekki-to for a long time.

A study confirmed the effectiveness of a short treatment with it for the restoration of low sperm concentration and sperm motility: Hochu-ekki-to was given orally to nine healthy male volunteers for 3 or 4 days. The number of sperm and the sperm motility rate were compared before and after treatment. The sperm concentration and the sperm motility increased in seven out of nine subjects and in six out of nine subjects, respectively. According to the results, Hochu-ekki-to could be expected to improve the concentration of sperm and the sperm motility even by short-term treatment and not by the long period of treatment for patients with asthnospermia and/or oligospermia (Nagao et al. 1996).

Nakayama et al. (1994) studied the efficacy of Hochu-ekki-to on spermatogenesis in hamster epididymal cells. In order to elucidate mechanisms of the Hochu-ekki-to spermatogenesis action, they investigated the effect of the sera from male mice, which were treated with Hochu-ekki-to orally, on protein synthesis in cultured hamster epidydimal cells. When the epididymal cells were cultured at 32°C under 5% CO_2 atmosphere for 9 to 10 days, the cultured cell exhibited the characteristic "paving stone" appearance. They assessed the activity of the sera on protein synthesis by measuring [^{3}H]-leucine incorporated in the cells.

The sera treated with Hochu-ekki-to showed a 29.1% increase in [^{3}H]-leucine uptake by the cultured cells. It is assumed that the maturation of spermatozoa involves protein synthesis in, and secretion from, the epithelium. These researchers suggested that Hochu-ekki-to could promote the synthesis of proteins involved in the functional maturation of spermatogenesis in the epididymal cells, although it is still unclear whether it has direct effective substances or whether it induces or produces some active substances in the body.

Nagao et al. (1996) reported that when serum from patients treated with Hochu-ekki-to was added to a culture medium of sperm, the motility and viability of sperm were enhanced and prolonged. Hochu-ekki-to in the body resulted in compounds that directly stimulated sperm motility. However, they were unable to find the potent compounds.

Amano et al. (1995, 1996) tried to identify the active chemicals in Hochu-ekki-to by an analysis of fluorescence spectra from it. First, they examined fluorescence spectra of semen, seminal plasma, and spermatozoa from 500 to 700 nm when excited at 488 nm by an Xe lamp and revealed that intensities of the emission peak were at 622 nm. This peak strongly correlated with the concentration of spermatozoa and the motility of sperm and semen plasma. The fluorescence analysis of Hochu-ekki-to showed an emission peak at 622 nm. These researchers concluded that compounds with an emission peak at 622 nm could influence the restoration of male infertility. Chemical analysis of active substances in this herbal medicine is needed to better understand its efficacy in this beneficial therapy.

REFERENCES

Amano, T., Kunimi, K. and Ohkawa, M. (1995) Analysis of fluorescence spectra from Chinese herbal medicine for male infertility. *Am. J. Chin. Med.,* 23, 213–221.

Amano, T., Hirata, A. and Namiki, M. (1996) Effects of Chinese herbal medicine on sperm motility and fluorescence spectra parameters. *Arch. Androl.,* 37, 219–224.

Egashira, N., Iizuka, S., Ishige, A., Komatsu, Y., Okada, M. and Maruno, M. (1996) Effects of various "Hozai" (formulation with tonic effects) on scopolamine induced amnesia in passive avoidance task. *J. Trad. Med.,* 13, 476–477.

Egashira, T., Takayam, F. and Yamanaka, Y. (1999a) Studies on pharmacological properties of Ninjin-yoei-to, report 1, Free radical scavenging activity of Ninnjin-yoei-to. *J. Trad. Med.,* 16, 108–115.

Egashira, T., Ikebe, A., Inoue, M., Ohyama, M., Nariai, M. Narusako, K., Tanaka, S., Takayama, F. and Yamanaka, Y. (1999b) Studies on pharmacological properties of Ninnjin-yoei-to, report 2. Effects of long-term treatment with Ninnjin-yoei-to on free radical and lipid peroxidation in mouse. *J. Trad. Med.,* 16, 116–122.

Inoue, T., Kawata, A., Hayashi, E., Morimoto, Y., Nagao, K., Fujiwara, Y., Koyama, R., Kagehisa, H., Yumioka, E., Miyata, H., Yamasaki, M., Tohnaka, M., Kawagishi, N., Morimoto, T. and Kanzaki, H. (1997) Influence on the sperm by administration of Hochu-ekki-to potion. *J. Fertilization Implantation,* 14, 73–76.

Kazama, T. (1998) Male sterility. *Curr. Therapy,* 6, 1683–1686.

Matsui, K., Uechi, Y., Horiguchi, A., Yang, G.-Y., Kitada, Y., Ono, Y, Ogata, Y., Wang, X.-X., Li, N., Komatsu, Y., Shimizu, Y. and Yamaguchi, N. (1997) Antibacterial effect of the Kampo herbal medicine, Hochu-ekki-to on methicillin-resistant *Staphylococcus aureus* positive mice. *Jpn. J. Orient. Med.,* 48, 357–367.

Mori, K., Kido, T., Dikuhara, H., Sakakibara, I., Sakata, T., Shimizu, K., Amagaya, S., Sasaki, H. and Komatsu, K. (1999) Effect of Hochu-ekki-to (TJ-41), a Japanese herbal medicine, on the survival of mice infected with influenza virus. *Antiviral Res.,* 44, 103–111.

Nagao, K., Morimoto, Y., Kawata, A., Fujiwara, Y., Koyama, R., Inoue, T., Miyata, H., Yamasaki, M., Tohnaka, M., Kawagishi, N., Kagehisa, H., Morimoto, T., Horikoshi, Y. and Kanzaki H. (1996) Direct effect of traditional Chinese medicines on sperm. *J. Fertilization Implantation,* 13, 174–177.

Nakai, S., Kawakita, T., Nagasawa, H., Himeno, K. and Nomoto, K. (1996) Thymus-dependent effects of a traditional Chinese medicine, Ren-Shen-Yang-Rong-Tang (Japanese name: Ninjin-Youei-To), in autoimmune MRL/MP-lpr/lpr mice. *Int. J. Immunopharmacol.,* 18, 271–279.

Nakayama, T., Goto, Y., Natsuyama, S. and Mori, T. (1994) Accelerating effects of Kampo medicines on protein synthesis in cultured hamster epidermal cells. *Jpn. J. Fertility Sterility,* 39, 278–282.

Tamaya, T., Ohno, Y. and Okada, H. (1987) Treatment of oligospermia with Hochyu-ekki-to and the guidance for the therapy based on Kampo diagnosis. *Jpn. J. Fertility Sterility,* 32, 385–390.

Wang, S.S., Kozuka, K., Asito, K. and Kano, Y. (1998) Pharmacological properties of galenical preparations (XVII): active compounds in blood and bile of rats after oral administrations of extracts of Polygalae Radix. *J. Trad. Med.,* 11, 44–49.

Yabe, T., Toriizuka, K. and Yamada, H. (1995) Choline acetyltransferase activity enhancing effect of Kami-untan-to(KUT) on basal forebrain cultured neurons and lesioned rats. *Phytomedicine,* 2, 41–46.

Yabe, T., Toriizuka, K. and Yamada, H. (1996) Kami-untan-to (KUT) improves cholinergic deficit in aged rats. *Phytomedicine,* 3, 253–258.

Yabe, T. and Yamada, H. (1996) Kami-untan-to enhances cholinacetyltransferase and nerve growth factor mRNA levels in brain cultured cells. *Phytomedicine,* 3, 361–367.

Yabe, T., Iizuka, S., Komatsu, Y. and Yamada, H. (1997) Enhancements of choline acetyltransferase activity and nerve growth factor secretion by Polygalae Radix extract containing active ingredients in Kami-untan-to. *Phytomedicine,* 4, 199–205.

Yabe, T., Tsuchida, H., Kiyohara, H., Takeda, T. and Yamada, H. (2003) Introduction of NGF synthesis in astrocytes by onjisaponins of *Polygala tenuifolia*, constituents of Kampo (Japanese herbal) medicine, Ninjin-yoei-to. *Phytomedicine,* 10, 106–114.

Yamamoto, T. (1995) Chinese medicine for dementia of Alzheimer's type. *J. Trad. Med.,* 12, 382–383.

Yamaoka, Y., Kawakita, T., Kishihara, K. and Nomoto, K. (1998) Effect of traditional Chinese medicine, Bu-Zhong-Yi-Qi-Tang, on the protection against an oral infection with *Listeria monocytogenes*. *Immunopharmacology,* 39, 215–223.

Appendix 1
Component Herbs of Juzen-taiho-to (Shi-Quan-Da-Bu-Tang in Chinese)

Herb	Decoction	Pharmaceutical Preparation*	
Astragali Radix	3.0 g	3.0 g	
(root of *Astragalus membranaceus* Bunge)			
Cinnamomi Cortex	3.0 g	3.0 g	
(bark of *Cinnamomum cassia* Blume)			
Rehmanniae Radix	4.0 g	3.0 g	
(root of *Rehmannia glutinosa* Liboschitz var. *purpurea* Makino)			
Paeoniae Radix	3.0 g	3.0 g	
(roots of *Paeonia lactiflora* Pallas)			
Cnidii Rhizoma	3.0 g	3.0 g	
(rhizome of *Cnidium officinale* Makino)			
Atractylodis Lanceae Rhizoma*		3.0 g	
(rhizome of *Atractylodes lancea* De Candolle)			
Atractylodis Rhizoma	4.0 g		
(rhizome of *Atractylodes japonica* Koidzumi ex Kitamura)			
Angelicae Radix	4.0 g	3.0 g	
(root of *Angelica acutiloba* Kitagawa)			
Ginseng Radix	3.0 g	3.0 g	
(root of *Panax ginseng* C.A. Meyer)			
Poria	4.0 g	3.0 g	
(sclerotium of *Poria cocos* Wolf)			
Glycyrrhizae Radix	1.5 g	1.5 g	
(root of *Glycyrrhiza uralensis* Fischer)			
	29.0 g	28.5 g	(total)

* Granules, TJ-48, Tsumura & Co.

Appendix 2
Composition of Kampo Formulas

Boi-bukuryo-to (Fang-Yi-Fu-Ling-Tang in Chinese)

Herb	
Poria	5.0 g
Glycyrrhizae Radix	2.0 g
Cinnamomi Cortex	3.0 g
Astragali Radix	5.0 g
Sinomeni Caulis et Rhizoma	5.0 g
	20.0 g (total)

Boi-ogi-to (Fang-Yi-Huang-Qi-Tang in Chinese)

	Decoction	Pharmaceutical Preparation (TJ-20)*
Herb		
Glycyrrhizae Radix	2.0 g	1.5 g
Astragali Radix	5.0 g	5.0 g
Atractylodis Lanceae Rhizoma or Atractylodis Rhizoma	3.0 g	3.0 g
Zizyphi Fructus	3.0 g	3.0 g
Zingiberis Rhizoma	3.0 g	1.0 g
Sinomeni Caulis et Rhizoma	5.0 g	5.0 g
	21.0 g	18.5 g (total)

Bukuryo-shigyaku-to (Fu-Ling-Si-Ni-Tang in Chinese)

Herb	
Poria	4.0 g
Ginseng Radix	2.0 g
Glycyrrhizae Radix	2.0 g
Zingiberis Siccatum Rhizoma	1.0 g
Aconiti Tuber	1.0 g
	10.0 g (total)

* Granule by Tsumura & Co.
** Decoction and pharmaceutical preparation composed of same crude drugs with same ratios.

Chorei-to (Zhu-Ling-Tang in Chinese) (TJ-40)**

Herb	
Poria	3.0 g
Polyporus	3.0 g
Alismatis Rhizoma	3.0 g
Talcum	3.0 g
Asini Gelatinum	3.0 g
	15.0 g (total)

Gorei-san (Wu-Ling-San in Chinese)

	Decoction	Pharmaceutical Preparation (TJ-17)*
Herb		
Poria	5.0 g	3.0 g
Polyporus	5.0 g	3.0 g
Alismatis Rhizoma	6.0 g	4.0 g
Cinnamomi Cortex	3.0 g	1.5 g
Atractylodis Lanceae Rhizoma or Atractylodis Rhizoma	5.0 g	3.0 g
	24.0 g	14.5 g (total)

Hacchin-to (Pa-Chen-Tang in Chinese)

Herb	
Astragali Radix	3.0 g
Cnidii Rhizoma	3.0 g
Paeoniae Radix	3.0 g
Rehmanniae Radix	3.0 g
Ginseng Radix	3.0 g
Atractylodis Lanceae Rhizoma or Atractylodis Rhizoma	3.0 g
Poria	3.0 g
Glycyrrhizae Radix	1.5 g
Zizyphi Fructus	1.5 g
Zingiberis Rhizoma	0.5 g
	24.5 g (total)

Hachimi-jio-gan (Ba-Wei-Di-Huang-Wan in Chinese)

	Decoction	Pharmaceutical Preparation (TJ-7)*
Herb		
Poria	3.0 g	3.0 g
Cinnamomi Cortex	1.0 g	1.0 g
Rehmanniae Radix	5.0 g	6.0 g
Dioscoreae Rhizoma	3.0 g	3.0 g
Alismatis Rhizoma	3.0 g	3.0 g
Moutan Cortex	3.0 g	2.5 g
Aconiti Tuber	1.0 g	0.5 g
Corni Fructus	3.0 g	3.0 g
	22.0 g	22.0 g (total)

* Granule by Tsumura & Co.

** Decoction and pharmaceutical preparation composed of same crude drugs with same ratios.

Hange-shashin-to (Ban-Xia-Xie-Xin-Tang in Chinese)

	Decoction	Pharmaceutical Preparation (TJ-14)*
Herb		
Pinelliae Tuber	5.0 g	5.0 g
Scutellariae Radix	2.5 g	2.5 g
Ginseng Radix	2.5 g	2.5 g
Zizyphi Fructus	2.5 g	2.5 g
Glycyrrhizae Radix	2.0 g	2.5 g
Coptidis Rhizoma	1.0 g	1.0 g
Zingiberis Rhizoma	1.0 g	2.5 g
	16.5 g	18.5 g (total)

Hochu-ekki-to (Bu-Zhong-Yi-Qi-Tang in Chinese) (TJ-41)

Herb	
Astragali Radix	4.0 g
Ginseng Radix	4.0 g
Atractylodis Lanceae Rhizoma	4.0 g
or Atractylodis Rhizoma	
Angelicae Radix	3.0 g
Aurantii Nobilis Pericarpium	2.0 g
Zizyphi Fructus	2.0 g
Bupleuri Radix	2.0g
Glycyrrhizae Radix	1.5 g
Cimicifugae Rhizoma	1.0 g
Zingiberis Rhizoma	0.5 g
	24.0 g (total)

Kakkon-to (Ge-gen-tang in Chinese)

	Decoction	Pharmaceutical Preparation (TJ-1)
Herb		
Puerariae Radix	8.0 g	4.0 g
Ephedrae Herba	4.0 g	3.0 g
Zizyphi Fructus	4.0 g	3.0 g
Cinnamomi Cortex	3.0 g	2.0 g
Paeoniae Radix	3.0 g	2.0 g
Glycyrrhizae Radix	2.0 g	2.0 g
Zingiberis Rhizoma	0.5 g	2.0 g
	24.5 g	18.0 g (total)

* Granule by Tsumura & Co.

Kami-untan-to (Jia-Wei-Wen-Dan-Tang in Chinese)

Herb	
Pinelliae Tuber	5.0 g
Phyllostachysis Folium	3.0 g
Aurantil Fructus Immaturus	3.0 g
Poria	3.0 g
Aurantii Nobilis Pericarpium	3.0 g
Glycyrrhizae Radix	2.0 g
Polygalae Radix	2.0 g
Scrophulariae Radix	2.0 g
Ginseng Radix	2.0 g
Rehmanniae Radix	2.0 g
Zizyphi Fructus	2.0 g
Zizyphi Spinosi Semen	2.0 g
Zingiberis Rhizoma	0.5 g
	31.5 g (total)

Makyo-kanseki-to (Ma-Xing-Gan-Shi-Tang in Chinese)(TJ-55)**

Herb	
Ephedra Herba	4.0 g
Armeniacae Semen	4.0 g
Glycyrrhizae Radix	2.0 g
Gypsum Fibrosum	10.0 g
	18.0 g (total)

Ninjin-yoei-to (Ren-Shen-Yang-Rong-Tang in Chinese)

Herb	Decoction	Pharmaceutical preparation (TJ-108)*
Rehmanniae Radix	4.0 g	4.0 g
Angelicae Radix	4.0 g	4.0 g
Atractylodis Lanceae Rhizoma or Atractylodis Rhizoma	4.0 g	4.0 g
Poria	4.0 g	4.0 g
Ginseng Radix	3.0 g	3.0 g
Cinnamomi Cortex	2.5 g	2.5 g
Paeoniae Radix	2.0 g	2.0 g
Polygalae Radix	2.0 g	2.0 g
Aurantii Nobilis Pericarpium	2.0 g	2.0 g
Astragali Radix	3.0 g	1.5 g
Glycyrrhizae Radix	1.0 g	1.0 g
Schisandrae Fructus	1.0 g	1.0 g
	32.5 g	31.0 g (total)

* Granule by Tsumura & Co.
** Decoction and pharmaceutical preparation composed of same crude drugs with same ratios.

Rikkunshi-to (Liu-Jun-Zi-Tang in Chinese)

	Decoction	Pharmaceutical Preparation (TJ-43)*
Herb		
Poria	4.0 g	4.0 g
Glycyrrhizae Radix	1.0 g	1.0 g
Ginseng Radix	4.0 g	4.0 g
Atractylodis Lanceae Rhizoma or Atractylodis Rhizoma	4.0 g	4.0 g
Zingiberis Rhizoma	1.0 g	0.5 g
Aurantii Nobilis Pericarpium	2.0 g	2.0 g
Zizyphi Fructus	2.0 g	2.0 g
Pinelliae Tuber	4.0 g	4.0 g
	22.0 g	21.5 g (total)

Ryo-kei-jutsu-kan-to (Ling-Gui-Shu-Gan-Tang in Chinese) (TJ-39)**

Herb	
Poria	6.0 g
Glycyrrhizae Radix	2.0 g
Cinnamomi Cortex	4.0 g
Atractylodis Lanceae Rhizoma or Atractylodis Rhizoma	3.0 g
	15.0 g (total)

Ryu-tan-shakan-to (Lung-Tan-Hsieh-Kan-Tang in Chinese)

	Decoction	Pharmaceutical preparation (TJ-76)*
Herb		
Angelicae Radix	5.0 g	5.0 g
Rehmanniae Radix	5.0 g	5.0 g
Akebiae Caulis	5.0 g	5.0 g
Scutellariae Radix	3.0 g	3.0 g
Alisma Rhizome	3.0 g	3.0 g
Plantaginis Semen	3.0 g	3.0 g
Gentianae Scabrae Radix	1.5 g	1.0 g
Gardeniae Fructus	1.5 g	1.0 g
Glycyrrhizae Radix	1.5 g	1.0 g
	28.5 g	27.0 g (total)

* Granule by Tsumura & Co.

** Decoction and pharmaceutical preparation composed of same crude drugs with same ratios.

Senkin-naitaku-sanryo (Chien-Chin-Nei-Tuo-San-Liao in Chinese)

Herb	
Angelicae Radix	3.0 g
Ginseng Radix	2.5 g
Astragali Radix	2.0 g
Cnidii Rhizoma	2.0 g
Saposhnikoviae Radix	2.0 g
Platycodi Radix	2.0 g
Magnoliae Cortex	2.0 g
Cinnamomi Cortex	2.0 g
Angelicae Dahuricae Radix	1.0 g
Glycyrrhizae Radix	1.0 g
	19.5 g (total)

Shikunshi-to (Si-Jun-Zi-Tang in Chinese)

Herb	Decoction	Pharmaceutical Preparation (TJ-75)*
Ginseng Radix	4.0 g	4.0 g
Atractylodis Lanceae Rhizoma or Atractylodis Rhizoma	4.0 g	4.0 g
Poria	4.0 g	4.0 g
Zizyphi Fructus	2.0 g	1.0 g
Glycyrrhizae Radix	2.0 g	1.0 g
Zingiberis Rhizoma	0.5 g	1.0 g
	16.5 g	15.0 g (total)

Shimbu-to (Chen-Wu-Tang in Chinese)

Herb	Decoction	Pharmaceutical preparation (TJ-30)*
Poria	5.0 g	4.0 g
Atractylodis Lanceae Rhizoma or Atractylodis Rhizoma	3.0 g	3.0 g
Paeoniae Radix	3.0 g	3.0 g
Zingiberis Rhizoma	0.5 g	1.5 g
Aconiti Tuber	1.0 g	0.5 g
	12.5 g	12.0 g (total)

Shimotsu-to (Si-Wu-Tang in Chinese) (TJ-71)**

Herb	
Angelicae Radix	3.0 g
Cnidii Rhizoma	3.0 g
Paeoniae Radix	3.0 g
Rehmanniae Radix	3.0 g
	12.0 g (total)

* Granule by Tsumura & Co.

** Decoction and pharmaceutical preparation composed of same crude drugs with same ratios.

Sho-saiko-to (Xiao-Chai-Hu-Tang in Chinese)

	Decoction	Pharmaceutical preparation (TJ-9)[*]
Herb		
Bupleuri Radix	7.0 g	7.0 g
Pinelliae Tuber	5.0 g	5.0 g
Scutellariae Radix	3.0 g	3.0 g
Ginseng Radix	3.0 g	3.0 g
Zizyphi Fructus	3.0 g	3.0 g
Glycyrrhizae Radix	2.0 g	2.0 g
Zingiberis Rhizoma	0.5 g	1.0 g
	23.5 g	24.0 g (total)

Sho-seiryu-to (Xia-Qing-Long-Tang in Chinese)

	Decoction	Pharmaceutical Preparation (TJ-19)[*]
Herb		
Pinelliae Tuber	6.0 g	6.0 g
Ephedrae Herba	3.0 g	3.0 g
Cinnamomi Cortex	3.0 g	3.0 g
Paeoniae Radix	3.0 g	3.0 g
Schisandrae Fructus	3.0 g	3.0 g
Asiasari Radix	3.0 g	3.0 g
Glycyrrhizae Radix	2.0 g	3.0 g
	23.0 g	24.0 g (total)

Toki-shakuyaku-san (Dang-Gui-Shao-Yao-San in Chinese) (TJ-23)

Herb	
Angelicae Radix	3.0 g
Poria	4.0 g
Atractylodis Lanceae Rhizoma or Atractylodis Rhizoma	4.0 g
Paeoniae Radix	4.0 g
Cnidii Rhizoma	3.0 g
Alismatis Rhizoma	4.0 g
	22.0 g (total)

Toki-shigyaku-san (Tang-Kuei-Szu-Ni-San in Chinese)

Herb	
Zizyphi Fructus	5.0 g
Angelicae Radix	3.0 g
Cinnamomi Cortex	3.0 g
Paeoniae Radix	3.0 g
Akebiae Caulis	3.0 g
Asiasari Radix	2.0 g
Glycyrrhizae Radix	2.0 g
	21.0 g (total)

[*] Granule by Tsumura & Co.

Tsu-do-san (Tong-Dao-San in Chinese)

	Decoction	Pharmaceutical Preparation (TJ-105)*
Herb		
Angelicae Radix	3.0 g	3.0 g
Aurantii Fructus Immaturus	3.0 g	3.0 g
Magnoliae Cortex	2.0 g	2.0 g
Aurantii Nobilis Pericarpium	2.0 g	2.0 g
Akebiae Caulis	2.0 g	2.0 g
Carthami Flos	2.0 g	3.0 g
Glycyrrhizae Radix	2.0 g	3.0 g
Rhei Rhizoma	2.0 g	1.8 g
	18.0 g	19.8 g (total)

* Granule by Tsumura & Co.

** Decoction and pharmaceutical preparation composed of same crude drugs with same ratios.

Appendix 3
Chinese and Japanese Herbs in Kampo Medicines

Name of Crude Drugs	Scientific Plant Namesand Part Used
Aconiti Tuber	root of *Aconitum carmichaeli* Debx.
Akebiae Caulis	stem of *Akebia quinata* Decaisne
Alismatis Rhizoma	rhizome of *Alisma orientale* Juzepczuk
Angelicae Radix	root of *Angelica acutiloba* Kitagawa
Angelicae Dahuricae Radix	root of *Angelica dahurica* Bentham et Hooker
Armeniacae Semen	seed of *Prunus armeniaca* Linne
Asiasari Radix	root of *Asiasarum sieboldii* F. Maekawa
Asini Gelatinum	aqueous extract of raw collagen obtained from bone, skin, ligament, or tendon of animals
Astragali Radix	root of *Astragalus membranaceus* Bunge
Atractylodis Lanceae Rhizoma	rhizome of *Atractylodes lancea* De Candolle
Atractylodis Rhizoma	rhizome of *Atractylodes japonica* Koidzumi ex Kitamura
Aurantii Nobilis Pericarpium	peel of *Citrus unshiu* Markovich
Aurantii Fructus Immaturus	fruit of *Citrus aurantium* Linne var. *daidai* Makino
Bupleuri Radix	root of *Bupleurum falcatum* Linne
Carthami Flos	tubulous flower of *Carthamus tinctorius* Linne
Cimicifugae Rhizoma	rhizome of *Cimicifuga simplex* Wormskjord
Cinnamomi Cortex	bark of *Cinnamomum cassia* Blume
Cnidii Rhizoma	rhizome of *Cnidium officinale* Makino
Corni Fructus	fruit of *Cornus officinalis* Siebold et Zuccarini
Coptidis Rhizoma	rhizome of *Coptis japonica* Makino
Dioscoreae Rhizoma	rhizome of *Dioscorea japonica* Thunberg
Ephedrae Herba	tuber of *Ephedra sinica* Stapf
Gardeniae Fructus	fruit of *Gardenia jasminoides* Ellis
Gentianae Scabrae Radix	rhizome of *Gentiana scabra* Bunge
Ginseng Radix	root of *Panax ginseng* C.A. Meyer
Glycyrrhizae Radix	root of *Glycyrrhiza uralensis* Fischer
Gypsum Fibrosum	Gypsum
Magnoliae Cortex	bark of *Magnolia obovata* Thunberg
Moutan Cortex	bark of *Paeonia suffruticosa* Andrews
Paeoniae Radix	root of *Paeonia lactiflora* Pallas
Phellodendri Cortex	bark of *Phellodendron amurense* Rupr.
Pinelliae Tuber	tuber of *Pinellia ternata* Treitenbach
Plantaginis Semen	seed of *Plantago asiatica* Linne
Platycodi Radix	root of *Platycodon grandiflorum* A. De Candolle
Polygalae Radix	root of *Polygala tenuifolia* Willdenow
Polyporus	sclerotium of *Polyporus umbellatus* Fries
Poria	sclerotium of *Poria cocos* Wolf
Puerariae Radix	root of *Pueraria lobata* Ohwi
Rehmanniae Radix	root of *Rehmannia glutinosa* Liboschitz var. *purpurea* Makino

(Continued)

Name of crude drugs	Scientific plant namesand part used
Rhei Rizoma	rhizome of *Rheum palmatum* Linne
Saposhnikoviae Radix	root of *Saposhnikovia divaricata* Schischkin
Schisandrae Fructus	fruit of *Schisandra chinensis* Baillon
Scrophulariae Radix	root of *Scrophularia ningpoensis* Hemsl.
Scutellariae Radix	root of *Scutellaria baicalensis* Georgi
Sinomeni Caulis et Rhizoma	rhizome of *Sinomenium acutum* Rehder et Wilson
Talcum	natural hydrous alminum silicate, silicon dioxide, etc.
Zingiberis Rhizoma	rhizome of *Zingiber officinale* Roscoe
Zingiberis Siccatum Rhizoma	rhizome of *Zingiber officinale* Roscoe
Zizyphi Fructus	fruit of *Zizyphus jujuba* Miller var. *inermis* Rehder
Zizyphi Spinosi Semen	seed of *Zizyphus jujuba* Miller

Index